DIFFUSION OF CONTAMINANTS IN THE OCEAN

OCEANOGRAPHIC SCIENCES LIBRARY

VOLUME 4

R. V. OZMIDOV

P. P. Shirshov Institute of Oceanology, Moscow, U.S.S.R.

Diffusion of Contaminants in the Ocean

Translated from the Russian by

I. LEIKIN

KLUWER ACADEMIC PUBLISHERS
DORDRECHT / BOSTON / LONDON

Library of Congress Cataloging-in-Publication Data

CIP

Ozmidov, R. V. (Rostislav Vsevolodovich)
[Diffuziia primeseĭ v okeane. English]
Diffusion of contaminants in the ocean / by R.V. Ozmidov.
p. cm. -- (Oceanographic sciences library ; 4)
Translation of: Diffuziia primeseĭ v okeane.
Includes bibliographical references.

ISBN-13: 978-94-010-6721-8 e-ISBN-13: 978-94-009-0523-8

DOI: 10.1007/978-94-009-0523-8

1. Marine pollution. I. Title. II. Series.
GC1085.09513 1990
628.1'68'09162--dc20 89-71648

Published by Kluwer Academic Publishers,
P.O. Box 17, 3300 AA Dordrecht, The Netherlands

Kluwer Academic Publishers incorporates
the publishing programmes of
D. Reidel, Martinus Nijhoff, Dr W. Junk and MTP Press.

Sold and distributed in the U.S.A. and Canada
by Kluwer Academic Publishers,
101 Philip Drive, Norwell, MA 02061, U.S.A.

In all countries, sold and distributed
by Kluwer Academic Publishers Group,
P.O. Box 322, 3300 AH Dordrecht, The Netherlands

Translated from the 1986 edition of

ДИФФУЗИЯ ПРИМЕСЕЙ В ОКЕАНЕ

Published by Gidrometeoizdat, Leningrad, U.S.S.R.

Softcover reprint of the hardcover 1st edition 1990

printed on acid-free paper

Contents

Preface

Diffusion of contaminants in the ocean is a major factor forming fields of both natural and man-mad substances introduced into the water medium. Without proper understanding of the laws of contaminant diffusion in the ocean it is impossible to choose the correct methods of calculating the transport of biogenic elements, dissolved gases and pollutants. Diffusion processes in the ocean are very complicated because of the larger number of factors influencing the distribution of a substance. In this regard, progress in studying the regularities of contaminant diffusion in the ocean can only be reached when combining theoretical and experimental methods.

Both theoretical and experimental studies on the diffusion of contaminants in the ocean have been devoted much attention in many countries of the world. The results are being published in a larger number of journals and collected works; however, until now there have been no comprehensive publications on the diffusion of contaminants in the ocean.

This monograph summarizes the results obtained on the problem by the author and other researchers. The presentation of theoretical results is combined with the data obtained in diffusion experiments with artificial tracers. Also given are practical recommendations on how to compute the behavior of contaminants in various hydrometeorological conditions.

Several parts of the monograph are based on the studies I have made in collaboration with A.N. Gerentsway, V.I. Zats, G.S. Karabashev, S.S. Muravyev, A. Okubo and others whose names are listed in References. I would like to thank my co-authors for the permission to use our papers in the book. I would like to express my sincere gratitude to A.S. Monin for his continuous attention and invaluable advice that helped to clarify the presentation.

Introduction

Diffusion, i.e., transport of substances by chaotic molecular or turbulent motions of the media, is of primary importance to the ocean. Diffusion is one of the most important factors influencing the formation of salinity fields, the oxygen content, biogenic elements and various pollutants in the ocean waters. In this regard, knowledge of the regularities of diffusion in the ocean is important for understanding the mechanism of formation of both natural hydrophysical fields and those of man-made pollutants. The transport of a substance of any origin as well as its concentration in a given point of the ocean certainly depend not only on the diffusion but on a large number of other factors. Among them there are biological (accumulation and transport of substances by living organisms), chemical (chemical decay of substances, their compounding with other substances resulting in transition to gaseous state or precipitation, etc.), physical (transition between states, coagulation, adsorption, nuclear decay) and, finally, mechanical ones (transport by the moving waters of the ocean). Mechanical transport, in turn, may be further presented as consisting of two parts: the transport by mean current velocities (advection) and the transport caused by the presence of random chaotic fluctuations in the velocity field (diffusion). The present monograph will, naturally, place most attention on the latter process; however, in some diffusion models, as will be shown further, it is possible to take into account some of the other factors listed above as well.

The prediction of the mechanical transport of substances in the ocean is a rather difficult task since it requires knowledge of the regularities of a larger number of factors influencing the process. For instance, for calculating advective transport of substances, data on the vector field of mean current velocities for the ocean region in question are needed as well those on the variability of this field in time. These data can be taken from mapes or obtained through calculations using the water density fields or the wind fields and heat fluxes across the ocean surface. Here, the choice of space-time scale of averaging is quite significant since it influences the level of non-stationary of the velocity field and the relative intensity of its fluctuating (turbulent) components. The setting of the turbulent components themselves, or their parameterization, is another difficult problem. The characteristics of ocean turbulence depend on the dynamical state of the ocean (mean current velocity gradients, surface and internal waves, density stratification of ocean waters). For example, under a stable stratification of the ocean layers, vertical turbulence and, for that matter, vertical diffusion become considerably retarded. Under an unstable stratification, in the case of convection, the contaminant, introduced into the surface layers of the ocean, can spread quickly down to large ocean depths.

The main difference between turbulent and molecular diffusion is the multiscale character of turbulent motions and the thus inherent correlation between the motions of individual particles of the diffusing contaminant at different points of the flow. This explains the dependence of the density of turbulent diffusion on the scale of the phenomenon considered.

The great importance of diffusion processes in the ocean to all the aspects of its 'life' was recognized as early as at the beginning of this century. For instance, Jacobsen in 1913 (see ref. [91]) was studying turbulent transport of salt of the Great Belt and Kattegat Straits. However, diffusion studies in the high seas were started much later. The first stage of thes investigations were attempts to find a relation between the diffusion parameters and the distributions of easy-to-observe sea-water properties: salinity S and temperature T. Better agreement of the observed distributions of S and T with the solutions of certain simplified semi-empirical equations for S and T transports was achieved through selection of the values of diffusion (or thermal diffusivity) coefficients. Until the late 1930's it was believed that, owing to the relatively small horizontal variations in the ocean of S and T, their horizontal diffusion (which is proportional to the gradient of the property) may be neglected. And only Rossby in 1936 and Parr in 1938 (see ref. [91]) came to a conclusion that the horizontal diffusion coefficients can be several orders higher than the vertical ones; that is why horizontal diffusion often appears quite significant. Soon, the studies of horizontal diffusion revealed the great variability of horizontal-diffusion coefficients as well as its relation with the scale of a diffusion process studied. It has become evident that such a phenomenon is related to the complicated structure of the turbulent current velocity field in the ocean. Without studying the detailed structure of ocean turbulence, further progress in the research of diffusion processes in the ocean would be rather difficult to ensure.

The next stage of the diffusion studies, based on direct measurements of turbulent velocities, began in the 1940's. The pioneering works by V.B. Shtokman [108, 109], carried in that direction, were developed later by A.G. Kolesnikov [44] and other authors. Adjoining to this direction of studies were the diffusion studies using discrete indicators (floats). Observing their chaotic motions in the ocean it was possible to determine the diffusion parameters (Richardson and Stommel [153], Ozmidov [72]). The use of dyes and radioactive isotopes for studying the diffusion regularities in the ocean turned out to be very fruitful as well. In these experiments on diffusion, emission of known amounts of such indicators enabled the carrying out of controlled experiments to obtain the necessary data under various conditions of the process. Such experiments, widespread in oceanology in the late 1950's and 1960's are being conducted now with the use of the newest contact and remote (air and space, for instance) methods of indicator registration.

As the data on the contaminant diffusion process in the ocean accrued, theoretical models for the phenomena also developed. As the first stage of investigations, semi-empirical diffusion equations served, as was mentioned above, as the basic tools of interpreting the observed contaminant distributions, with the main goal being the evaluation of diffusion coefficients in these equations. When these evaluations considerably accrued in number a question arose of how to find the dependence of diffusion coefficients on their determining factors such as the phenomenon scale, current-velocity and water-density gradients, parameters of wave motions, etc. In the subsequent years, not only semi-empirical diffusion equations, which were used to calculate averaged contaminant fields, but also relations for computing more complex characteristics of contaminant concentration fields – for instance, of spectral or correlation functions of concentration fluctuations – came into use for the description of diffusion processes in the ocean. The advent of sophisticated computers also made it possible to solve numerically systems of equations for simultaneous determination of current velocity and contaminant concentration fields.

A large number of journal publications have been devoted to the results of experimental and theoretical studies of the diffusion of contaminants in the ocean. However, there are no comprehensive monographs on the problem either in the U.S.S.R. or abroad. Admittedly, as early as 1975 a monograph on the diffusion of contaminants in the atmosphere was issued [8]. A number of regularities of atmospheric diffusion are common to the similar ones in the ocean; however, the ocean specifics, naturally, bring about diffusion pecularities that need special consideration. Certain issues of the diffusion of contaminants in the ocean are dealt with in the book [121]. That edition, however, places emphasis on the general theory of diffusion with its applications as regards various media but without an adequate analysis of the specific features of the process in the ocean. One paragraph in the chapter "Turbulence in the ocean" in [91] rather briefly presents the results of the studies on contaminant diffusion regularities that were carried out prior to 1976 – 1977.

Representative forums have met a number of times to discuss the problem of the diffusion of contaminants in the ocean. In 1970, for example, an international symposium [115] was held at Warnemünde, GDR, where the results of national and joint efforts of the Council for Mutual Economic Assistance (COMECON) member states in studying turbulent diffusion were discussed. In 1976 Moscow hosted the next International Symposium on similar subjects [83], and in 1982 as International Meeting on Contaminant Diffusion Models was held, also in Moscow [84].

Much experience of both experimental and theoretical studies on the diffusion of contaminants in the ocean has bee acquired at the P.P. Shirshov Institute of Oceanology of the Academy of Sciences of the USSR. The Institute has created the first fluorimeters for determining concentrations of fluorescent dyes, used in experiments on diffusion, and elaborated methods for conducting experiments with instantaneous and continuous dye sources. It has been widely using discrete indicators of various types in experiments carried out both in coastal waters and in the high seas. Back in the early 1970's, the researchers of the institute, together with other Soviet and COMECON oceanologists, started studying statistical characteristics of contaminant fields in ocean, with experiments conducted in the Baltic and Caspian seas, Lake Baikal and in the high seas.

This monograph is an attempt to summarize the information available on the regularities of the diffusion of contaminants in the ocean that has been obtained both by the Institute of Oceanology and other Soviet and foreign organizations. Special attention in the monograph is devoted to the regularities of small-scale and medium-scale (from hundreds of meters to dozens of kilometers) diffusion processes, which are important for working out measures to control the ocean pollution. Naturally, it is necessary, first and foremost, to try to partially or completely eliminate the discharge of pollutants into the ocean. Such measures, however, are usually very costly, so it turns out to be difficult to totally eliminate the discharge. Therefore, the question of predicting the behavior of contaminants in the ocean arises. Knowing the regularities of the diffusion of contaminants in the ocean, we can optimize discharge locations and normalize the discharge rates at which allowable contaminant concentrations in ocean waters are not exceeded. In this regard, this book devotes much attention to the methods of calculating the concentrations of diffusing contaminants in the ocean under various hydrometeorological conditions.

CHAPTER 1

Elements of Contaminant-Diffusion Theory

1.1. Molecular Diffusion

Concentration c is the main quantitative characteristic of a contaminant field in a fluid. It is represented by the mass Δm of contaminant in reference to the volume ΔV of fluid in which it is contained, i.e.,

$$c = \Delta m/\Delta V. \tag{1.1}$$

When the variation of the concentration c over a small distance Δx is not large, it can be assumed with a fair degree of approximation that

$$c(x + \Delta x) - c(x) = \Delta x \, \partial c/\partial x. \tag{1.2}$$

The contaminant flux q through a surface element $\Delta x \Delta y$ over the time Δt, may, evidently, be defined by the expression

$$q = \Delta m/(\Delta x \Delta y \Delta t), \tag{1.3}$$

where Δm is the total mass of the contaminant passing through the surface element in question over the time Δt. When contaminant concentration on one side of the surface element is greater than that on its other side, random molecular motion in the fluid results, on the average, in a contaminant flux through that surface in the direction of lower concentration. At the same time, this flux may be assumed to be proportional to the gradient of concentration, i.e.,

$$q = -D \, \partial c/\partial x, \tag{1.4}$$

where the quantity D is called the molecular diffusion coefficient. Expressions similar to (1.4) can also be written (provided, naturally, that the coefficient D is isotropic) for the fluxes along the other two of the Cartesian coordinates. The total contaminant flux (vector quantity), in that case, can obviously be written as

$$\mathbf{q} = -D \, \nabla c, \tag{1.5}$$

where ∇ denotes the three-dimensional gradient.

(1.5) is the principal relation in the theory of contaminant diffusion in fluids at rest or in laminar motions.

In the kinetic theory of gases, it has been demonstrated that the molecular diffusion coefficient D can be expressed through the mean travel velocity U_m of gas (fluid) molecules and its mean free path length l_m by means of the simple formula

$$D = u_m l_m. \tag{1.6}$$

The molecular diffusion coefficient D is dependent on the type of fluid and contaminant, temperature and pressure, as well as on a number of other factors. Its value for ocean waters is of the order of 10^{-9} m^2/s and it varies, approximately, twofold over the range of sea water temperatures.

Now, let us derive an equation for the contaminant balance in a cubic element $\Delta x \Delta y \Delta z$ of a fluid at rest (see Figure 1.1). According to (1.4), the quantity of contaminant entering this cubic element through the bottom surface $\Delta x \Delta y$, which is perpendicular to the axis z, over the time Δt is

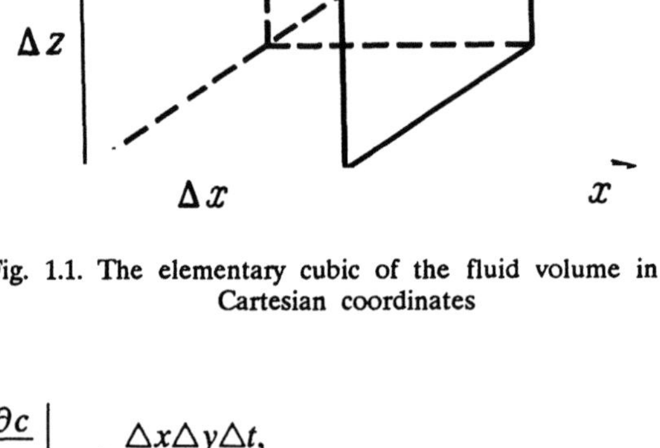

Fig. 1.1. The elementary cubic of the fluid volume in the Cartesian coordinates

$$m_{z1} = -D \left.\frac{\partial c}{\partial z}\right|_{z=0} \Delta x \Delta y \Delta t,$$

while the simultaneous loss of the contaminant through the top surface is

$$m_{z2} = -D \left.\frac{\partial c}{\partial z}\right|_{z=\Delta z} \Delta x \Delta y \Delta t;$$

i.e., the net increase in the quantity of contaminant in the cubic element on account of the above two fluxes will be equal to

$$\Delta_z m = D\left(\frac{\partial c}{\partial z}\bigg|_{z=\Delta z} - \frac{\partial c}{\partial z}\bigg|_{z=0}\right)\Delta x\Delta y\Delta t = D\Delta_z\left(\frac{\partial c}{\partial z}\right)\Delta x\Delta y\Delta t, \tag{1.7}$$

where Δz denotes the increment along the axis z. Similarly, by analyzing the fluxes of the substance through the cube's lateral surfaces perpendicular to the Cartesian coordinates x and y, we find that the total change in the quantity of the contaminant in the cubic element over the time Δt is

$$\Delta m = D\left[\Delta_x\left(\frac{\partial c}{\partial x}\right)\Delta y\Delta z + \Delta_y\left(\frac{\partial c}{\partial y}\right)\Delta x\Delta z + \Delta_z\left(\frac{\partial c}{\partial z}\right)\Delta x\Delta y\right]\Delta t, \tag{1.8}$$

where Δx and Δy are increments along the axes x and y. Then, by dividing the quantity Δm by the volumes of the cube $\Delta x\Delta y\Delta z$ and by the time Δt, and proceeding from finite differences to infinitesimal increments, we obtain an expression for the variation of contaminant concentration in time of the following form:

$$\frac{\partial c}{\partial t} = D\left(\frac{\partial^2 c}{\partial x^2} + \frac{\partial^2 c}{\partial y^2} + \frac{\partial^2 c}{\partial z^2}\right) = D\nabla^2 c, \tag{1.9}$$

where ∇^2 is the Laplace operator.

The relation (1.9) is known as the molecular diffusion equation for a medium at rest. When the fluid is in motion relative to the coordinate system, there emerge additional contaminant fluxes due to advection through the sides of the considered cubic volume. For example, such fluxes through the surfaces perpendicular to the axis z are, respectively, $(wc)_{z=0}\Delta x\Delta y\Delta t$ and $(wc)_{z=\Delta z}\Delta x\Delta y\Delta z$, where w is the velocity component of fluid motion along the axis z. Because of these fluxes, the contaminant diffusion equation has to be completed by an additional term of the form $\nabla(\mathbf{u}\cdot c)$, where $\mathbf{u}$ denotes the flow velocity vector. If the fluid may be considered incompressible, then $\nabla(\mathbf{u}c) = \mathbf{u}\nabla\mathbf{c}$ and, allowing for advection, the diffusion equation takes the form

$$\partial c/\partial t + \mathbf{u}\nabla c = D\nabla^2 c. \tag{1.10}$$

It was already mentioned that in a given volume of fluid diffusing contaminant may decay or, on the contrary, arise. To allow for this phenomenon, the diffusion equation must be supplemented by introducing a certain term Q describing the specific (per unit volume and time) intensity of sources (sinks). Often it may be assumed that the specific increase or decrease in the quantity of contaminant in a body of water is proportional to its concentration. In this case, the diffusion equation is complemented by a term of the form γc where γ is a coefficient of proportionality whose sign depends on whether the contaminant decays or arises.

In order to solve the diffusion equation (1.10), the latter must be supplemented by initial and boundary conditions. The initial condition means assigning a function $c(x, y, z, t)$ at a certain moment of time $t = t_0$, generally considered as zero time, i.e., at $t = 0$. Boundary condition for equation (1.10) may be of various types. Concentration values at the boundaries Γ of the studied domain G may be pre-assigned, i.e., the function $c(t)|_{\Gamma}$ is assumed to be known. There also exist boundary conditions of a more complex type such as, for instance, pre-assigned fluxes at Γ, proportional to certain powers of the contaminant concentration at the boundary. When considering a case of contaminant diffusion occurring far from the boundaries or if the researcher is only interested in the initial moments of diffusion from localized sources, the boundary effect may usually be disregarded, and the diffusion problem is solved as one with the boundary condition of zero concentration at infinitely large distances from the source.

Sources of contaminants, naturally, may vary substantially in type. As a rule, to facilitate the solution of a diffusion equation, researchers tend to consider some idealized types of sources such as, for instance, an instantaneous point source from which the contaminant is assumed to be released into the ocean at a fixed moment of time at a point (x_0, y_0, z_0). A more complicated type of idealized source is a continuous point source with specific intensity q/s which acts from the moment t_0 at a point (x_0, y_0, z_0). Even more complicated are such sources as instantaneous sources distributed in space, sources continuously distributed but acting at a constant intensity at each point in space, sources distributed in space and in time according to a given law, etc.

In some cases, the diffusion equation (1.10) can be simplified. If the contaminant transport in two directions can be neglected the equation is reduced to a one-dimensional form. In ocean conditions, such situations occur at lower values of horizontal contaminant gradients (for instance, in case of a uniform atmospheric fall-out over a large sea area). Provided that, at the same time, it is also possible to neglect the vertical advection of the contaminant, the diffusion equation is reduced to the form

$$\partial c/\partial t = D\partial^2 c/\partial z^2. \tag{1.11}$$

The solution of the one-dimensional equation (1.11) under the boundary conditions of zero concentration c at $|z| \to \infty$, for the case of an instantaneous point source of unit intensity placed at the point $z = 0$ at the moment of time $t = 0$, has the form*

* For methods of the solution of equation (1.11) under various boundary and initial conditions, see, e.g., book [101].

$$c_1(z, t) = \frac{1}{2\sqrt{\pi D t}} \exp\left(-\frac{z^2}{4Dt}\right). \tag{1.12}$$

The expression (1.12) is known as the fundamental solution of one-dimensional equations of diffusion. From this solution it may be seen that at $t \to 0$ $c \to 0$ for all $|z| > 0$ and $c \to \infty$ at $z = 0$. The latter peculiarity is tantamount to the assertion that at $t \to 0$ the function c tends to the delta function $\delta(z)$. With the increasing t the contaminant concentration at the point $z = 0$ decreases in inverse proportion to the square root of t, and the contaminant is then propagated in both directions from the source. Naturally, in the process of diffusion the total amount of contaminant remains unchanged, which may be easily ascertained by integrating the expression (1.12) with respect to z from $+\infty$ to $-\infty$.

Distributions of contaminant concentration along the axis z for certain values of Dt are displayed in Figure 1.2. The curves shown there are known as plots of Gaussian or normal functions, tabulated in a number of handbooks and manuals on probability theory. A normal curve is symmetrical about the origin of coordinates and its 'width' is determined by the expression $\sigma^2 = 2Dt$, named as dispersion. The square root of dispersion is known as 'standard deviation'. Since the latter obviously has the dimension of length it may serve as a spatial scale of the diffusion process.

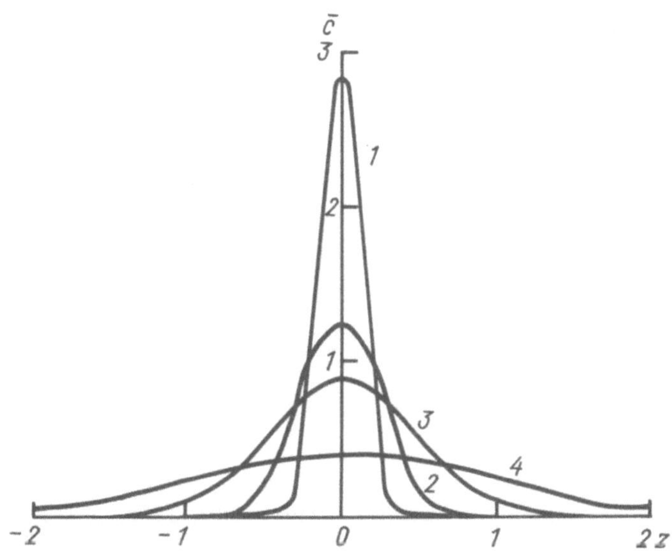

Fig. 1.2. Curves of normalized one-dimensional distribution of the concentration of a contaminant diffusion from an instantaneous point source of unit intensity. Curves 2, 3 and 4 refer to the contaminant distribution at diffusion times exceeding that for Curves 1 5-, 10- and 50-fold, respectively.

As follows from the fundamental solution (1.12), contaminant concentration differs from zero for all values of z even when diffusion occurs over very short periods of time t. This particular feature may be interpreted as a result of infinitely fast propagation of contaminant in the process of diffusion. Such an inference, naturally, does not correspond to the physical reality and is attributable to the fact that the assumptions used in deriving the diffusion equation make no allowance either for molecular inertia or for the finiteness of molecular velocity. However, errors in concentration values due to these assumptions are usually negligible and may be disregarded in practical computations. Besides, as a rule, far greater errors arise because of the idealized boundary conditions at large distances from the point where the contaminant is released.

For a two-dimensional diffusion equation, the fundamental solution, for instance on the plane (x, y), has the form

$$c_2(x, y, t) = \left(\frac{1}{2\sqrt{\pi D t}}\right)^2 \exp\left(-\frac{x^2+y^2}{4Dt}\right). \tag{1.13}$$

According to the above formula, the contaminant concentration, in this case, at the point $x = 0$, $y = 0$, decreases in time inversely to the first power of the diffusion time.

In the case of three-dimensional diffusion in an infinite space, the solution of the diffusion equation for an instantaneous point source of unit intensity placed at the moment $t = 0$ at the origin of coordinates is written in the form

$$c_3(x, y, z, t) = \left(\frac{1}{2\sqrt{\pi D t}}\right)^3 \exp\left(-\frac{x^2+y^2+z^2}{4\,Dt}\right). \tag{1.14}$$

Here, the contaminant concentration at the origin of coordinates decreases in inverse proportion to the diffusion time in the power of 3/2.

The fundamental solutions (1.12), (1.13) and (1.14) also make it possible to construct the solutions of diffusion equations for more complicated types of contaminant sources. Due to the linearity of equation (1.11), the diffusion of contaminant from individual sources occurs independently, and therefore, the total contaminant concentration created by all these sources is a sum of the concentrations created by each of them. In the diffusion theory, the possibility of this summation of concentration values is known as the superposition principle. The latter considerably simplifies the construction of solutions for complicated contaminant sources. In fact, each source distributed in space and in time may be presented as a sum of elementary sources acting in a small volume of space during short periods of time. In such a case, the solution of the diffusion equation can be obtained simply by integrating the fundamental solution over an appropriate volume and time interval. Thus, for instance, when the intensity of an instantaneous point source is not a unit one but equal to Q, the solution of the problem is reduced to simple multiplication of the fundamental solutions by the quantity Q. If in the time moment $t = 0$ a contaminant is released according to the space distribution law $\varphi(x, y, z)$, then the solution of the diffusion equation (1.11) for an infinite space takes the form

$$c(x, y, z, t) = \left(\frac{1}{2\sqrt{\pi D t}}\right)^3 \iiint_{-\infty}^{\infty} \exp\left(-\frac{(x-\xi)^2+(y-\eta)^2+(z-\zeta)^2}{4Dt}\right) \times$$

$$\times \varphi(\xi, \eta, \zeta)\, d\xi d\eta d\zeta, \tag{1.15}$$

where ξ, η and ζ stand for integration variables.

Solutions of diffusion equations with more complicated initial and boundary conditions, however, may not be that trivial, especially a case in which the boundaries of the area, where the diffusion of a contaminant occurs, have complicated outlines. Solutions of diffusion equations with advective terms may be even more complicated. Here, analytical solution can only be found, as a rule, for the simplest velocity fields. In other cases, numerical methods needs to be used.

1.2. Contaminant Diffusion in Turbulent Flows

The flow velocity undergoes, in the case of turbulent motion, chaotic fluctuations due to the numerous eddies of various sizes existing there. It is not only the flow velocity that fluctuates in a turbulent flow. Also fluctuating are all the other hydrodynamical and thermodynamical characteristics, including the concentration of contaminants present there. Determining contaminant concentration values for each point in the turbulent flow at any given moment of time is, obviously, a hopeless and usually unnecessary task. What may be of interest in such a flow are data on concentration fields, averaged in one way or another, as well as on statistical characteristics of random fluctuating field $c(x, y, z, t)$. Naturally, values of velocity components in equation (1.10) cannot be explicitly defined as functions of coordinates and time for a turbulent flow. In this regard, the diffusion equation (1.10) has to be averaged. Generally speaking, such averaging should be carried out over an ensemble of realizations of fields in question. Some set of rules, known as the Reynolds rules of averaging, must be held:

$$\overline{\varphi + \psi} = \bar{\varphi} + \bar{\psi}; \tag{1.16}$$

$$\overline{a\varphi} = a\bar{\varphi}, \quad a = \text{const}; \tag{1.17}$$

$$\overline{\bar{\varphi}\psi} = \bar{\varphi}\bar{\psi} \tag{1.18}$$

$$\overline{\lim f_n} = \lim \bar{f}_n, \tag{1.19}$$

where φ and ψ are some random functions (such as velocity components and contaminant concentration), f_n is the functional sequence, and the bar denotes averaging.

The Reynolds rules say, *inter alea*, that repeated averaging must not change the newly obtained values. However, as will be shown below, this rule in the ocean conditions is not always satisfied if space or time averaging is used instead of ensemble averaging.

Let us represent the values of instantaneous contaminant concentrations and of fluid velocity components as sums of values and fluctuations: $c = \bar{c} + c'$, $u = \bar{u} + u'$, $v = \bar{v} + v'$, $w = \bar{w} + w'$, and substitute these expressions into equation (1.10). On averaging the expressions thus obtained we arrive at the following diffusion equation:

$$\frac{\partial \bar{c}}{\partial t} + \bar{u}\frac{\partial \bar{c}}{\partial x} + \bar{v}\frac{\partial \bar{c}}{\partial y} + \bar{w}\frac{\partial \bar{c}}{\partial z} = D\nabla^2\bar{c} - \frac{\partial}{\partial x}(\overline{u'c'}) - \frac{\partial}{\partial y}(\overline{v'c'}) - \frac{\partial}{\partial z}(\overline{w'c'}). \quad (1.20)$$

The last three terms on the right-hand side of equation (1.20) account for the turbulent nature of the flow. They describe the contribution to the contaminant balance made by transport due to fluctuation velocities. Unlike the molecular diffusion equation, equation (1.20) remains unclosed since the contaminant turbulent fluxes $\overline{u'c'}$, $\overline{v'c'}$, and $\overline{w'c'}$ are unknown. Hence, there arises the closure problem for equation (1.20). To closure the equation the new unknown values need, this way or another, to be expressed through the function $\bar{c}$ we search for. It would be most natural to assume (by analogy with the expression for a contaminant flux in the case of molecular diffusion) that turbulent fluxes are proportional to the mean contaminant-concentration gradients. For a one-dimensional case, such expressions were offered by Taylor as long ago as in 1915 [160] and by Schmidt in 1917 [155]. Considering a general case of three-dimensional diffusion, the hypothesis of proportionality between contaminant turbulent fluxes and the gradient of the average contaminant concentration may obviously be written down as follows:

$$\overline{u_i'c'} = -K_{ij}\partial\bar{c}/\partial x_j, \quad (1.21)$$

where u_i' is any of the fluctuation velocity components ($i = 1, 2, 3$); x_j is any of the Cartesian coordinates ($j = 1, 2, 3$); and K_{ij} is a second-rank tensor called the turbulent diffusion (or transport) coefficient tensor. Since the value of K_{ij} is not, unlike the coefficient of molecular diffusion, a physical constant, the components of the tensor K_{ij} may, generally speaking, be functions both of coordinates and time.

Substituting the expression (1.21) into equation (1.20) we arrive at what is called the semi-empirical equation of turbulent diffusion:

$$\frac{\partial \bar{c}}{\partial t} + \bar{u}_i \frac{\partial \bar{c}}{\partial x_i} = D\nabla^2 \bar{c} + \frac{\partial}{\partial x_i}\left(K_{ij}\frac{\partial \bar{c}}{\partial x_j}\right). \tag{1.22}$$

In most cases, turbulent transport of contaminants in the ocean is, as will be shown later, by many orders higher than the molecular one. That is why the terms with molecular diffusion are, as a rule, omitted from equation (1.22). Diffusion using full or simplified versions of semi-empirical equation (1.22) are called K-theories and are widely used for solving diffusion problems.

1.3. Turbulent Diffusion Coefficients

The second-rank tensor of turbulent diffusion coefficients has nine components and is written in detailed form as follows:

$$K = \begin{pmatrix} K_{11} & K_{12} & K_{13} \\ K_{21} & K_{22} & K_{23} \\ K_{31} & K_{32} & K_{33} \end{pmatrix}. \tag{1.23}$$

Difficulties in setting all the nine tensor components and in solving the semi-empirical diffusion equation makes us seek ways of simplifying the tensor (1.23). Primarily, it would be natural to assume that (see [68], Part I) the turbulent diffusion coefficient tensor is symmetrical, i.e., $K_{ij} = K_{ji}$. However, in that case there are as many as six components remaining, and it does not seem realistic to set all of them in order to solve the diffusion equation. Really, even if for a given point of the flow all three components of the contaminant turbulent flux are known as well as all three derivatives of the average contaminant concentration field ($\overline{u_i'c'}$ and $\delta\bar{c}/\delta x_j$, respectively), it is impossible to determine six components of the tensor K_{ij} based on the relations between $\overline{u_i'c'}$ and $\partial\bar{c}/x_j$. Naturally, this is accounted for by the fact that coefficients K_{ij} are not physical constants and, in turn, are dependent on the flow characteristics. Further simplification of the tensor (1.23) can be done if assuming the hypothesis on the symmetry of the tensor K_{ij} in the ocean relative to the vertical direction z [42]. This hypothesis is based on the preferred role of the vertical axis in the ocean due to the density stratification of its waters. Stratification, generally a stable one, provides drastic differences in the diffusion process along the vertical and horizontal directions x and y, with the origin of the coordinates placed on the ocean surface and the axis z directed vertically downward. Differences in the diffusion process in the horizontal plane along the x and y axes in the cases of large-scale processes may apparently be neglected. This is not true of small-scale processes, which may be essentially influenced by horizontal anisotropy of smaller-scale processes, i.e., ocean waves. It should be noted that the symmetry of the tensor K_{ij} must not necessarily give rise to the symmetry of $\overline{u_i'c'}$ fluxes.

The general expression for an axially symmetrical second-rank tensor has the form [118]

$$K_{ij} = K_L \sigma_{ij} + (K_z - K_L)\lambda_i\lambda_j. \tag{1.24}$$

where values K_L ('horizontal' transport coefficient of turbulent diffusion) and K_z ('vertical' transport coefficient of turbulent diffusion) are already scalar values; σ_{ij} is a unit tensor; and λ_i and λ_j are components of the unit vector which sets the symmetry axis.

Provided the homogeneity of diffusion in the horizontal plane and the assumptions made above, we arrive at the following expressions for turbulent contaminant fluxes in a system of (x, y, z) coordinates:

$$\overline{u'c'} = -K_L \partial\bar{c}/\partial x; \quad \overline{v'c'} = -K_L \partial\bar{c}/\partial y; \quad \overline{w'c'} = -K_z \partial\bar{c}/\partial z, \tag{1.25}$$

and the semi-empirical equation for turbulent diffusion or turbulent-transport equation (disregarding molecular diffusion) takes the following form:

$$\frac{\partial\bar{c}}{\partial t} + \bar{u}\frac{\partial\bar{c}}{\partial x} + \bar{v}\frac{\partial\bar{c}}{\partial y} + \bar{w}\frac{\partial\bar{c}}{\partial x} = K_L\left(\frac{\partial^2 c}{\partial x^2} + \frac{\partial^2 c}{\partial y^2}\right) + \frac{\partial}{\partial z}\left(K_z\frac{\partial\bar{c}}{\partial z}\right), \tag{1.26}$$

where coefficient K_z is supposed to be dependent on the vertical coordinate and, generally speaking, on time. Coefficient K_L in equation (1.26) may also depend upon z and t.

Diffusion equation (1.26) as well as its modernized or simplified versions are widely used in oceanology for solving diffusion problems. In order to solve the equation, it naturally should be supplemented with boundary conditions on the surfaces that contour the ocean area in question, and with corresponding initial conditions as well. Moreover, to solve a diffusion problem thus formulated it is necessary to set the field of averaged velocity components $\bar{u}(x, y, z, t)$, $\bar{v}(x, y, z, t)$ and $\bar{w}(x, y, z, t)$. These components are, generally speaking, functions of coordinates and time. And finally, we have to make assumptions for functional forms and values of the coefficients of turbulent transport K_L and K_z.

Some time ago, estimations of the orders of K_L and K_z coefficient values were made by comparing the observed distributions of salinity (and temperature) in the ocean with the solutions of the simplified versions of equation (1.26). For example, in Jacobsen's paper on turbulent transport in Kattegat, mentioned in the Introduction, the following diffusion equation was used:

$$\bar{u}\frac{\partial\bar{c}}{\partial x} = \frac{\partial}{\partial z}\left(K_z\frac{\partial\bar{c}}{\partial z}\right), \tag{1.27}$$

i.e., the process of salt transport in the strait was considered stationary, with the averaged field velocity vector directed along the axis x.

Integrating expression (1.27) from 0 to z_1 and assuming that the flux of salt across the sea surface equals zero we arrive at the following formula for the coefficient of vertical turbulent diffusion at horizon z_1:

$$K_z(z_1) = \left(\frac{\partial \bar{c}}{\partial z}\right)^{-1}_{z=z_1} \int_0^{z_1} \bar{u}\, \frac{\partial \bar{c}}{\partial x}\, dz. \tag{1.28}$$

Using the data on the salinity field and the currents in Kattegat, Jacobsen has calculated values of K_z for several horizons. The values are given in Table I.I.

TABLE I.I.

Values of the vertical turbulent diffusion coefficients in Kattegat (after Jacobsen)

Depth, m	3.5	5.0	7.5	10.0	12.5	15.0	17.5	20.0
K_z cm^2/s	0.3	0.4	0.18	0.05	0.04	0.2	0.74	2.1

The calculations made by Jacobson are, certainly, quite approximate, since the assumptions used for the derivation of equation (1.28) are not strict enough. However, these calculations convincingly and for the first time showed that vertical turbulent exchange in the sea exceeds molecular diffusion by at least two to three orders (in upper sea waters anyway). This conclusion was later proved by V.B. Satokman [107] who, using equation (1.28), obtained K_z values for the Caspian Sea (the area of the Agrakhan Peninsula) reaching 30 cm^2/s.

An even simpler relation for evaluating K_z was used by Kremser and Brosin [134] who believed that the coefficient K_z in their area of study (the Baltic Sea, to the north-east of Gotland Island) could be considered a constant value. However, velocity of the flow $\bar{u}$ in the region studied often remains unknown. In such cases, we can, from the simplified equation of turbulent diffusion, find the following expression only given data on the salinity field:

$$\frac{K_z}{\bar{u}} = \frac{\partial \bar{c}}{\partial x} \Big/ \frac{\partial^2 c}{\partial z^2}\,. \tag{1.29}$$

Expression (1.29) was used in 1939 by Montgomery (see ref. [91]) for evaluating the $K_z/\bar{u}$ ratio in the Equatorial Counter-Current in the Atlantic.

Non-stationary diffusion equations were also used for estimating the vertical turbulent diffusion coefficients. For example, Jeffreys (see ref. [81]), assuming there was no mean current, neglecting the horizontal diffusion and considering the coefficient K_z a constant, transformed the diffusion equation (1.26) to the form

$$\partial\bar{c}/\partial t = K_z \partial^2\bar{c}/\partial z^2. \tag{1.30}$$

Further on, using data on the annual variations of salinity in the Segne fiord area, Jeffreys obtained K_z values within the range from 8 to 35 cm^2/s. Similarly, V.B. Shtokman [108] found that the value K_z for the Caspian Sea were 0.88 ... 2.55 cm^2/s. Equation (1.30) was also used by Ozmidov and Popov [90] who, basing their calculations on the strontium-90 distribution data for the Atlantic, evaluated K_z at 30 cm^2/s.

In 1927 Jacobsen (see ref. [81]) introduced a graphical method of determining K_z based on the properties of the so-called T, S-diagrams (T for temperature and S for sea-water salinity). The theory of T, S-diagrams has been elaborated by Shtokman [110] and other authors. It enables the estimation of coefficients of turbulent transport between water masses of different temperatures and salinities.

Provided that the salt content in the surface layer of the ocean changes periodically, wave-like disturbances of the salinity field will spread deeper into the ocean, with their speed of propagation and amplitude depending on K_z. Actually, the solution of Equation (1.30) with the boundary condition on the surface of the ocean

$$\bar{c}\big|_{z=0} = \bar{c}_0 \cos(\omega t) \tag{1.31}$$

and $\bar{c}$ approaching zero at $z \longrightarrow \infty$ has the form

$$\bar{c}(z, t) = \bar{c}_0 \exp\left(-\sqrt{\frac{\omega}{2K_z}}\, z\right) \cos\left(\omega t - \sqrt{\frac{\omega}{2K_z}}\, z\right), \tag{1.32}$$

where $\bar{c}_0$ is the amplitude of salinity fluctuations at the surface of the ocean and ω is the angular oscillation frequency ($\omega = 2\pi/T_0$, where T_0 is the oscillation period).

It would be interesting to note that it is necessary to set initial conditions for equation (1.30) for the present problem provided there is a sufficiently long time elapsed since the origination of oscillations. At a multiple, periodic repetition of $\bar{c}$ values at the sea surface the influence of the initial contaminant-concentration distribution becomes insignificant and may be disregarded. It follows from equation (1.32) that the amplitude $A(z)$ of oscillations of contaminant-concentration values for the thick layers of the ocean is described by

$$A(z) = \bar{c}_0 \exp\left(-\sqrt{\frac{\omega}{2K_z}}\, z\right), \tag{1.33}$$

i.e., it decreases with the depth according to the exponential law while the phase shift of oscillation φ is related with z in a linear way:

$$\varphi = \sqrt{\frac{\omega}{2K_z}}\, z. \tag{1.34}$$

Given the oscillation amplitudes of contaminant concentrations at depths z_1 and z_2 (A_1 and A_2, respectively), the coefficient K_z can be determined through a formula derived from (1.33):

$$K_z = \frac{\omega}{2} \left(\frac{z_2 - z_1}{\ln \dfrac{A_1}{A_1}} \right)^2 . \tag{1.35}$$

K_z can also be calculated from the data on the phase shift of oscillations at levels z_1 and z_2 through a formula derived from (1.34):

$$K_z = \frac{\omega}{2} \left(\frac{z_2 - z_1}{\varphi_2 - \varphi_1} \right)^2 . \tag{1.36}$$

Estimation of the coefficients of vertical turbulent transport in the ocean has been made by a good many authors using the formulas (1.35) and (1.36). For example, Shtokman [103] used both these formulas to determine K_z in the Sargossa Sea. However, Shtokman, like Kremser and Matthäus [135], Pivovarov [95] and others did later, used data on heat fluxes rather than on vertical fluxes of diffusing contaminants for their calculations. Seemingly, the orders of the turbulent-transport coefficients for thermal diffusivity do not differ that much from those for turbulent diffusion. The point is that the equation of turbulent thermoconductivity is completely identical to that of turbulent diffusion, and the introduction of the turbulent thermal diffusivity coefficients is based on the assumptions and hypotheses that heat is transported by turbulent-velocity fluctuations, which are similar to the assumptions and hypotheses used for introducing the turbulent-diffusion coefficients. Nevertheless, there may be no complete identity between the turbulent-diffusion coefficient values and those for thermal diffusivity because changing temperature of a fluid considerably effects its dynamics since the density of a fluid depends on the temperature and, in turn, determines considerably the character of motion of a fluid. As for the concentrations of diffusing contaminants, they usually appear to be small and cannot significantly influence the fluid density. Therefore, the question of relationship between the coefficients of turbulent thermal diffusivity and those of turbulent diffusion is rather complicated and needs special consideration. When the orders of the values of these coefficients are assessed at relatively small temperature variations of the medium, however, they may be considered approximately the same.

Using the formulas (1.35) and (1.36), Shtokman found that, in the case of the Sargasso Sea, the vertical turbulent transport coefficients were within the ranges of 2.3 – 9.5 cm^2/s for the 0 – 50 m layer, 1.4 – 9 cm^2/s for the 50 – 100 m layer and 14 – 22 cm^2/s for the 100 – 200 m layer. Using data on the daily variations of temperature in the Kattegat and the above formulas, K_z values were obtained that varied from several units to 24.2 cm^2/s. The variability of the obtained K_z values with depth led Shtokman to the idea of taking into consideration the dependence of K_z on the vertical coordinate z in the transport equation itself, i.e., of not placing the coefficient K_z outside the derivative with respect to z. Obviously, for an analytical solution of such an equation it is necessary to set some dependence of K_z on z. It was suggested that K_z be regarded as an exponentially decreasing function of depth

$$K_z = K_z^0 \exp(-\beta z), \tag{1.37}$$

where K_z^0 is the K_z coefficient value at the level $z = 0$ and β is a constant coefficient.

The solution of the non-stationary diffusion equation with the coefficient K_z depending on the vertical coordinate according to the formula (1.37) and with a periodic boundary condition has the form

$$\bar{c}(z, t) = \bar{c}_0 \sqrt{\exp(\beta z)} \sqrt{\frac{hei_1^2\, \xi + her_1^2\, \xi}{her_1^2\, \xi_0 + hei_1^2\, \xi_0}} \cos(\omega t + \varphi), \tag{1.38}$$

where hei_1 and her_1 are real coefficients of the real and imaginary parts of the Hankel function of the first kind and of the first order,

$$\xi = \frac{2}{\beta} \sqrt{\frac{\omega \exp(\beta z)}{K_z}}, \quad \xi_0 = \frac{2}{\beta} \sqrt{\frac{\omega}{K_z}},$$

and the phase shift φ is, in turn, written down as follows:

$$\varphi = \operatorname{arctg} \left(\frac{hei_1 \xi_0 her_1 \xi - her_1 \xi_0 hei_1 \xi}{her_1 \xi_0 her_1 \xi + hei_1 \xi_0 hei_1 \xi} \right). \tag{1.39}$$

Again, formulas (1.38) and (1.39) can be used for obtaining the expression for the transport coefficient through the ratio of the amplitudes and phase shifts of temperature waves at the two horizons of observation. After making calculations with the use of these same data on water temperatures in the Sargossa Sea, Shtokman found that the variances of the transport coefficients calculated with assumptions of constancy and dependency of the coefficient on z can reach 50 – 100%.

A somewhat different method of evaluating the depth variable coefficient K_z under a periodic boundary condition for the equation of the form

$$\frac{\partial \bar{c}}{\partial t} = \frac{\partial}{\partial z}\left(K_z \frac{\partial \bar{c}}{\partial z}\right) \tag{1.40}$$

was suggested by Fjeldstad in 1933 [123]. Let us assume that variations of the contaminant concentration (or water temperature) at the depth z may be presented by the Foufier series

$$\bar{c}(z, t) = \sum_{n=1}^{\infty} A_n(z) \cos(n\omega t - \varphi_n), \tag{1.41}$$

where A_n and φ_n are amplitudes and phase shifts of corresponding harmonics of $\bar{c}(z, t)$ oscillations. By substituting the expression (1.41) into equation (1.40), differentiating and grouping the terms we then yield an identity which must be valid for any harmonic:

$$\left[\frac{\mathrm{d}}{\mathrm{d}z}\left(K_z A_n \frac{\mathrm{d}\varphi_n}{\mathrm{d}z}\right) + K_z \frac{\mathrm{d}A_n}{\mathrm{d}z}\frac{\mathrm{d}\varphi_n}{\mathrm{d}z} + A_n n\omega\right] \sin(n\omega t - \varphi_n) +$$

$$+ \left[\frac{\mathrm{d}}{\mathrm{d}z}\left(K_z \frac{\mathrm{d}A_n}{\mathrm{d}z}\right) - K_z A_n \left(\frac{\mathrm{d}\varphi_n}{\mathrm{d}z}\right)^2\right] + \cos(n\omega t - \varphi_n) = 0 \tag{1.42}$$

For the validity of (1.42), it is necessary that

$$\frac{\mathrm{d}}{\mathrm{d}z}\left(K_z A_n \frac{\mathrm{d}\varphi_n}{\mathrm{d}z}\right) + K_z \frac{\mathrm{d}A_n}{\mathrm{d}z}\frac{\mathrm{d}\varphi_n}{\mathrm{d}z} + A_n n\omega = 0 \tag{1.43}$$

and

$$\frac{\mathrm{d}}{\mathrm{d}z}\left(K_z \frac{\mathrm{d}A_n}{\mathrm{d}z}\right) - K_z A_n \left(\frac{\mathrm{d}\varphi_n}{\mathrm{d}z}\right)^2 = 0. \tag{1.44}$$

Multiplying (1.43) by A_n we may place this expression in a more compact form:

$$\frac{\mathrm{d}}{\mathrm{d}z}\left(K_z A_n^2 \frac{\mathrm{d}\varphi_n}{\mathrm{d}z}\right) + A_n^2 n\omega = 0. \tag{1.45}$$

Further integrating (1.45) from z to h we arrive at

$$\left(K_z A_n^2 \frac{\mathrm{d}\varphi_n}{\mathrm{d}z}\right)_h - \left(K_z A_n^2 \frac{\mathrm{d}\varphi_n}{\mathrm{d}z}\right)_z = -n\omega \int_z^h A_n^2 \,\mathrm{d}z. \tag{1.46}$$

Now, if to choose the level h at such a depth where the n-th harmonic of the $\bar{c}(z,t)$ oscillations gets damped we will arrive at the Fjeldstad formula for determining the coefficient $K_z(z)$:

$$K_z(z) = \frac{n\omega}{A_n^2 \dfrac{d\varphi_n}{dz}} \int_z^h A_n^2 \, dz. \tag{1.47}$$

If the coefficient K_z depends not only on the vertical coordinate z but changes in time as well its estimation can be done using data on the spacial and time variations of the field $\bar{c}(z,t)$. Actually, by writing down equation (1.30) in its finite-difference form we can come to an expression for the coefficient $K_z(z,t)$ at a point z_0 and in a moment of time t_0 of the following form:

$$K_z(z_0, t_0) = \frac{\bar{c}(z_0, t'') - \bar{c}(z_0, t')}{2\Delta t} \; \frac{h^2}{\bar{c}(z'', t_0) + \bar{c}(z', t_0) - 2\bar{c}(z_0, t_0)}, \tag{1.48}$$

where t'' and t' are moments of time separated from t_0 by Δt and $-\Delta t$, respectively, and z'' and z' are points that lie at distances h and $-h$ from z_0, respectively.

Formula (1.48) was used by Shtokman [103] for evaluating the seasonal variability of K_z in the Caspian, Barents and Sargasso seas. The K_z values appeared to be within less than one square centimeter per second and $12-16$ cm^2/s for different regions studied, with pronounced seasonal changes of K_z. Differences in the character of such seasonal changes for each of the seas in question could be accounted for by the specific features of their hydrometeorological regimes.

As was mentioned in the Introduction, prior to, approximately, the late thirties, oceanologists believed that the orders of the horizontal and vertical transport coefficients were about the same. But in 1939 Montgomery (see ref. [91]) came to the conclusion that the distribution of salinity in the Equatorial Counter-Current in the Atlantic can only be accounted for by assuming that for that area $K_L = 4 \cdot 10^7$ cm^2/s. In the same year, Sverdrup [158] showed that in the South Atlantic $K_L = 10^8$ cm^2/s, and the ratio of the vertical to horizontal transport coefficients should be on the order of 10^6. Shtokman attempted to explain such difference between the coefficients K_z and K_L [103]. He suggested presenting fluctuations of contaminant (temperature) concentrations c' as a product of variations of averaged concentration and a certain value l having the dimension of length. This assumption is, obviously, similar to the widely known method of introducing the so-called 'mixing length', the authorship of which in the theory of turbulence belongs to Prandtl. Given the above assumption, the expressions for turbulent flows of contaminants along the horizontal and vertical axes x and z are as follows:

$$\overline{u'l_x}\ \partial\bar{c}/\partial x = -K_L\ \partial\bar{c}/\partial x; \quad \overline{w'l_z}\ \partial\bar{c}/\partial z = -K_Z\ \partial\bar{c}/\partial z \tag{1.49}$$

where l_x and l_z are projections of the mixing length on the directions x and z, respectively.

It would be interesting to note that, according to (1.49), the turbulent-transport coefficients become, under the above assumptions, equal to the product of fluctuation velocities and some value with the dimension of length which is completely analogous to the formula (1.6) for the molecular-diffusion coefficient. But, unlike the strictly defined mean free path length l_m, for the turbulent type of fluid motion the mixing length cannot be precisely defined. The fluctuation turbulent velocity also cannot be defined as clearly as the velocity u_m. Nevertheless, presenting turbulent-transport coefficients as products of some characteristic speeds and characteristic length scales appears to be quite helpful, and we will use it a number of times henceforward.

From (1.49) we obtain the following coefficient ratio:

$$K_x/K_z = \overline{u'l_x}/\overline{w'l_z}.$$

If, according to Prandtl, we assume further that the projections of the mixing length are proportional to the corresponding components of fluctuation velocity, it would not be difficult to arrive at the following ratios:

$$\frac{K_x}{K_z} = \frac{l_x^2}{l_z^2} = \left(\frac{\partial\bar{c}}{\partial z}\Big/\frac{\partial\bar{c}}{\partial x}\right)^2. \tag{1.50}$$

The field data show that the values of mean horizontal and vertical temperature gradients in the ocean are on the order of 10^{-7} and 10^{-4} °C/cm, respectively. Deriving from here, we find the value of 10^6 which Sverdrup has obtained from field observations for the K_x/K_z ratio. We should, however, promptly note that this ratio may, certainly, vary considerably in time, from place to place and, most importantly, depending on the scale of averaging of those horizontal and vertical processes for which transport coefficients are to be determined. Besides the ratio (1.50) cannot serve as an *explanation* for such strong anisotropy of the transport process in the ocean. The differences in mean temperature gradients along the horizontal and vertical axes is more likely to be a consequence of the transport anisotropy rather than a reason for it. Reasons for it are the vertical density stratification of waters and the differences in the characteristic vertical and horizontal scales of the ocean. The role of these factors in the formation of the peculiarities of turbulent transport in the ocean will be dealt with below in a more detailed way.

The dependence of the horizontal turbulent transport coefficients on the phenomenon scale was noted soon after the determination of the K_L values in the ocean had begun. For example, Sverdrup and Fleming [159] evaluated K_L at $2 \cdot 10^6$ cm^2/s for the Californian Current area, and McEwen [138], while investigating the dynamics of large-scale (synoptic, as they say today) eddies off California, found $K_L = 5 \cdot 10^5$ cm^2/s. At the same time, Munk, Ewing and Revelle [140] found the value for horizontal turbulent diffusion coefficient for the Bikini Atoll area to be as low as $1.5 \cdot 10^5$ cm^2/s, based on the contaminant diffusion data (with a radius of about 3 km). On the other hand, Miyadzaki (see ref. [81]) came to the conclusion that, when calculating the general circulation scheme for the Pacific ocean, one has to operate with the value for the horizontal coefficient of the order of 10^{12} cm^2/s for better agreement with the field data. This and other figures given above may, of course, be argued about because the authors made many insufficiently grounded assumptions for calculating the K_L values. However, the fact that K_L is dependent upon the phenomenon scale could raise no doubt.

1.4. Dependence of Transport Coefficients upon the Determining Factors

The vertical turbulent diffusion and, therefore, the vertical-diffusion coefficients must be significantly influenced by the density stratification of oceanic waters. As a matter of fact, vertical turbulent fluctuations are to be 'suppressed' in a stably stratified fluid since the transport of particles of the denser fluid up into the above layer or of lighter particles down into the below layer requires considerably energy. Actually, if the vertical gradient of water density $\partial\rho/\partial z$ equals γ, in order to mix a fluid column with the cross-section of 1 cm^2 and height l

$$E = g\gamma l^3/12 \tag{1.51}$$

energy units will be required. For mixing a column 1 cm high and of unit mass the energy required is equal to

$$E_0 = g\gamma l^2/(12\rho_0) = l^2N^2/12, \tag{1.52}$$

where ρ_0 is the mean fluid density in the column, g is the acceleration due to gravity and the value

$$(g\gamma/\rho_0)^{1/2} = N \tag{1.53}$$

has the dimension of a reciprocal second and is called the Väisälä – Brunt frequency* . In the stratified ocean waters, particles taken out of their equilib-

* Strictly speaking, the expression for the Väisälä–Brunt frequency should involve the potential fluid density, i.e., a density adiabatically reduced to standard pressure. So when using density *in situ* in the expression (1.53) it is necessary to introduce the correction $-g\rho_0/c^2$ to the density gradient $\partial\rho/\delta z$, where c is the velocity of sound in sea water. However, for most practical purposes this correction may be neglected.

rium positions by small vertical displacements make free oscillations with the frequency N.

The damping influence of the stable water-density gradient on turbulence is able to deny turbulence in flow in question while it could have existed given all other conditions. As a criterion for the existence of turbulence in a stratified medium the non-dimensional number Ri, introduced into hydrodynamics by Richardson in 1921, is used:

$$Ri = \frac{g}{\rho_0} \frac{\partial \rho / \partial z}{(\partial \bar{u} / \partial z)^2} = N^2 (\partial \bar{u} / \partial z)^{-2}. \tag{1.54}$$

The Richardson number (1.54) is called a gradient or kinematic number, as distinct from the dynamical number Rf which is expressed as follows:

$$Rf = \frac{K_z^\rho}{K_z^u} Ri, \tag{1.55}$$

where K_z^ρ is the vertical turbulent-transport coefficient for mass, and K_z^u is the vertical-transport coefficient for momentum (the eddy viscosity coefficient).

Generally, when both horizontal components of velocity vector are presented in a flow the expression for Ri (1.54) is generalized into:

$$Ri = N^2 [(\partial \bar{u} / \partial z)^2 + (\partial \bar{v} / \partial z)^2]^{-1}. \tag{1.56}$$

The Richardson number is a ration of the work of Archimedean forces (with a minus) to the work of turbulent strength and, thus, reflects the ratio of the work of the turbulence damping mechanism (without counting viscosity forces) to that of its generation. So it is only natural that the criterion for the existence of turbulence in a stably stratified flow must be the condition $Rf < 1$ (the condition that $Rf = 1$ can be assumed only if fluid viscosity approaches zero). The Richardson criterion can be more specific for different types of motion. Particularly, the stability criterion for a stationary, horizontally homogeneous flow with stable vertical density gradient, as Miles and Howard reported [127, 139] is

$$Ri \geqslant 1/4. \tag{1.57}$$

If there is some turbulence in a stably stratified fluid its vertical observance and parameters (including the vertical transport coefficient) should essentially depend on the Richardson number. Such a dependence can, for instance, be derived from the equation of turbulent energy balance in a stratified flow or from field data. Pursuing this path, some authors have arrived at formulas of the following type:

$$K_z = K_z^0(1 + aRi)^n, \tag{1.58}$$

where K_z is the turbulent vertical-transport coefficient in a stratified fluid, K_z^0 is that for a non-stratified fluid and a and n are some constant values. In meteorology expressions correlating K_z with the height of the observation point above the earth's surface are widespread. There have been formulas suggested according to which K_z linearly grows up to a certain height and then remains constant, as well as power and exponential expressions.

Other concepts, more grounded from the physical standpoint, on the influence of density stratification of a medium on the turbulent transport characteristics can be obtained from the Monin – Obuhov theory of the turbulent regime in the near-ground atmospheric layer (see ref. [68] and/or [67]). The theory says that the characteristics of a stationary, horizontally homogeneous turbulent flow are determined by the five external governing parameters: the depth of layer h, the height of boundary roughness z_0, the buoyancy parameters g/ρ_0, the velocity difference δU and the potential density $\delta\rho$ across the layer. When determining such turbulent field characteristics as temperature T and salinity S, corresponding temperature gradients δT and salinity gradients δS should be used instead of $\delta\rho$, and parameters αg and βg instead of g/ρ_0, where α and β are sea-water compressibility-coefficients accounted for by changing the temperature and salinity, respectively. We can, from the above external parameters, determine the so-called velocity friction (square root of the turbulent momentum flux divided by the fluid density) u_*, vertical turbulent heat fluxes q_T, those of salt q_S and mass q_ρ on the rough boundary of the layer. Also we can introduce, in a uniquely possible way, the scale of length (the sublayer thickness) L_* with constant turbulent fluxes over the depth as well as the ρ density variation scale R_* and corresponding scales T_* and S_* for temperature and salinity:

$$L_* = \frac{u_*^3}{\kappa q_\rho g/\rho_0}; \quad R_* = -\frac{q_\rho}{\kappa u_*}; \quad T_* = -\frac{q_T}{\kappa c_p \rho u_*};$$

$$S_* = -\frac{q_S}{\kappa \rho u_*}, \tag{1.59}$$

where c_p is water heat capacity at constant pressure and κ is the so-called von Kármán's constant introduced into the formulas to facilitate further calculations.

At heights significantly exceeding the average size of irregularities of the underlying surface, the value z is not supposed to influence the motion characteristics in the layer. Therefore, for every point of the flow, all the non-dimensional characteristics of the processes there (with scales exceeding the characteristic ones for molecular effects) should be universal functions of the non-dimensional ratio h/L_*. With decreasing density stratification and its transition into an unstable one the turbulent velocity fluctuations will increase (under otherwise identical conditions). The density fluctuations are small at very large and stable gradients since the velocity turbulence fluctuations will be damped very much in that case. Turbulence gets stronger with the reducing stratification; however, the mean density gradient gets lower. That is why the rate of growth ρ' will retard and the density fluctuations will subside to their minimum thereafter, at a neutral stratification of the flow (we would like to draw your attention to the fact that fluctuations of scalar value are proportional to the product of the velocity fluctuations and the gradient of the mean field of substance in question). Under unstable stratification and the development of convective motions, the fluctuations of ρ will grow again up to the regimes with strong convective mixing when the fluid density evens off and the density fluctuations subside again. The fluctuations of temperature, salinity and contaminant concentration can behave in a much more complicated way with the changing density stratification, depending on the contributions of the gradients of T, S and c to the density gradient as well as on the types of the sources of these substances and the boundary conditions.

The length scales of the L_* type, beyond which the influence of the density stratification is low, can be drawn from another (but equivalent) set of governing parameters. For example, Panchev [93] suggested using the following parameters for a temperature-stratified flow: $\partial\bar{u}/\partial z$, ε, ε_T and αg, where ε and ε_T are the turbulent energy dissipation rate and the rate of smoothing of temperature irregularities, respectively. The values ε and ε_T are determined by the following expressions:

$$\varepsilon = \frac{\nu}{2} \sum_{i,j=1}^{3} \overline{\left(\frac{\partial u_i}{\partial x_j} + \frac{\partial u_j}{\partial x_i}\right)^2} ; \tag{1.60}$$

$$\varepsilon_T = 2\chi \sum_{i=1}^{3} \overline{\left(\frac{\partial T}{\partial x_i}\right)^2}, \tag{1.61}$$

where ν and χ are kinematic molecular coefficients of viscosity and thermal diffusivity, respectively.

From the above parameters and on dimensional grounds the following general expression for the scale of length can be derived:

$$L = \varepsilon^{1/2} \left(\alpha g \frac{\partial \bar{T}}{\partial z}\right)^{-3/4} \varphi \left[\frac{\alpha g \varepsilon_T}{\varepsilon} \left(\frac{\partial \bar{T}}{\partial z}\right)^{-1}, \left(\alpha g \frac{\partial \bar{T}}{\partial z}\right)^{-1/2} \frac{\partial \bar{u}}{\partial z}\right], \tag{1.62}$$

where φ is some universal function of two non-dimensional arguments.

If we postulate a power form for φ and drop one by one each of the five governing parameters we can arrive at seven different expressions for the scale of length. In oceanology, the most wide-spread scale has been one introduced by Ozmidov in 1965 [78]. In the general case, not only for heat stratification but for density as well, the expression for this scale has the following form:

$$L_0 = a\left[\bar{\rho}\varepsilon^{2/3}\Big/\left(g\,\frac{\partial\bar{\rho}}{\partial z}\right)\right]^{3/4} = a\varepsilon^{1/2}N^{-3/2}, \tag{1.63}$$

where a is some constant whose form will be presented in more detail further.

The expression for L_0 can be derived not only on dimensional grounds but also from certain physical reasonings. Only turbulent eddies of the characteristic size L and with definite velocity gradient $\Delta\bar{u}$ are, apparently, capable of overcoming the Archimedean forces in a stratified flow. If the density 'threshold' is overcome for such eddies, turbulence under the scale L should be three-dimensionally isotropic; then $\Delta\bar{u}$ is related with L by a well-known relationship of the Kolmogorov theory of locally isotropic turbulence [46]:

$$\Delta\bar{u} \sim \varepsilon^{1/3}L^{1/3}, \tag{1.64}$$

and the characteristic velocity gradient in that case will be equal to

$$d\bar{u}/dz = c\varepsilon^{1/3}L^{-2/3}, \tag{1.65}$$

where c is the universal non-dimensional constant of the order of unity.

However, only such motions can overcome the density threshold for which the Richardson number has reached its critical value Ri_{cr}. Substituting the velocity gradient in the expression for Ri_{cr} in accordance with the relation (1.65) we will arrive at

$$Ri_{cr} = \frac{g}{\bar{\rho}}\,\frac{\partial\bar{\rho}/\partial z}{c^2\varepsilon^{2/3}L^{-4/3}}, \tag{1.66}$$

where it follows for L that

$$L = L_0 = Ri_{cr}^{3/4}c^{3/2}\varepsilon^{1/2}N^{-3/2}, \tag{1.67}$$

which coincides with (1.63) at $a = Ri_{cr}^{3/4}c^{3/2}$.

The considerations given right now may be used for clarifying the form of dependence of the coefficient of vertical turbulent transport K_z upon the density gradient $\partial\bar{\rho}/\partial z = \gamma$. If for scales under L_0 the turbulence is three-dimensionally isotropic then the only governing parameters for processes not exceeding the scale L_0 but large enough to be influenced by molecular effects would be the values ε and l, where l is the scale of a process under consideration (Kolmogorov's hypothesis [46]). In that case, based on dimensional grounds for the isotropic transport coefficient K we obtain

$$K = c_1 \varepsilon^{1/3} l^{4/3}, \tag{1.68}$$

where c_1 is a universal non-dimensional constant.

According to (1.68), the coefficient of vertical turbulent transport increases with the scale of the phenomenon in the power 4/3. But up to what limits may this growth occur in a density-stratified medium? It is obvious that the process will go on until the scale L_0 is achieved; afterward the coefficient K_z should stop increasing in the vertical direction (or at least, the increase should be retarded abruptly) since larger vertically oriented eddies should be absent in a stratified flow. Therefore, the 'maximum' value for the coefficient K_z in a stratified flow has the form

$$K_z^{\max} = c_1 \varepsilon^{1/3} L_0^{4/3} = c_1 c^3 Ri_{cr} \varepsilon N^{-2}. \tag{1.69}$$

In the light of the above-considered mechanism of turbulent transport in a stratified turbulent medium, it seems interesting to watch how the diffusion of a contaminant occurs if the contaminant is placed inside the flow at a certain moment of time and into a small ('point') volume of space. At smaller sizes of the contaminant cloud, its diffusion will be predominantly influenced by molecular diffusion effects with the constant coefficient D. Therefore, for this period the contaminant diffusion will be described by the expression (1.14). However, on reaching some characteristic scale l_0, small turbulent eddies will become important, and the contaminant diffusion will get somewhat accelerated. l_0 is called the Kolmogorov scale and, on dimensional grounds, has the form

$$l_0 = \sqrt[4]{\nu^3/\varepsilon} . \tag{1.70}$$

As the spot grows further in size, its diffusion will become influenced by turbulent eddies of larger sizes, thus making its diffusion accelerate isotropically more and more in all directions. The contaminant spot at this stage of diffusion will, naturally, have a spherical form (in this or another method of averaging the turbulent fluctuations in the contaminant concentration field). But as soon as the size of the spot reaches the value of L_0 it will get 'flattened', as the vertical growth of the spot will be no more accelerated (the coefficient K_z will approach its 'saturation'), while in the horizontal directions the acceleration will still continue since the increase in the horizontal transport coefficient will not be hampered. Therefore, beginning from that moment, the diffusion process will become anisotropic with a constant coefficient of vertical diffusion and an increasing, depending on the scale of the phenomenon (and thus on the time of the diffusion), coefficient K_L, isotropic in the horizontal plane. The increase in the coefficient K_L may continue up to the scale at which horizontal isotropy of processes gets disturbed, for instance, because of the anisotropy of boundary conditions at the ocean surface, the presence of mean velocity field gradients and so on (see Figure 1.3).

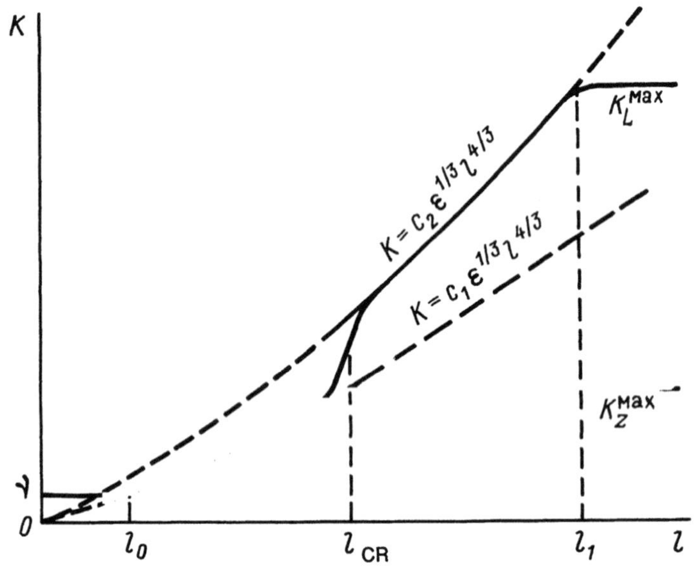

Fig. 1.3. Dependence of the vertical and horizontal turbulent transport coefficient on the phenomenon scale in a density-stratified ocean (after [78]).

Also of interest is the question of what happens to the horizontal transport coefficient when the scale of the phenomenon reaches the value L_0. There is no consensus on that among oceanologists. Some authors believe that the vertical density stratification does not in any way influence the horizontal transport. Another point of view is that the horizontal exchange is bound to be more intensive in a stratified fluid as compared with a homogeneous medium (under otherwise equal conditions). This conclusion is based on the fact that at the phenomenon scales exceeding L_0 turbulent eddies 'operate' only in the horizontal plane, consuming no energy on vertical mixing. The properties of such a quasi-two-dimensional turbulence has, naturally, to be different from the regularities of a three-dimensional phenomenon. The basic difference of a two-dimensional turbulence from a three-dimensional one is the existence, in the two-dimensional case, of two inertial scale ranges.

One of them is where the transport of kinetic energy at the rate ε from small-scale to large-scale eddies occurs (unlike the opposite flux in the three-dimensional case), while the other interval is where the transport of entrophy (a half of the root-mean-square vorticity) at the rate ε_ω in the direction of smaller eddies (larger wave numbers k) occurs. If there is an influx of energy and entrophy into the ranges of wave numbers k_0 then the energy distribution for turbulent fluctuations $E(k)$ in the wave-numbers space for smaller k, to which the energy flux is directed, will be proportional to k in the power ≪-5/3≫ (the 'Kolmogorov' spectrum), while for larger k (up to the scale $k_d = (\varepsilon_\omega/\nu)^{1/6}$, where the entrophy flux exists, we, on dimensional grounds, arrive at $E(k) \sim k^{-3}$. If the influxes of energy and entrophy occur not only "at the point" k_0 or if the range for the values k is limited, the laws of energy distribution in the wave-number space will, naturally, become more complicated.

If ε and ε_ω are governing parameters for the two-dimensional turbulence in the inertial range then, on dimensional grounds, we can, for this range, also yield the expressions for the distribution in the wave-number space of the fluctuation 'energy' for passive contaminant concentrations $E_c(k)$, diffusing in the flow. $E_c(k)$, briefly defined as the 'spectrum' of fluctuations of the concentration c', is related to c' by the expression

$$\overline{c'^2} = \int_0^\infty E_c(k)\, dk. \tag{1.71}$$

In the general case we must include in the set of dimensional parameters governing $E_c(k)$ the rate of transport over the spectrum of inhomogeneity measure of contaminant-concentration fluctuations ε_c. Then, on dimensional grounds, for $E_c(k)$ in the inertial range we obtain

$$E_c(k) = \varepsilon_c \varepsilon^{-1/3} k^{-5/3} f(l/L), \tag{1.72}$$

where $f(l/L)$ is some non-dimensional function which has the value of 1 at its zero argument; $L = (\varepsilon/\varepsilon_\omega)^{1/2}$ is the constant scale of length; and $l = 1/k$ is the current scale of length. At l large enough to neglect the forces of viscosity and, therefore, the energy dissipation (inertially convective range of scales), ε must not be present in the expression (1.72). In that case, $f(l/L) \sim k^{2/3}\varepsilon^{1/3}\varepsilon_\omega^{-1/3}$ and for $E_c(k)$ we obtain

$$E_c(k) = c_2 \varepsilon_\omega^{-1/3} \varepsilon_c k^{-1}, \tag{1.73}$$

where c_2 is a numerical constant.

We can also arrive at the expression for the coefficient of diffusion $K_L(l)$ in the inertial range of a two-dimensional turbulence on similar dimensional grounds. For the general case, we have

$$K_L(l) = \varepsilon^{1/3} l^{4/3} \psi(l/L), \tag{1.74}$$

where $\psi(l/L)$ is a universal function.

If the scale l belongs to the range of entrophy transport then the parameter ε must fall out from the formula (1.74), for which it is necessary that $\psi(l/L) \sim k^{2/3} \varepsilon^{1/3} \varepsilon_\omega^{-1/3}$. In that case,

$$K_L(l) = c_3 \varepsilon_\omega^{1/3} l^2, \tag{1.75}$$

where c_3 is a non-dimensional constant.

The expression (1.75) was obtained by Lin [137] in 1972 and critically analyzed thereafter by Mirabel and Monin [64]. The point is that the two-dimensional semi-empirical equation, presented in polar coordinated,

$$\frac{\partial \bar{c}}{\partial t} = r^{-1} \frac{\partial}{\partial r} \left[r K(r) \frac{\partial \bar{c}}{\partial r} \right] \tag{1.76}$$

(r is the radius vector), has no solution for an instantaneous point source of a contaminant just in the case when $K(r) \sim r^2$. There exist solutions for all other power functions $K(r)$ on r, with some of these solutions being used in oceanology, as will be mentioned below. The above peculiarity of the case $K(r) \sim r^2$ is certainly related to the specifics of the spectrum (1.73) at which, as is easy to see from formula (1.71), logarithmic divergence of the mean square of fluctuation concentration $\overline{c'^2}$ occurs at small k, i.e, a strong overestimation of the contribution to $\overline{c'^2}$ from the concentration field disturbances with very small k is observed. In [64] a spectral equation for $E_c(k)$ is used for the derivation of formulas for the spectrum $E_c(k)$ and coefficient $K(l)$, and, in order to close it, the hypothesis of the non-local character of entrophy transport in the spectrum is applied. As a result of this theory, the following generalizations of the formulas (1.73) and (1.75) were obtained in [64]:

$$E_c(k) = (4)^{2/3} \varepsilon_c \varepsilon_\omega^{-1/3} k^{-1} \left(\ln \frac{k}{k_0} \right)^{-1/3}, \tag{1.77}$$

$$K_L(l) = (2^{2/3}/3)\, \varepsilon_\omega^{1/3} l^2 \left(\ln \frac{l}{l_0} \right)^{-2/3}, \tag{1.78}$$

where k_0 is the wave number limiting the inertial range and l_0 is the scale at which the energy supply of the flow occurs.

Additional logarithmic factors in these two formulas provide the spectrum with an effective 'out-off' boundary at small k, thus eliminating the divergence of the integral (1.71), and the transport coefficient (1.78) guarantees the solution of equation (1.76), since for $l \longrightarrow \infty$ we have $l^2K(l) \longrightarrow 0$.

The question of dependence of transport coefficients upon the scale of a phenomenon under the ocean conditions can certainly not be resolved through the above formulas. The influxes of energy and entrophy to the motions of the ocean waters may occur, generally speaking, over the entire range of motion scales. Besides, these motions themselves (even the most large-scale ones) are not purely two-dimensional. That is why it is hardly possible to expect the exact fulfillment of the formulas (1.68) or (1.75) for the transport coefficients. In order to facilitate, to a certain extent, the options for the dependence of transport coefficients upon the phenomena scale, Ozmidov [79] postulated a hypothesis of quasi-discrete supplies of energy and entrophy for the ocean. According to this theory, the influxes of energy and entrophy to the ocean from external sources are primarily concentrated in the neighborhood of some specific values of the wave numbers k_i, whereas in the intervals between them such influxes, in a number of cases, can be neglected. Under this hypothesis, the spectral distribution of the water-motion energy as well as the dependence of transport coefficients on the scale of the phenomenon must be rather specific. The energy and entrophy transferred to the ocean motions in the scales $L_i = 1/k_i$ by external sources will be redistributed further on among other scales of motions. Because of the chaotic mechanism of breaking the turbulent eddies we can assume that in scales significantly different from L_i the eddies will lose their prevailing orientation determined by the external forces and will become close to the isotropic ones (either three-dimensionally or in the horizontal plane only). At a sufficiently large interval between the 'neighboring' values L_i there may be ranges with the turbulent characteristics universally dependent upon the scale of the phenomenon and other factors. Naturally, the parameters for such dependences may be different for each of these ranges as in the energy-supply zones there are 'additions' to the fluxes of energy ε and of entrophy ε_ω. Therefore, each interval where the universal laws are to be held true should have, generally speaking, its own values of ε_i and $\varepsilon_{i\omega}$. If the influx of energy and entrophy into the given energy-supply zone is high as compared with those in the other zones, the turbulent characteristics in the range of scales considered will be determined by the 'local' values of the parameters ε_i, $\varepsilon_{i\omega}$, i.e., the processes in this case will not depend on phenomena with other scales. It is a main feature that distinguishes the ocean from many other turbulent flows of fluids where there is one governing scale at which the flow energy supply occurs. In that case, the influx of energy (and entrophy) totally determines the dynamics of all motions in the flow, beginning from L and ending with smaller motions that already get damped by the molecular viscosity. Such a governing scale, for example, for the motions of a fluid running in pipes or in canals is the diameter (radius) of the pipe or the width of the canal. The type of fluid motion also depends significantly on the scale L. The transition from the laminar type of motion to the turbulent one is known to occur under certain values of the Reynolds number

$$Re = LU/\nu, \tag{1.79}$$

where L is the characteristic space scale, U is the typical velocity of the fluid motion, and ν is the molecular viscosity coefficient for the fluid.

For the flows in pipes, canals or in other devices of this type, where the hydraulic head or the force of gravity serves as the only external motive force, L in (1.79) coincides with the typical size of the flow; therefore the number Re is determined with the use of that value. If there are several energy-supply zones and characteristic scales L_i there are, generally speaking, several values of the Reynolds number Re_i. This circumstance explains the possibility of occurrence in the ocean of regions with laminar or close to laminar regimes of the water's motions. In such regions, the local values of the Reynolds number Re_i are comparatively small, and the influence of other scale motions can be neglected if the 'neighboring' energy-supply scales L_{i-1} and L_{i+1} are variant enough from L_i and their energy and entrophy fluxes are not large.

The fulfillment in each of the ranges of the universal laws of the two-dimensional or three-dimensional turbulent regimes is determined by a number of factors. In most cases, not quite 'clear-cut' regularities of three-dimensional or two-dimensional turbulence are, apparently, typical for the ocean except for some intermediate cases. If, for example, the range of wave-numbers is limited (and this, naturally, always occurs in the ocean), then instead of the ≪-3 power law≫ for the energy spectrum in the case of two-dimensional turbulence one can arrive at [94]

$$E(k) \sim \varepsilon_{\omega 1}^{2/3} k^{-3} (1 - k^2/k_1^2)^{2/3}, \tag{1.80}$$

where $k_1 = (\varepsilon_{\omega 1}/\varepsilon_1)^{1/2}$; $\varepsilon_{\omega 1}$ is the entrophy flux in the direction usual for two-dimensional turbulence and ε_1 is the energy flux in the direction usual for three-dimensional case.

In the case being considered, where the range of k is limited and where influxes of energy and entrophy occur in the neighborhood of k_0, on both sides of k_0 there will be a decrease in the level of $E(k)$, with the steepness of this decrease gradually increasing. More complicated pictures for the functions $E(k)$ and, naturally, $K(k)$ will be observed at several values of wave numbers k_i at which there is a pronounced influx of energy and entrophy. For instance, if one of the sources is located in the range of baroclinic instability of the large-scale ocean circulations and other sources are in the scales of inertial and tidal motions, then in the range of synoptic scale motions (about 100 kilometers) the fluxes ε and ε_ω from these sources will become overlapped. The forms of dependence of the spectra $E(k)$, $E_c(k)$, as well as of the transport coefficient $K(k)$ on the wave number (scale) will be significantly influenced in that case by the relative intensity of the sources, their number, mutual location, etc.

A simple scheme of the kinetic-energy distribution in the ocean among motions of different scales was suggested, as was noted above, by Ozmidov [79]. The scheme had three specific scales L_i of energy influx (the global circulation scale, the scale of tidal and inertial oscillations, the scale of surface waves) and three ranges of wave numbers where the laws of locally isotropic turbulence are held (but with different values of the governing parameter ε). Thus, according to the scheme, the function $E(k)$ in the ranges between the energy-supply zones must be described by the "–5/3 power law" and the transport coefficient by the "4/3 power law". The graph of dependence of the transport coefficient on the phenomenon scale, following from the scheme, is given in Figure 1.4. The values of ε_i for each region where the "4/3 power law" is held are chosen equal to 10^{-5}, 10^{-3}, and 10^{-1} cm^2/s^3.

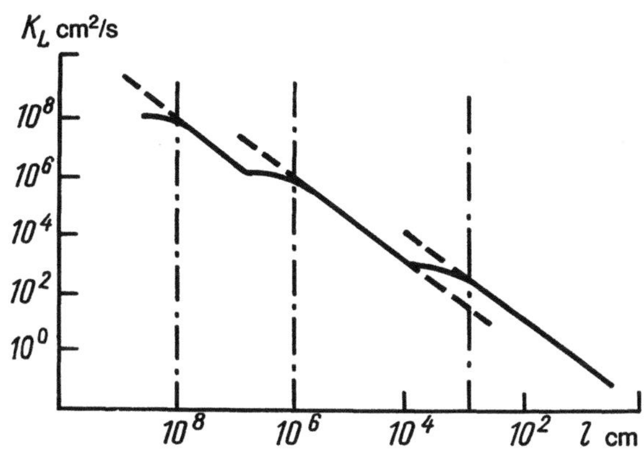

Fig. 1.4. A pattern of dependence of the horizontal turbulent transport coefficient K_L on the phenomenon l in the ocean (after [80]) y-axis: K_L cm^2/s; x-axis: l cm.

The fulfillment of the "–5/3 power law" for the function $E(k)$ and of the "4/3 power law" for the transport coefficient for different scale ranges has been proven a number of times by the field observation data. The review of these results can be found in monographs [81] and [67]. Here we will only reproduce the dependence for the horizontal diffusion coefficient in the ocean obtained by

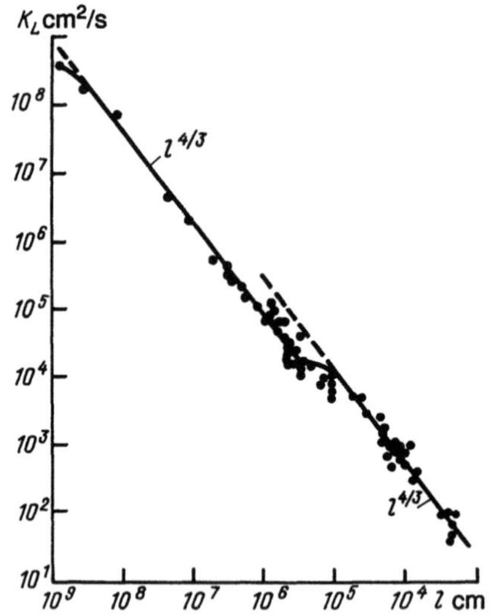

Fig. 1.5. Dependence of the coefficient K_L on the scale of the phenomenon l, based on dye-spot diffusion experiments (after [92]).
y-axis: K_L cm^2/s; x-axis: l cm.

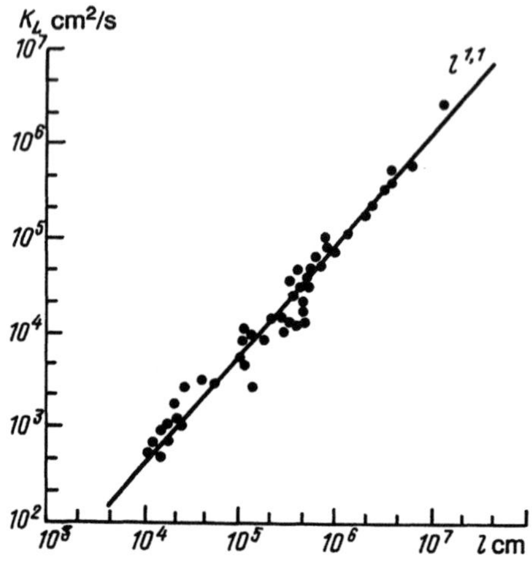

Fig. 1.6. Dependence of the coefficient K_L on the scale of the phenomenon within a wide range of variations of l (after [146]).
y-axis: K_L cm^2/s; x-axis: l cm.

Okubo and Ozmidov [92] and Okubo [146] using diffusion measurements in various regions of the ocean (Figures 1.5 and 1.6). As is seen from Figure 1.6, the dependence $K(l)$ is, on the average, easily approximated rather well for a broad range of scales from about 10^4 to 10^7 cm by a power law with the exponent 1.1. For some narrower scale range, the dependence $K(l) \sim l^{4/3}$ is fulfilled better (Figure 1.5) with different proportionality factors, as it follows from the general diagram in Figure 1.4. For some intermediate scale range, approximation of experimental points by power dependences with other exponents is possible. For example, Joseph and Sendner [129] suggested using a linear relation of K with l. They showed that in a number of cases a better agreement between calculation results and experimental data is achieved for this type of dependence than when other power expressions for $K(l)$ are used. It should also be mentioned that there is much agreement between the theory and experiment if assuming that K = const. Such an 'irregularity' in functional forms of $K(l)$ in the ocean, obviously, illustrates the complicated picture of energy supply of the ocean from external sources (wind, solar heat, tidal forces, etc.) as well as the complicated mechanisms of the redistribution of this energy among the motions of ocean waters of different scales which lead to the variety of the forms of the functions $K(l)$.

1.5. Diffusion Models with Transport Coefficients Depending on Time and the Phenomenon Scale

It was shown in the previous section that the transport coefficients, introduced into semi-empirical diffusion equations, significantly depend in the turbulent flow on the scale l of the phenomenon under consideration. If diffusion from localized sources is being studied then the dependence $K(l)$ can be presented as the variability of the coefficient K in time. Actually, at small sizes of the diffusing contaminant spot the effective transport coefficient must be small, too, since only smaller eddies participate in 'dragging apart' of the spot. But with the growth of the spot, i.e., with the increasing diffusion time, the diffusion rate (and, therefore, the diffusion coefficient) must increase. These phenomena are, naturally, most pronounced for the horizontal diffusion of the contaminant because the scale range in the horizontal directions in the ocean are several orders higher than in the vertical directions.

Attempts to take into consideration the dependences of diffusion coefficients on the phenomenon scale and time within the framework of the semi-empirical diffusion equation have been made by a number of authors. In 1958 Ozmidov [74] suggested using the horizontal coefficient of diffusion in the form (1.68) for calculating diffusion in the ocean from point sources of contaminants, with the radius vector of the observation point r (the contaminant source was placed in the origin of the coordinates) being considered the phenomenon scale l. With such a functional form of the coefficient $K_L(l)$ the increasing values of K_L were automatically taken into account along with the growth of the diffusing contaminant spot. A similar method of substituting the radius vector of the observation point for the phenomenon scale l was used in the same year by Joseph and Sendner [129], who presented $K_L(r)$ in the form of a linear function of r with the proportionality factor u_1 having the dimension of velocity (which was called the 'diffusion rate'). The work [143] considered the cases $K_L(r, t) = a_1 r^{4/3} t$ and $K_L(r, t) = a_2 r t$ and in [142] the formula $K_L(r, t) = a_3 t^2 r$, where a_1, a_2 and a_3 are some constants with different dimensions, was suggested. A more general case of using the horizontal coefficient of diffusion in the form

$$K_L(r, t) = a_4 r^n f(t) \tag{1.81}$$

was considered in [143], where a_4 is a constant, n is the exponent such that $0 \leqslant n \leqslant 2$ and $f(t)$ is some function of time. The solution of the horizontal-diffusion equation (1.76) with the horizontal-transport coefficient used in the form $K_L(r) = c_1 \varepsilon^{1/3} r^{4/3}$ was found in [74]. Substituting the expression for $K_L(r)$ into (1.76) and differentiating it we will arrive at

$$\frac{\partial \bar{c}}{\partial t} = c_1 \varepsilon^{1/3} \left[\frac{7}{3} r^{1/3} \frac{\partial \bar{c}}{\partial r} + r^{4/3} \frac{\partial^2 \bar{c}}{\partial r^2} \right]. \tag{1.82}$$

Changing the variables according to the formula

$$r^{1/3} = \rho, \tag{1.83}$$

we transform equation (1.82) to the form

$$\frac{\partial \bar{c}}{\partial t} = \frac{c_1 \varepsilon^{1/3}}{9} \left(\frac{5}{\rho} \frac{\partial \bar{c}}{\partial \rho} + \frac{\partial^2 \bar{c}}{\partial \rho^2} \right). \tag{1.84}$$

The solution of (1.84) is sought for in the form

$$\bar{c}(\rho, t) = m t^n \exp\left[-\frac{\rho^2}{4} \Big/ \left(\frac{c_1 \varepsilon^{1/3}}{9} t \right) \right], \tag{1.85}$$

where m and n are yet unknown constants.

Differentiating (1.85) and substituting it into (1.84) we find that $n = -3$. Therefore, the solution of equation (1.84) has the form

$$\bar{c}(\rho, t) = \frac{m}{t^3} \exp\left[-\frac{\rho^2}{4} \Big/ \left(\frac{c_1 \varepsilon^{1/3}}{9} t \right) \right]. \tag{1.86}$$

The constant m is defined from the condition of contaminant conservation in the diffusion process

$$Q = \int_0^\infty 2\pi r \bar{c}(r, t)\, dr, \tag{1.87}$$

where Q is the intensity of the instantaneous point source of the contaminant.

Substituting (1.86) into (1.87) and integrating it we find that

$$m = Q \Big/ \left[6\pi \left(\frac{4c_1\varepsilon^{1/3}}{9} \right)^3 \right]. \tag{1.88}$$

So the solution of equation (1.76) for the instantaneous point source with intensity Q, placed in the moment of time $t = 0$ into an infinite two-dimensional space at the point $x = y = 0$, has, in this case, the form

$$\bar{c}(r, t) = \frac{Q}{6\pi(4c_1\varepsilon^{1/3}/9)^3 t^3} \exp\left[-\frac{r^{2/3}}{4(c_1\varepsilon^{1/3}/9)\, t} \right]. \tag{1.89}$$

A substantial difference between the solution of (1.89) and that of the two-dimensional diffusion equation a constant coefficient (see the formula (1.13)) is the fast (proportional to t^{-3}) decrease in the contaminant concentration in the center of the spot $\bar{c}(0, t)$, that occurs using (1.84), while with a constant coefficient in the two-dimensional case $\bar{c}(0, t)$ subsides only proportionally to the first power of t. At the same time, the lowering concentration with the distance r from the origin of the coordinates occurs, according to (1.89), at a lower rate (the exponential factor is proportional to $r^{2/3}$) than follows from the solution of the diffusion equation with constant coefficient (here the argument of the exponential function is proportional to r^2).

The solution of the three-dimensional equation with the diffusion coefficient being proportional to the spacial radius vector of the observation point R in the power 4/3 can be obtained in a similar way [77]. For the three-dimensionally isotropic process, the diffusion equation in the spherical coordinates has the form

$$\frac{\partial \bar{c}}{\partial t} = \frac{1}{R^2} \left[\frac{\partial}{\partial R} K_L(R)\, R^2 \frac{\partial \bar{c}}{\partial R} \right]. \tag{1.90}$$

Substituting the expression for the transport coefficient into (1.90) we arrive at

$$\frac{\partial \bar{c}}{\partial t} = c_2\varepsilon^{1/3} \left(\frac{10}{3} R^{1/3} \frac{\partial \bar{c}}{\partial t} + R^{4/3} \frac{\partial^2 \bar{c}}{\partial R^2} \right). \tag{1.91}$$

Then, changing the variables according to (1.83) and assuming that the solution has the form (1.85), we similarly find that for the three-dimensional case,

$n = -9/2$. The condition of conservation of the contaminant in the diffusion process will have the following form:

$$Q = \int_0^\infty 4\pi R^2 \bar{c}(R, t)\, dR, \tag{1.92}$$

where Q is, again, the intensity of the instantaneous point source.

The final solution of the three-dimensional diffusion equation, with the coefficient given by (1.68), for the instantaneous point source of the contaminant placed in the moment $t = 0$ into the origin of the coordinates will be of the following form:

$$\bar{c}(R, t) = \frac{Q}{192\cdot 7!\pi^{3/2}\varepsilon^{3/2}\left(\frac{c_2 t}{9}\right)^{9/2}} \exp\left[-\frac{R^{2/3}}{4(c_2/9)\varepsilon^{1/3}t}\right], \tag{1.93}$$

where the constant c_2 may, generally speaking, differ from the constant c_1 for the two-dimensional case.

The contaminant concentration decreases in the case of three-dimensional diffusion rather quickly in the center of the spot: according to the formula (1.93), proportionally to the time in the power 9/2 (with a constant diffusion coefficient only proportionally to t in the power 3/2). At the same time, the decrease in the concentration with the increasing distance from the contaminant release point is again slower in this case than in the case of classical diffusion with a constant coefficient (the arguments of the exponential functions are proportional to $R^{2/3}$ and R^2, respectively).

The solution of the two-dimensional diffusion equation (1.76) with the transport coefficient in a more general form (1.81) was found later in [143]. It has the following form:

$$\bar{c}(r, t) = \frac{(2-n)Q}{2\pi(2-n)^{4/(2-n)}\Gamma[2/(2-n)]\, a_4^{2/(2-n)}[\psi(t)]^{2/(2-n)}} \times$$
$$\times \exp\left[-\frac{r^{2-n}}{(2-n)^2 a_4 \psi(t)}\right], \tag{1.94}$$

where Γ is the gamma function and $\psi(t) = \int_0^t f(t)\, dt$.

It is easy to see that the solutions of two-dimensional diffusion equation for different particular forms of the coefficient $K_L(r, t)$, given above, are included into the general solution (1.94). Table I.II presents certain characteristics of the solution of the equation for horizontal diffusion of the contaminant from instantaneous sources at various horizontal-diffusion coefficients $K_L(r, t)$.

Okubo [143] made a comparison of the relations presented in Table I.II with experimental results of diffusion observations in the ocean that were at his disposal. However, he failed to state a preference for any of the tabulated formulas based on such a comparison. This once again reflects the diversity of horizontal diffusion regimes in the ocean in different scales of space and time and under different conditions of the occurring processes.

TABLE I.II

Characteristics of two-dimensional contaminant diffusion from the point source with intensity Q at various forms of the diffusion coefficient $K_L(r, t)$

Author of solution	Dependence of contaminant concentration in the center of the spot $\bar{c}(0, t)$ on time t	Variations of the logarithms of normalized contaminant concentration along the radius vector of the spot $\ln[\bar{c}(r, t)/\bar{c}(0, t)]$
Classical diffusion equation with constant coefficient	$\frac{Q}{4\pi K_L} t^{-1}$	$-\frac{1}{4K_L t} r^2$
Joseph and Sendner [129]	$\frac{Q}{2\pi u_1^2} t^{-2}$	$-\frac{1}{u_1 t} r$
Okubo [143]	$\frac{Q}{\frac{3}{4}\pi^{3/2} a_1^3} t^{-3}$	$-\frac{1}{a_1^2 t^2} r^{4/3}$
Ozmidov [74]	$\frac{Q}{6\pi(c_1 \varepsilon^{1/3})^3} t^{-3}$	$-\frac{1}{c_1 \varepsilon^{1/3} t} r^{2/3}$
Okubo and Pritchard (see [143])	$\frac{Q}{\pi a_2^2} t^{-2}$	$-\frac{1}{a_2^2 t^2} r^2$
Obukhov [142]	$\frac{Q}{\pi a_3^3} t^{-3}$	$-\frac{1}{a_3^3 t^3} r^2$

Naturally, in the case of horizontal diffusion, diffusion in the vertical direction must significantly influence the contaminant concentration field. Admittedly, because of the substantial differences in the values of the vertical and horizontal diffusion coefficients these two processes can often be considered separately. This may be possible when the studied process has got a large scale in the horizontal direction as well as in the cases when the contaminant concentration gradients in the vertical direction or relative to the horizontal coordinates (e.g., at a uniform contaminant fall-out from the atmosphere over a large ocean area) are certainly small. Under large vertical density stratification of waters, when vertical transport is considerably retarded, it is possible in a number of cases to neglect the contaminant diffusion in the vertical direction.

In a more detailed form the question of interaction of horizontal and vertical diffusion processes was considered by Monin [66]. He suggested singling out four regimes of vertical diffusion: under the first one the vertical transport may be neglected completely; under the second one, the vertical-transport coefficient K_z is constant; under the third, this coefficient is linearly related to the vertical coordinate; and, finally, under the fourth regime, K_z is described by the "4/3 power law", i.e.,

$$(1)\ K_z = 0; \tag{1.95}$$

$$(2)\ K_z = \text{const}; \tag{1.96}$$

$$(3)\ K_z = \kappa u_* z; \tag{1.97}$$

$$(4)\ K_z \sim \varepsilon^{1/3} z^{4/3}, \tag{1.98}$$

where u_* is the friction velocity, and κ is Kármán's constant.

The first regime can be realized in the above cases; the second one, under considerable vertical density stratification of waters when the stratification scale L_0 is small; the third, under stratification of ocean waters close to neutral and with turbulence of the wall type; the fourth, under developed three-dimensional turbulence which, for instance, exists in the case of density convection in the upper layer of the ocean. As for the horizontal diffusion, it is always considered to be described by an equation with the coefficient K_L related to the scale of the phenomenon by the "4/3 power law". The case of three-dimensionally isotropic turbulence with such a diffusion coefficient as well as the case of purely horizontal diffusion were already considered above. For the second regime of diffusion, the fundamental solution for an instantaneous point source of contaminant with the intensity Q, placed at the moment $t = 0$ into the origin of the coordinates, must depend on the coordinates, time and the parameter ε. If there is an ocean current with the constant velocity V directed along the axis x then the solution should depend on x and y only in the combination $(x-Vt)^2+y^2$. In this case, the solution can be presented in the following form (under the condition of impenetrability of the ocean surface for the contaminant):

$$\bar{c}(x, y, z, t) = \frac{Q}{\varepsilon t^3} f_0\left(\frac{(x-Vt)^2+y^2}{\varepsilon t^3}\right) \frac{1}{\sqrt{\pi K_z t}} \cdot \exp\left(-\frac{z^2}{4K_z t}\right), \tag{1.99}$$

where f_0 is some universal function.

Under the third regime, the corresponding solution will have the form

$$\bar{c}(x, y, z, t) = \frac{Q}{\varepsilon t^3} f_0\left(\frac{(x-Vt)^2+y^2}{\varepsilon t^3}\right) \frac{1}{\kappa u_* t} \cdot \exp\left(-\frac{z}{\kappa u_* t}\right), \tag{1.100}$$

where f_0 is a universal function different, generally speaking, from the f_0 in the formula (1.99).

Interestingly, according to the formulas (1.89), (1.93), (1.99) and (1.100), the contaminant concentration in the center of the diffusing spot lowers with time for the four diffusion regimes above in accordance with the laws t^{-3}, $t^{-3.5}$, t^{-4} and $t^{-4.5}$; in other words, the more intensified the vertical mixing is, the faster it occurs.

Using the superposition principle it is not difficult to find general expressions for the contaminant concentration under the four diffusion regimes if the stationary contaminant source with the intensity q is placed into the point $x = y = z = 0$. We, correspondingly, arrive at

$$\bar{c}(x, y, z, t) = \frac{q\varepsilon}{V^4} f_1\left(\frac{\varepsilon r}{V^3}, \frac{x}{r}\right) \delta(z) \tag{1.101}$$

$$\bar{c}(x, y, z, t) = \frac{q\varepsilon^{3/2}}{K_z^{1/2} V^5} f_1\left(\frac{\varepsilon r}{V^3}, \frac{x}{r}, \frac{\varepsilon^{1/2} z}{K_z^{1/2} V}\right); \tag{1.102}$$

$$\bar{c}(x, y, z, t) = \frac{q\varepsilon^2}{u_* V^6} f_1\left(\frac{\varepsilon r}{V^3}, \frac{x}{r}, \frac{\varepsilon z}{u_* V}\right); \tag{1.103}$$

$$\bar{c}(x, y, z, t) = \frac{q\varepsilon^2}{V^7} f_1\left(\frac{\varepsilon R}{V^3}, \frac{x}{R}\right), \tag{1.104}$$

where f_1 are universal functions (all different, generally speaking), $\delta(z)$ is a delta function, $r = \sqrt{x^2 + y^2}$ and $R = \sqrt{x^2 + y^2 + z^2}$.

In this case, the lowering of the contaminant concentration in the point where the contaminant source is located is related, naturally, not to the diffusion time t but to the current velocity V, with the lowering of the concentration being proportional to V^{-4}, V^{-5}, V^{-6} and V^{-7}; i.e., its rate is increasing with the intensification of the vertical transport.

When there is no mean flow the continuous point source of contaminant will create a concentration field described under the above four diffusion regimes by the following expressions:

$$\bar{c}(x, y, z, t) = \frac{q}{\varepsilon^{1/3} r^{4/3}} f_2(0)\ \delta(z); \tag{1.105}$$

$$\bar{c}(x, y, z, t) = \frac{q}{K_z^{1/2} \varepsilon^{1/6} r^{5/3}} f_2(0)\ \delta(z); \tag{1.106}$$

$$\bar{c}(x, y, z, t) = \frac{q}{u_* r^2} f_2\left(\frac{\varepsilon^{1/3} z}{u_* r^{2/3}}\right); \tag{1.107}$$

$$\bar{c}(x,\ y,\ z,\ t) \ = \ \frac{q}{\varepsilon^{1/3}R^{7/3}}\ f_2(0), \tag{1.108}$$

where f_2 are universal functions (all different, generally speaking).

In this case, the contaminant concentration on the ocean surface will decrease with the distance from the source according to the laws $r^{-4/3}$, $r^{-5/3}$, $r^{-6/3}$ and $r^{-7/3}$, respectively; i.e., the more intensified the vertical transport is the faster it will occur. It should be mentioned that in the case of two-dimensional diffusion with a constant transport coefficient the stationary point source of the contaminant creates, after a long period of activity, an infinite contaminant concentration, while in the case of the diffusion with the variable coefficient K_L this phenomenon does not occur – the source may act as long as possible without causing infinite concentrations (except for the source point itself) since in this case diffusion is capable of 'dragging' the contaminant away from the source fast enough and for large distances.

It is not difficult to derive from formulas (1.106) – (1.108) the expressions for the vertical distribution of the contaminant concentrations under the stationary source. Suffice it only to come in (1.106) – (1.108) to the limit $r \longrightarrow 0$ and we will arrive at the expressions

$$\bar{c}(0,\ 0,\ z) \ = \ \text{const}\ \frac{qK_z^2}{\varepsilon z^5}\ ; \tag{1.109}$$

$$\bar{c}(0,\ 0,\ z) \ = \ \text{const}\ \frac{qu_*^2}{\varepsilon z^3}\ ; \tag{1.110}$$

$$\bar{c}(0,\ 0,\ z) \ = \ \text{const}\ \frac{q}{\varepsilon^{1/3}z^{7/3}}\ . \tag{1.111}$$

Here the contaminant concentration falls with the depth according to laws z^{-5}, z^{-3} and $z^{-7/3}$; i.e., the more intensified the vertical transport is the slower it declines.

All the formulas given above right now naturally can and do find their use when calculating the contaminant-concentration fields in the ocean. However, as was truly noted in [66], there are, as a rule, more complicated conditions in nature for the diffusion processes. In particular, the field of the mean current velocity in the ocean can very rarely be considered homogeneous and stationary. And the current velocity gradients and the current velocity variability in time can significantly complicate the diffusion picture. The next part of the book is devoted to the influence of mean velocity field space-time variability on the contaminant diffusion.

1.6. Contaminant Diffusion in a Non-Homogeneous and Non-Stationary Mean Current Velocity Field

The mean components of current velocity u, v and w in the turbulent diffusion equation with advective terms (1.26) can, generally speaking, be functions of coordinates and time. Obviously, such a variability of the velocity field must significantly influence the distribution of the concentration of the contaminant diffusing in the flow. Taylor, studying the contaminant transport in pipes [162, 163], was the first to draw attention to the effect of the velocity gradient (shear) on contaminant diffusion. The joint effect of vertical diffusion and velocity shear leads to a quick elongation of the contaminant spot in the direction of the flow. This phenomenon can be described by introducing some effective coefficient of horizontal diffusion K_L^{Eff}. The analysis is based on the stationary equation with the terms describing vertical diffusion and horizontal advection, with the horizontal velocity of the current (directed along the axis x) being assumed as a function of the vertical coordinate z. If we average such a diffusion equation from the surface to the seabed ($z = H$), then a new term will appear in the equation which can be regarded as an additional one to the horizontal transport of the contaminant. It turns out that the coefficient K_L^{Eff}, which is introduced in this way, is inversely proportional to the vertical diffusion coefficient K_z. The expression for K_L^{Eff} for the simple forms of the function $\bar{u}(z)$ was in [154], and the one for the arbitrary profile of $\bar{u}(z)$ in [116]. A rather simple and clear-cut model of the phenomenon was suggested in [148], where the sequence of deformation events for the vertical 'column' of the contaminant by the velocity gradient is considered followed by the 'smearing' of such a deformed column over the thickness of the flow by the vertical diffusion. The contaminant concentration after an n number of these events is distributed in the elementary unit cells along the axis x according to a binominal law. When n approaches infinity, the binominal distribution tends to the normal one with the variance $nV\tau$, where V is the current velocity averaged over depth and τ is the characteristic time of the vertical diffusion of the contaminant in a layer of thickness H. The derivative of the variance with respect to time appears to be equal to $V^2\tau/2$ as n approaches infinity. τ, in turn, has the form $\tau = H^2/(\pi^2 K_z)$ (this expression follows from the solution of an equation like (1.12) having, for the limited space, the following form: $\bar{c}(z, t) = \cos(\pi z/H)\exp(-\pi^2 K_z t/H^2)$). Further, considering that the transport coefficient can be defined as 1/2 of the derivative of the variance of the contaminant concentration distribution with respect to time, we finally arrive at

$$K_L^{\mathrm{Eff}} = V^2 H^2/2\pi^2 K_z. \tag{1.112}$$

For other models of the phenomenon, the expression (1.122) may be somewhat altered (for example, in [154] it was found that $K_L^{\text{Eff}} = \overline{V}^2H^2/(30K_z)$); however, the inverse dependence of K_L^{Eff} on K_z remains valid.

In the cases of rather easy forms of the functions $\bar{u}(x, y, z)$, $\bar{v}(x, y, z)$ and $\bar{w}(x, y, z)$ the diffusion equation (1.26) can be solved analytically. For example, Okubo [144] gave consideration to the case of the two-dimensional diffusion equation where horizontal components of the current velocity were dependent on the coordinates in the following way:

$$\bar{u}(x, y) = \alpha x + (h - \eta)y; \tag{1.113}$$

$$\bar{v}(x, y) = (h + \eta)x - \alpha y, \tag{1.114}$$

where

$$\alpha = \left(\frac{\partial \bar{u}}{\partial x}\right)_0 = -\left(\frac{\partial \bar{v}}{\partial y}\right)_0; \; h = \frac{1}{2}\left[\left(\frac{\partial \bar{v}}{\partial x}\right)_0 + \left(\frac{\partial \bar{u}}{\partial y}\right)_0\right];$$

$$\eta = \frac{1}{2}\left[\left(\frac{\partial \bar{v}}{\partial x}\right)_0 - \left(\frac{\partial \bar{u}}{\partial y}\right)_0\right], \tag{1.115}$$

and the lower index indicates the constancy of the corresponding derivatives for the entire flow. In that case, the diffusion equation had the form

$$\frac{\partial \bar{c}}{\partial t} + [\alpha x(h - \eta)y]\frac{\partial \bar{c}}{\partial y} + [(h + \eta)x - \alpha y]\frac{\partial \bar{c}}{\partial y} = K_L\left(\frac{\partial^2 \bar{c}}{\partial x^2} + \frac{\partial^2 \bar{c}}{\partial y^2}\right). \tag{1.116}$$

The solution of equation (1.116) for an instantaneous point source with the intensity Q placed at the moment $t = 0$ into the origin of the coordinates was obtained in the following form:

$$\bar{c}(x, y, t) = \frac{Q}{4\pi K_L P(t)} \exp\left[-\frac{M_1(t)x^2 + 2M_2(t)xy + M_3(t)y^2}{4 K_L P^2(t)}\right], \tag{1.117}$$

where

$$P(t) = \left\{\frac{h^2 + \alpha^2}{(h^2 + \alpha^2 - \eta^2)^2} \text{sh}^2[(h^2 + \alpha^2 - \eta^2)^{1/2}t] - \right.$$

$$\left. - \frac{\eta^2}{h^2 + \alpha^2 - \eta^2} t^2\right\}^{1/2}; \tag{1.118}$$

$$M_1(t) = \frac{h^2 + \alpha^2 + h\eta}{2(h^2 + \alpha^2 - \eta^2)^{3/2}} \operatorname{sh}[2(h^2 + \alpha^2 - \eta^2)^{1/2}t] - $$
$$- \frac{h\eta + \eta^2}{h^2 + \alpha^2 - \eta^2} t - \frac{\alpha}{h^2 + \alpha^2 - \eta^2} \operatorname{sh}(h^2 + \alpha^2 - \eta^2)^{1/2}t; \quad (1.119)$$

$$M_2(t) = \frac{\alpha\eta}{2(h^2 + \alpha^2 + \eta^2)^{3/2}} \operatorname{sh}[2(h^2 + \alpha^2 - \eta^2)^{1/2}t] - $$
$$- \frac{\alpha\eta}{h^2 + \alpha^2 - \eta^2} t + \frac{h}{h^2 + \alpha^2 - \eta^2} \operatorname{sh}^2(h^2 + \alpha^2 - \eta^2)^{1/2}t; \quad (1.120)$$

$$M_3(t) = \frac{h^2 + \alpha^2 - h\eta}{2(h^2 + \alpha^2 - \eta^2)^{3/2}} \operatorname{sh}[2(h^2 + \alpha^2 - \eta^2)^{1/2}t] + $$
$$+ \frac{h\eta - \eta^2}{h^2 + \alpha^2 - \eta^2} t + \frac{\alpha}{h^2 + \alpha^2 - \eta^2} \operatorname{sh}^2(h^2 + \alpha^2 - \eta^2)^{1/2}t. \quad (1.121)$$

The solution (1.117) can be reduced to simpler fields of the mean current velocity. If, for example, $\alpha = h = 0$ and $\eta \neq 0$ (the fluid rotates as a solid body), then (1.117) is reduced to the classical solution (1.13) (certainly, substituting K_L for D). But if, however, $h = \eta = 0$ and $\alpha > 0$ (extensional deformation), then the solution takes the form

$$\bar{c}(x, y, t) = \frac{Q}{2\pi \frac{K_L}{\alpha}[(e^{2\alpha t} - 1)(1 - e^{-2\alpha t})]^{1/2}} \times$$
$$\times \exp\left[- \frac{x^2}{2\pi \frac{K_L}{\alpha}[(e^{2\alpha t} - 1)} - \frac{1}{2\pi \frac{K_L}{\alpha}[(e^{-2\alpha t} - 1)} \right]. \quad (1.122)$$

The expression (1.122) was obtained earlier by Townsend [164].

The third case, analyzed by Okubo, when $\alpha = 0$ and $h = -\eta > 0$, is close to the one considered earlier by Novikov [71]. He found the solution to the three-dimensional diffusion equation of the form

$$\frac{\partial \bar{c}}{\partial t} = (u_0 + a_1 z)\frac{\partial \bar{c}}{\partial x} + v_0 \frac{\partial \bar{c}}{\partial y} - w_0 \frac{\partial \bar{c}}{\partial z}$$
$$= K_z \frac{\partial^2 \bar{c}}{\partial z^2} + K_L\left(\frac{\partial^2 \bar{c}}{\partial x^2} + \frac{\partial^2 \bar{c}}{\partial y^2}\right), \quad (1.123)$$

where u_0, v_0, w_0, K_z, K_L and a_1 are constants.

The fundamental solution of equation (1.123), found by Novikov, can be presented in the following form:

$$\bar{c}(x, y, z, t) = \frac{Q}{(4\pi t)^{3/2}\sqrt{K^* K_L^2}} \exp\left[-\frac{(x - u_0 t - \frac{1}{2}a_1 t z)^2}{4K^* t} - \frac{(y - v_0 t)^2 + (z - w_0 t)^2}{4K_L t}\right], \tag{1.124}$$

where

$$K^* = K_z + \tfrac{1}{12}K_L a_1^2 t_2^2. \tag{1.125}$$

All the solutions given right above of diffusion equations with variable velocity components include, as is easy to see, quadratic forms relative to the variables x, y and z. This means that isolines of contaminant concentrations in the diffusing spots have an elliptical form. The main axes of such ellipses do not, generally speaking, coincide with the coordinate axes, with the ellipses being capable of turning in relation to the coordinate system with time. Variances of concentration distributions in spots also change with time in the directions of the main axes of the ellipses (σ_x^2, σ_y^2, σ_z^2). Therefore, the spots not only turn in the process of diffusion, but they may get 'flattened' or elongated in various directions. The intensity of these processes, naturally, depends on the governing parameters of the solution. The characteristic time t_{cr}, from which the influence of velocity gradients on the process becomes significant, also depends on them. For example, for an equation of the following form:

$$\frac{\partial \bar{c}}{\partial t} + (u_0 + a_1 z)\frac{\partial \bar{c}}{\partial x} + (v_0 + a_2 z)\frac{\partial \bar{c}}{\partial y} = K_z\frac{\partial^2 \bar{c}}{\partial z^2} + K_L\left(\frac{\partial^2 \bar{c}}{\partial x^2} + \frac{\partial^2 \bar{c}}{\partial y^2}\right), \tag{1.126}$$

time t_{cr} is determined by the equation:

$$t_{cr} = \sqrt{\frac{12}{\frac{K_z}{K_L}(a_1^2 + a_2^2)}}, \tag{1.127}$$

and the angle α between the larger axis of the ellipse of concentration isolines and the axis x by the formula

$$\text{tg}\,\alpha = \frac{2a_1 a_2}{a_1^2 + a_2^2}. \tag{1.128}$$

For the ellipse semi-axes ratio λ we in turn can obtain

$$\lambda = \sqrt{1 + \tfrac{1}{12}(a_1^2 + a_2^2)\frac{K_z}{K_L}t^2}. \tag{1.129}$$

Analytical solutions of the diffusion equation can also be found in the case of the dependence of the components of mean current velocity on time. For instance, Okubo [145] considered the case in which one of the velocity components is a harmonic function of time:

$$\frac{\partial \bar{c}}{\partial t} + \left[\bar{u}_s(z) + \bar{u}_t(z) \sin \frac{2\pi t}{T}\right] \frac{\partial \bar{c}}{\partial x} = K_L \frac{\partial^2 \bar{c}}{\partial x^2} + K_z \frac{\partial^2 \bar{c}}{\partial z^2}, \tag{1.130}$$

where $\bar{u}_s(z)$ and $\bar{u}_t(z)$ are linear functions of the coordinate z and T is the period of velocity variability.

The analysis of the solution of equation (1.130) showed that in an unlimited space and after a sufficiently long diffusion time the shear effect of the stationary current becomes the prevailing factor as compared with the shear in the periodical flow. For the depth-limited ocean, the relation of the time τ_0 necessary for the mixing of the contaminant throughout the ocean body to the oscillation period T becomes important. If $\tau_0 \gg T$, the velocity shear in the periodical current does not practically, influence the contaminant distribution. If τ_0 is comparable to T, the contribution of this shear to the transport of the contaminant becomes comparable to that of the velocity displacement in the stationary flow.

The stationary diffusion equation with the velocity component $\bar{u}$ and diffusion coefficients K_z and K_L, dependent on the coordinates, was studied by V.I. Goroshko [21]. He presented the velocity component as a power function of the following form:

$$\bar{u}(y, z) = \bar{u}_1 z^{\alpha} y^{\tau}, \quad \alpha, \tau > 0, \tag{1.131}$$

and the transport coefficients as the functions

$$K_z = k_1 y^{1-\beta} z^{\alpha}; \quad K_L = k_2 z^{1-\beta} y^{\tau}, \tag{1.132}$$

where $\bar{u}_1$, α, τ, β, k_1 and k_2 are some constants.

The diffusion equation

$$\bar{u} \frac{\partial \bar{c}}{\partial x} = \frac{\partial}{\partial y}\left(K_L \frac{\partial \bar{c}}{\partial y}\right) + \frac{\partial}{\partial z}\left(K_L \frac{\partial \bar{c}}{\partial z}\right) \tag{1.133}$$

was solved for a basin with depth h and a straight shore-line directed along the axis x. Contaminant fluxes through the free surface of the basin, the bottom and the side wall were neglected. The origin of the coordinates was placed on the shoreline at the water's edge, with the axis y directed perpendicular to the shoreline toward the basin. The boundary conditions of the problem were as follows:

$$\left(K_z(y,\ z)\ \frac{\partial \bar{c}(x,\ y,\ z)}{\partial z}\right)_{z \to 0} = 0; \tag{1.134}$$

$$K_z(y,\ h)\ \frac{\partial \bar{c}(x,\ y,\ h)}{\partial z} = 0; \tag{1.135}$$

$$\left(K_L(y,\ z)\ \frac{\partial \bar{c}(x,\ y,\ z)}{\partial y}\right)_{y \to 0} = 0; \tag{1.136}$$

$$\bar{c}(x,\ y,\ z)_{x, y \to \infty} = 0. \tag{1.137}$$

The continuous point source of the contaminant with the intensity q was placed on the axis y in the ocean at depth z_0 and at the distance y_0 from the shoreline.

The solution of equation (1.133) with the coefficients given by the formulas (1.131) – (1.132) and with the boundary conditions (1.134) – (1.137) was found in the following form:

$$\begin{aligned}\bar{c}(x,\ y,\ z) \ = \ & \frac{a y^{\beta/2}}{x} \exp\left(-\ \frac{y^{2\sigma} + y_0^{2\sigma}}{2bx}\right) I_{-\beta/(2\sigma)}\left[\frac{(y\ y_0)^{\sigma}}{ax}\right] \times \\ & \times \left\{1 + l z^{3/2} \sum_{k=1}^{\infty} J_{-\varepsilon/(2\lambda)}\left[\sigma_k\left(\frac{z_0}{h}\right)\right]^{\lambda} J^{-2}_{-\varepsilon/(2\lambda)}(\sigma_k)\ J_{-\varepsilon/(2\lambda)} \times \right. \\ & \left. \times \left[\sigma_k\left(\frac{z}{h}\right)\right]^{\lambda} \exp(-m\sigma_k^2 x)\right\},\end{aligned} \tag{1.138}$$

where

$$a = \frac{q\,(\alpha + 1) y^{\beta/2}}{(\beta + \tau + 1) k_1 h^{\alpha+1}}; \quad \sigma = \frac{\beta + \tau + 1}{2}; \quad b = \frac{2k_1\sigma^2}{\bar{u}_1};$$

$$l = \frac{(\varepsilon + \alpha + 1) z_0^{\varepsilon/2}}{(\alpha + 1) h^2}; \quad \lambda = \frac{\varepsilon + \alpha + 1}{2}; \quad m = \frac{\lambda k_2}{\bar{u}_1 h^{\varepsilon+\alpha+1}};$$

σ_k are positive roots of the equation $J(\alpha+1)/(\varepsilon+\alpha+1)(\delta) = 0$; J is the Bessel function; and I is a modified Bessel function of the first kind.

Figure 1.7 gives, as an example, normalized distribution curves for contaminant concentration, according to (1.138), for a contaminant source placed at a distance of 45 m from the shoreline. It follows from the figure that with an increasing longitudinal coordinate x the maximal contaminant concentration in the transverse direction gradually approaches the shore. This is related to the increasing transport coefficients with the enlargement of the coordinate y. It is also interesting that the shoreline ($y = 0$) does not, practically, get sub-

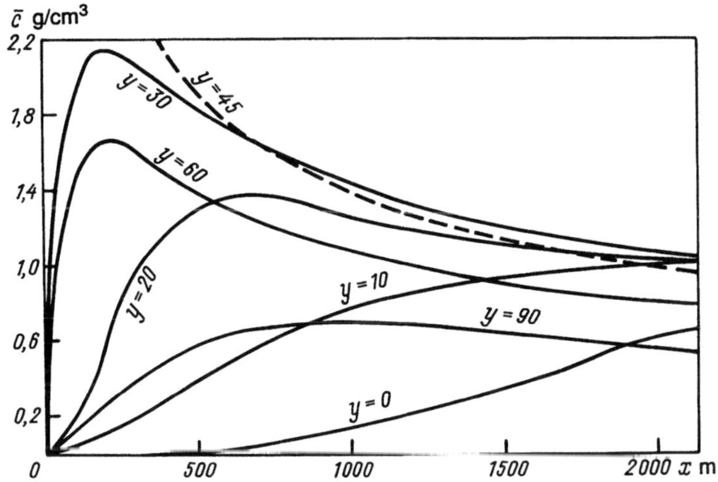

Fig. 1.7. Distribution of contaminant concentration along the longitudinal axis x at variations by the distance from the shoreline y (after [21]). y-axis: $\bar{c}$ g/cm^3; x-axis: x m.

jected to the pollution at distances up to 500 m downstream from the location of the source of a contaminant. Naturally, this specific behavior of the contaminant concentration can be used to optimize the location of sewage-water discharge systems in order, for instance, to prevent the pollution of certain seashore areas.

Despite the complexity of the problem as stated in (1.133) – (1.137), it still remains very much idealized. In reality, of course, the shoreline is rarely straight, the depth of the sea, even approximately, constant, and the flow velocity rarely invariable. Taking into account these and other factors influencing the process, naturally leads to diffusion equations that are impossible to solve analytically. In such cases, we have to use numerical methods which will be dealt with in more detail below.

1.7. Diffusion of Discrete Particles in the Turbulent Flow

A diffusion equation of the type (1.26) can be derived not only from the balance conditions for the diffusion contaminant but also from statistical considerations of the motion of diffusing particles in the flow. Let us denote the probability of a particle coming from the point $|M_0|$ at the moment t_0 to be present in the moment $t > t_0$ in the small vicinity of the point M as

$$P(M, t; M_0, t_0)\ \Delta V, \tag{1.139}$$

where ΔV is an element of space. Since the presence of the particle inside the entire space V is a reliable event the function (1.139) satisfies the following normalization condition:

$$\int_V p(M, t; M_0, t_0) \, dV = 1. \tag{1.140}$$

If a large number (N) of particles, not interacting with each other, come from the point M_0 over a short period of time Δt, then the concentration of such particles at $\Delta t \to 0$ at the point M at the moment t will be described by the expression (1.139), provided the mass of all the particles that have left the point M_0 equals one. This establishes a relation between the notion of contaminant concentration in a point of space and the probability for one particle, emitted from the source, of being in the vicinity of that point. If the initial concentration of particles in the moment of time t_0 equals $\varphi(M)$ then the concentration $c(M, t)$ of these particles in the moment $t > 0$ will be equal to

$$c(M, t) = \int_{V_p} p(M, t; P, t_0) \, \varphi(P) \, dV_p, \tag{1.141}$$

where P is the integration variable.

The following relation follows from equation (1.141):

$$p(M, t; M_0, t_0) = \int_{V_p} p(M, t; P, \theta) \, p(P, \theta; M_0, t_0) \, dV_p, \tag{1.142}$$

It is true for every moment of time $t_0 < \theta < t$. Equation (1.142) is called the Einstein – Kolmogorov equation.

Under certain conditions imposed on the function $p(M, t; M_0, t_0)$. equation (1.142) can be transformed into a partial differential equation of the diffusion type. The first condition is imposed on the mean velocity of the diffusing particles. If, for the sake of simplicity, we consider the case of one-dimensional motion of particles (along the axis x), this condition will have the form:

$$\lim_{\tau \to 0} \frac{\overline{x - \xi}}{\tau} = \lim \frac{1}{\tau} \int_L (x - \xi) \, p(x, t + \tau; \xi, t) \, d\xi = A(x, t), \tag{1.143}$$

where ξ is the coordinate of the particle at the initial moment of time; x is the coordinate on the expiration of the time τ; $A(x, t)$ is some bounded function; and L is the integration domain.

The second condition has the form

$$\lim_{\tau \to 0} \frac{\overline{(x - \xi)^2}}{\tau} = \lim \frac{1}{\tau} \int_L (x - \xi)^2 \, p(x, t + \tau; \xi, t) \, d\xi = 2K(x, t), \tag{1.144}$$

where $K(x,t)$ is some bounded function. The condition (1.144) means a linear dependence of the mean square of the particle displacement on time at small values of τ.

The third condition is imposed on the mean cube of the particle displacement and has the following form:

$$\lim_{\tau\to 0} \frac{\overline{|x-\zeta|^3}}{\tau} = \lim \frac{1}{\tau} \int_L |x-\xi|^3\, p(x,\, t+\tau;\, \xi,\, t)\mathrm{d}\xi = 0. \tag{1.145}$$

If we further multiply both parts of equation (1.142) by an arbitrary function $\psi(z)$, turning it into zero together with its derivative at the boundaries of the integration domain, integrate the obtained equation over the entire domain and then expand into the Taylor series with respect to the shear $(x-\xi)$, we can, using the above properties of the function $p(M,t;\,M_0t_0)$, arrive at the expression

$$\int_L \psi(x)\,[\partial p/\partial t + \partial(Ap)/\partial x - \partial^2(Kp)/\partial x^2]\,\mathrm{d}x = 0. \tag{1.146}$$

Since the function $\psi(x)$ is an arbitrary one the relation (1.146) must be held when the expressions in brackets in the integrand equals zero. This gives the differential form of the Einstein – Kolmogorov equation:

$$\partial p/\partial t = -\partial(Ap)/\partial x + \partial^2(Kp)/\partial x^2. \tag{1.147}$$

Equation (1.147) may be presented in the following form:

$$\frac{\partial p}{\partial t} = \frac{\partial}{\partial x}\left(K\frac{\partial p}{\partial x}\right) + \alpha\frac{\partial p}{\partial x} + \beta p, \tag{1.148}$$

where $\alpha = -A + \partial K/\partial x$ and $\beta = -\partial A/\partial x + \partial^2 K/\partial x^2 = \partial\alpha/\partial x$.

As is seen from (1.148), K has the physical meaning of the diffusion coefficient. If the diffusion process is homogeneous in space and time, i.e., the function p depends only on the differences $x-x_0$ and $t-t_0$, the coefficients A and K turn out to be constants. But if the function p depends only on the modulus of particle displacement $|x-\xi|$ then $A = 0$, and equation (1.148) is reduced to the simplest diffusion equation

$$\partial p/\partial t = K\partial^2 p/\partial x^2, \tag{1.149}$$

where the probability density for the presence of the particle at a given point of space and at given moment of time serves as an unknown quantity instead of the contaminant concentration, and the diffusion coefficient is determined according to the formula (1.144).

It follows from the formula (1.144) that the diffusion coefficient, determined according to it, is, generally speaking, a function of time. It is possible to specify this function if the variability of the particle-displacement variance $\sigma_x^2 = \overline{[x(t)-x(t_0)]}$ in the time is known. The displacement of the particle along the axis x over the time t can apparently be determined from the expression

$$x = \int_0^t u(\alpha)\, d\alpha, \tag{1.150}$$

where u is the particle-velocity component and α is the integration variable. If we assume that there is no mean velocity of the flow, the fluctuation velocity u' is substituted for the complete velocity u in the expression (1.150). After squaring and averaging (1.150) we arrive at the relation

$$\overline{x^2} = \overline{\left[\int_0^t u'(\alpha)\, d\alpha\right]^2} = \int_0^t\int_0^t \overline{u'(\alpha_1)\, u'(\alpha_2)}\, d\alpha_1\, d\alpha_2, \tag{1.151}$$

where α_1 and α_2 are integration variables.

If we divide $\overline{u'(\alpha_1)u'(\alpha_2)}$ by the mean square of the momentum velocity $\overline{u'^2}$ (it is assumed that $\overline{u'^2}$ does not depend either on the coordinate or time), in the integrand we will obtain the expression for the particle-velocity correlation coefficient (the Lagrangian correlation coefficient):

$$R_u(\alpha_2-\alpha_1) = \overline{u'(\alpha_1)\, u'(\alpha)}\,/\overline{u'^2}. \tag{1.152}$$

Considering (1.152), we transform (1.151) into the form

$$\overline{x^2} = \overline{u'^2} \int_0^t\int_0^t R_u(\alpha_2-\alpha_1)\, d\alpha_1\, d\alpha_2. \tag{1.153}$$

If the correlation coefficient depends only on the difference between times and not on the times themselves, then, considering that $R_u(\tau)$ is an even function, we will arrive at the formula obtained by Taylor [161]:

$$x^2 = \overline{2u'^2} \int_0^t d\alpha_2 \int_0^{\alpha_2} R_u(\alpha_1)\, d\alpha_1, \tag{1.154}$$

or at the one derived by Kampè de Fèriet (see [104]):

$$\overline{x^2} = \overline{2u'^2} \int_0^t (t-\alpha) R_u(\alpha)\, d\alpha. \tag{1.155}$$

The diffusion coefficient $K_x(t)$ now may be presented, according to the formula (1.144), as follows:

$$K_x(t) = \overline{u'^2} \int_0^t R_u(\alpha)\, d\alpha. \tag{1.156}$$

Similarly we can obtain expressions for the diffusion coefficients along the coordinate axes y and z.

The integral

$$T_u = \int_0^\infty R_u(\alpha)\, d\alpha \tag{1.157}$$

has the dimension of time and characterizes the time interval at which the velocities of particles are correlated. T_u is called the Lagrangian time scale of turbulence.

If the diffusion time t is large as compared with T_u, the formula (1.155) may, approximately, be presented in the form

$$\overline{x^2} \approx \overline{2u'^2}\, T_u t - \overline{2u'^2} \int_0^\infty \alpha R_u(\alpha)\, d\alpha. \tag{1.158}$$

In this equation, the second term on the right-hand side is a constant, thus at larger values of t the first term begins to prevail over the second one, and we can arrive at the approximation

$$\overline{x^2} = \overline{2u'^2}\, T_u t, \tag{1.159}$$

and for the diffusion coefficient at $t \gg T_u$ we obtain

$$K_x = \overline{u'^2}\, T_u; \tag{1.160}$$

i.e., in this case the diffusion coefficient becomes a constant, and the turbulent-diffusion equation becomes analogous to the molecular-diffusion equation but with the coefficient presented by the expression (1.160). There is also an analogy between the free path length l_m in the theory of molecular diffusion and the value with the dimension of length $T_u\sqrt{\overline{u'^2}}$ in the case of the turbulent flow of a fluid (see the formula (1.6)). In both cases, these values characterize the distance at which the correlation between the motions of the particle (molecule) becomes lost. But if the particle has not yet 'gone' outside the scale of correlation, then the diffusion coefficient must be a function of the diffusion time. At short diffusion times, the function $R_u(\tau)$ can be expanded into a series with respect to time. If we neglect all the terms of the orders higher than two we will arrive at

$$\overline{x^2} \approx \overline{u'^2}\, t^2,$$

and in that case the diffusion coefficient will be a linear function of time.

The dependence of the diffusion coefficient on time can generally be found if the form of the function $R_u(\tau)$ is known. For various types of turbulent flows the function $R_u(\tau)$ (as well as the coefficients of correlation along the axes y and z: $R_v(t)$ and $R_w(t)$) was determined based on the empirical data or on certain theoretical considerations. Various analytical expressions for $R_u(t)$ have been suggested. For example, Taylor [161] suggested that $R_u(t) = e^{-\tau/\tau_1}$, where τ_1 is some scale of time. Later Sutton [157] proposed to approximate the function $R_u(\tau)$ by the expression

$$R_u(t) = (1-\tau/\tau_0)^n,$$

where τ_0 is the scale of time and the exponent n may change depending, for example, on the density stratification of the fluid.

If we consider the diffusion in the turbulent flow not of individual particles but of pairs then a number of specific expressions for the variance of the distance between these particles as well as for the coefficient of their relative diffusion can be obtained. The problem of relative diffusion of particles is more complicated than that of an individual particle because the relative diffusion depends not only on the turbulence characteristics of the fluid but also on the initial separation of the particles. If the distance between the particles is not large then only small-scale turbulent eddies (small-scale velocity fluctuations) participate in their 'dragging apart' while larger eddy formations transport the particles as a whole. That is why relative diffusion of particles is an accelerating process, with this acceleration occurring until the particles get separated by such a distance at which the correlation between them becomes lost (i.e., there will be no more turbulent eddies of this scale in the flow). The relative diffusion parameters are defined, as Batchelor showed [114], generally as mutual correlations between the motions of the particles at different moments of time. At the final stage of diffusion, when the particles' motion become independent, their mutual variance will simply be equal to the doubled variance of the motion of the individual particles (i.e., to the doubled variance values given by the formulas (1.153 – 1.155). We can also specify the expressions for mutual variance of the pair of particles in the case in which the distance between the particles lies within the range of validity of the locally isotropic turbulence theory given by Kolmogorov. In this range, all the characteristics of the processes may, as is known, only depend on the dissipation rate of the turbulent energy ε, kinematic fluid viscosity ν and, naturally, on the particle separation distance l itself and on the diffusion time t. If the initial separation of the particles l_1 is much greater than the Kolmogorov scale (1.70), then ν must fall out of consideration. In that case, i.e., in the inertial scale range, the rate of variation of the mutual variance of the particles $\overline{dx^2}/dt$ (doubled relative diffusion coefficient) can be expressed by the following formulas:

(1) for short diffusion times, when the importance of the initial separation of particles is still insignificant,

$$\overline{dx^2}/dt \sim (\varepsilon l_1)^{2/3}\, t; \tag{1.161}$$

(2) for the case when the diffusion time is long enough and the particles have 'forgotten' about their initial separation (therefore, the parameter l_1 is omitted from the formula) we arrive at

$$\overline{dx^2}/dt \sim \varepsilon t^2. \tag{1.162}$$

The moment of time t^*, separating these two diffusion regimes, can also be determined from the dimensional considerations through the formula

$$t^* = l_1^{2/3}\varepsilon^{-1/3}. \tag{1.163}$$

In order to take into account the growth of the relative diffusion coefficient with the growth of l, Richardson [152] suggested presenting the coefficient of relative diffusion of particles F as the function of mutual separation of particles l. The diffusion equation in that case was presented by him for the density particle distribution function q according to their mutual distances in the following form:

$$\frac{\partial q(l,t)}{\partial t} = \frac{\partial}{\partial l}\left[F(l)\,\frac{\partial q(l,t)}{\partial l}\right]. \tag{1.164}$$

The transition from the function $q(l,t)$ to the usual concentration $\bar{c}(x,y,z,t)$ is rather complicated. In most cases, however, we can, as Richardson showed, let $F \approx 3K$. In the case of independent motion of diffusing particles $F = 2K$.

Having analyzed the atmospheric diffusion data, Richardson established that the coefficient $F(l)$ is related to the phenomenon scale l according to the formula

$$F(l) = kl^{4/3}, \tag{1.165}$$

which is identical to the expression for the '4/3 power law' (1.68) at $k = c_1\varepsilon^{1/3}$.

1.8. Concentration Fluctuations for the Diffusion Contaminant

It is not only the data on the mean concentration of the contaminant diffusion in the ocean or those on the dispersion of the contaminant spots and on the dispersions of the diffusing particles that may be of interest for practical purposes, but also the data on the fluctuating components of the contaminant concentration field themselves, $c'(x, y, z, t)$. For studying these fluctuations it is convenient to subtract the contaminant-transport equation written for the instantaneous concentrations and velocity components (1.10) from the averaged equation (1.20). It will result in

$$\frac{\partial c'}{\partial t} + u_i' \frac{\partial \bar{c}}{\partial x_i} + \overline{u_i} \frac{\partial c'}{\partial x_i} + \left(u_i' \frac{\partial c'}{\partial x_i} - \overline{u' \frac{\partial c'}{\partial x_i}} \right) = D\nabla^2 c', \tag{1.166}$$

where $i = 1, 2, 3$, as usual. Now, multiplying (1.166) by $2c'$, then averaging and reducing some terms with the help of the continuity equation for fluctuation velocities

$$\partial u_i'/\partial x_i = 0, \tag{1.167}$$

we will arrive at the equation for the variance of contaminant concentration fluctuations

$$\frac{\partial \overline{c'^2}}{\partial t} + \overline{u_i} \frac{\partial \overline{c'^2}}{\partial x_i} = 2\overline{u_i' c'} \frac{\partial \bar{c}}{\partial x_i} + \frac{\partial}{\partial x_i} \left(D \frac{\partial \overline{c'^2}}{\partial x_i} - \overline{u_i' c'^2} \right) - 2D \overline{\left(\frac{\partial c'}{\partial x_i} \right)^2}. \tag{1.168}$$

An equation, similar to (1.168), was obtained by Corrsin [119] for the variance of temperature fluctuations in turbulent flows. The form of equation (1.168) is also similar to that of the energy balance for velocity fluctuations in the turbulent flow. In this regard, it may be interpreted as a balance equation for the 'energy' of contaminant-concentration fluctuations. The variation of this 'energy' with time is stipulated, as follows from the equation, by its advection by the mean current (the second term in the left-hand side of the equation), by the transports due to molecular processes and turbulent-velocity fluctuations (the second-last term on the right), by the generation of $\overline{c'^2}$ under the influence of the turbulent flux of the contaminant in the field of its mean concentration gradient (the first term on the right) and, finally, by the 'dissipation' of $\overline{c'^2}$ (the term $2D \cdot \overline{(\partial c'/\partial x_i)^2} = \varepsilon_c$.

The components of the mean current velocity $\overline{u_i}$ as well as the mean contaminant concentration are considered to be known values in equation (1.168). But, as it is easy to see, in this case the equation remains unclosed since, besides the unknown quantity $\overline{c'^2}$, the unknown functions are the moments $\overline{u'c'}$ and $\overline{u'c'^2}$ (the fluxes of c' and c'^2) as well as ε_c. These values need, this way or another, to be expressed through the known functions. The most commonly used method, which was used a number of times above, is when the fluxes are considered proportional to the gradients of the mean concentration of the substance to be transported. Under this assumption,

$$\overline{u_i'c'} = -K\ \partial\bar{c}/\partial x_i; \tag{1.169}$$

$$\overline{u_i'c'^2} = -K_c \frac{\partial \overline{c'^2}}{\partial x_i}, \tag{1.170}$$

where K is a 'usual' turbulent diffusion coefficient, and K_c is the transport coefficient $\overline{c'^2}$, which is, generally speaking, different from K.

Under various simplifying assumptions about the velocity field (for example, in the case of its isotropy) ε_c may also be expressed through the unknown function $\overline{c'^2}$, and thus the closure of equation (1.168) is achieved. Naturally, some other methods can be used to close equation (1.168). Toompuu [102], for example, suggested expressing ε_0 in the following way:

$$\varepsilon_c = f(x)\ \overline{c'^2}, \tag{1.171}$$

where $f(x)$ is some universal function of coordinates.

Under the hypotheses (1.169) – (1.171) equation (1.168), for the case of cylindrical symmetrical diffusion and uniform current with velocity V in the direction of the axis x, will have the form

$$V\frac{\partial \overline{c'^2}}{\partial x} = 2K\left(\frac{\partial \bar{c}}{\partial r}\right)^2 + K_c\left(\frac{\partial^2 \overline{c'^2}}{\partial r^2} + \frac{1}{r}\frac{\partial \overline{c'^2}}{\partial r}\right) - f(x)\ \overline{c'^2}, \tag{1.172}$$

where $r = y^2 + z^2$. For a semi-space $z > 0$ and a source of unit intensity, the solution of Equation (1.172) has the form [102]

$$\overline{c'^2}(z, z_0, x) = \frac{1}{\pi\sigma_\sigma^2(x)}\frac{\varphi(x)}{\varphi(0)} \exp\left[-\frac{r^2 + r_0^2}{2\sigma_\sigma^2(x)}\right] J_0\left[\frac{rr_0}{\sigma_\sigma^2(x)}\right], \tag{1.173}$$

where $\varphi(x) = \exp[-1/V \int f(x)\mathrm{d}x]$; r_0 is the radius of the source; and σ_σ^2 is the spacial dispersion of σ_c^2 (with the dimension of cm^2), which is related to K_c in the following way:

$$K_c = \frac{V}{2}\frac{\mathrm{d}\sigma_\sigma^2}{\mathrm{d}x}. \tag{1.174}$$

If K_c is a constant then $\sigma_\sigma^2 = 2K_c x/V$. In that case, the value of the dispersion of the contaminant-concentration fluctuations will, according to (1.173), decrease in inverse proportion to the coordinate x and fall exponentially with the distance from axis x.

However, data on space-time variability of σ_c^2 is not enough for solving some of the diffusion problems. Often it is necessary to know the spectral composition of the contaminant concentration fluctuations. In that case, it is convenient to present $c'(x, y, z, t)$ as well as the fluctuations of velocity components $u_j'(x, y, z, t)$ in the form of the Fourier integrals:

$$C(\mathbf{k}, t) = \int c' \exp(i\mathbf{kx})\, \mathrm{d}\mathbf{x}; \tag{1.175}$$

$$U_j(\mathbf{k}, t) = \int u_j' \exp(i\mathbf{kx})\, \mathrm{d}\mathbf{x}; \tag{1.176}$$

where $\mathbf{k}$ is the wave vector, $\mathbf{x}$ is the space vector and $i = \sqrt{-1}$.

Now, performing the Fourier transformation of equation (1.166) it can be reduced to the following form:

$$\partial C(\mathbf{k}, t)/\partial t + i\int k_j' U_j(\mathbf{k}-\mathbf{k}')\, C(\mathbf{k}')\, \mathrm{d}k' = -Dk^2 C(\mathbf{k}, t). \tag{1.177}$$

Let us define the spectrum of the contaminant-concentration fluctuations $E_c(\mathbf{k}, t)$ as a mean product of the function $C(\mathbf{k}, t)$ and its complex-conjugate value $C^*(\mathbf{k}, t)$:

$$E_c(\mathbf{k}, t) = C(\mathbf{k}, t)C^*(\mathbf{k}, t); \tag{1.178}$$

and, correspondingly, let us define the mutual spectrum of velocity-component fluctuations as

$$E_{jk}(\mathbf{k}, t) = U_j(\mathbf{k}, t)U_k^*(\mathbf{k}, t). \tag{1.179}$$

An equation for the spectral function $E_c(\mathbf{k}, t)$ can be obtained from (1.177) by adding to it the complex-conjugate equation. Normally, the equation for $E_c(\mathbf{k}, t)$ appears unclosed since the non-linear term which describes the interaction between E_c and E_{jk} cannot, generally speaking, be expressed through these functions. (The mutual spectrum of velocity components E_{jk} is considered a known quantity in this case). The spectral equation for $E_c(\mathbf{k}, t)$ becomes closed if we assume that there is a small correlation between E_c and E_{jk}. In this case, the equation for the spectrum E_c takes the form

$$\partial E_c/\partial t + E_{jk} \int k_j k_k E_c \, \mathrm{d}\mathbf{k} = -2Dk^2 E_c. \tag{1.180}$$

However, the assumption of a small correlation between E_c and E_{jk} is, obviously, far from reality since the fluctuations of c must apparently be closely re lated with velocity-component fluctuations in the turbulent flow. More realistic is a hypothesis according to which the whole spectrum of turbulent eddies may be subdivided into two ranges, with smaller isotropic eddies participating in the diffusion of E_c and the larger ones transporting E_c as a whole ('advectively'). If we denote the wave number dividing these two ranges by k_0, the equation for the spectrum E_c will have the form

$$-\frac{\partial}{\partial t}\int_0^k E_c(k)\mathrm{d}k = \left[D + a\int_{k_0}^{\infty} (E(k)\, k^{-3})^{1/2}\, \mathrm{d}k\right] \int_0^{k_0} 2E_c(k)\, k^2 \mathrm{d}k, \tag{1.181}$$

where $E(k) = E_{ii}(k)$ and a is a numerical constant.

In order to solve equation (1.181) it is necessary to prescribe the energy spectrum of turbulent-velocity fluctuations $E(k)$. In the case of isotropic turbulence, the Kolmogorov spectrum can be used:

$$E(k) = c_1 \varepsilon^{2/3} k^{-5/3}, \tag{1.182}$$

where ε is the turbulent-energy-dissipation rate and c_1 is a universal constant.

A numerical solution of equation (1.181) with the use of $E(k)$ in the form of (1.182) was given by Ichiye [128]. Having differentiated equation (1.181) with respect to k he reduced it to the following non-dimensional form:

$$\frac{\partial E_c'(k', t')}{\partial t} = -(R^{-1} + ak^{-4/3})\, k^2 E_c'(k', t') +$$

$$+ \frac{4}{3} ak^{-7/3} \int_0^k E_c'(k', t')\, k'^2 \mathrm{d}k', \tag{1.183}$$

where k' and t' are the non-dimensional wave number and time, respectively; a and R are non-dimensional parameters: $a = 6/7 c_1 \varepsilon T' K'^{2/3}$, $R = (2DT'K'^2)^{-1}$, where T' and K' are, in turn, characteristic scales of time and wave number. Ichiye considered problems with the two initial distributions of the function $E_c'(k', t')$:

$$E_c'(k', 0) = \exp(-k'^2/b^2); \quad E_c'(k, 0) = m(-k'/b)^4 \exp(-k'^2/b^2), \tag{1.184}$$

where b and m are constants.

Numerical calculations of the initial value problems (1.183) – (1.184) were made for several values of the parameters R and a. The obtained graphs of the function $E_c(k, t)$ displayed its transformation in time and wave-number space. In some cases, there existed a flux toward larger wave numbers of the 'energy' of contaminant-concentration fluctuations, while at other values of the parameters this flux changed its direction. In a number of cases, however, there existed wave-number ranges where the function $E_c(k, t)$ became negative, which is not realistic from the physical standpoint. Ichiye believes that this is a result of the insufficiently grounded assumptions made in the derivation of equation (1.183). In this regard, as was pointed out by Ichiye, the experiment must play a decisive role in determining properties of the function $E_c(k, t)$.

A probability density distribution for concentration fluctuations $p_{c'}$ may give more complete information on fluctuating contaminant concentration fields in the turbulent flow. The value $p_{c'}\mathrm{d}c'$ determines the share of concentration fluctuations within the range from c' to $c' + \mathrm{d}c'$ relative to all possible fluctuations. It is then obvious that the function $p_{c'}$ satisfies the normalization condition

$$\int_0^\infty p_{c'}\, \mathrm{d}c' = 1. \tag{1.185}$$

An equation for the function $p_{c'}$ as well as one for the spectrum of concentration fluctuations can be derived from (1.166). Let φ be some function of the contaminant concentration fluctuation c'. Then, using the differentiation rule for a composite function, we can obtain an equation for $\varphi(c')$:

$$\frac{\partial \varphi}{\partial t} + \frac{\partial \varphi u_i}{\partial x_i} = \frac{\partial \varphi}{\partial c'} \left[\mathrm{D}\nabla^2 c' + \frac{\partial}{\partial x_i} (\overline{c'u_i'} - \bar{c}u_i') \right]. \tag{1.186}$$

Now, using the identity

$$\frac{\partial \varphi}{\partial c'} \nabla^2 c' = \nabla^2 \varphi - \frac{\partial^2 \varphi}{\partial c'^2} \left(\frac{\partial c'}{\partial x_i} \right)^2 \tag{1.187}$$

we transform (1.186) to the form

$$\frac{\partial \varphi}{\partial t} + \frac{\partial \varphi u_i}{\partial x_i} = D\nabla^2 \varphi - \frac{\partial^2 \varphi}{\partial c'^2} D \left(\frac{\partial c'}{\partial x_i} \right)^2 + \frac{\partial \varphi}{\partial c'} \frac{\partial}{\partial x_i} (\overline{c'u_i'}) - \\ - \frac{\partial \varphi}{\partial c'} \frac{\partial}{\partial x_i} (\bar{c}u_i'). \tag{1.188}$$

Then, averaging (1.188) and using the property (1.19), we obtain

$$\frac{\partial}{\partial t}\bar{\varphi} + \frac{\partial}{\partial x_i}(\overline{\varphi u_i}) = D\nabla^2\bar{\varphi} - D\,\overline{\frac{\partial^2\varphi}{\partial c'^2}\left(\frac{\partial c'}{\partial x_i}\right)^2} + $$

$$+ \overline{\frac{\partial\varphi}{\partial c'}}\,\frac{\partial}{\partial x_i}\,(\overline{c'u_i'}) - \overline{\frac{\partial\varphi}{\partial c'}\,u_i'}\,\frac{\partial\bar{c}}{\partial x_i}\,. \tag{1.189}$$

As is seen, this equation has a number of terms that are mean products of two functions. Let us consider in more detail the mean value $\overline{\psi(c')u_i'}$, where $\psi(c')$ is some function of c'. Based on the probability-statistical definition of averaging, it may be presented as

$$\overline{\psi(c')u_i'} = \iint \psi(c')\;u_i'p_{\mathbf{u}',c'}\;d\mathbf{u}'\;dc', \tag{1.190}$$

where $p_{\mathbf{u}',c'}$ is the joint probability density distribution function for the velocity and contaminant-concentration fluctuations. Further, if we use the conditional probability density distribution function for the velocity fluctuation at fixed concentration fluctuations $p_{\mathbf{u}'/c'}$, then (1.190) will have the form

$$\overline{\psi(c')u_i'} = \int \psi(c')\,v_i'(c')p_{c'}\;dc', \tag{1.191}$$

where $v_i(c') = \int u_i'p_{\mathbf{u}'/c}d\mathbf{u}'$.

For the term $\overline{\psi u_i}$ the expression (1.191) will be presented as

$$\overline{\varphi u_i} = \int \varphi(c')\{\bar{u}_i + v_i'\}\;p_{c'}\;dc', \tag{1.192}$$

and for the term $\overline{(\partial\varphi/\partial c')u_i'}$, correspondingly, as

$$\overline{\frac{\partial\varphi}{\partial c'}\,u_i'} = \int \frac{\partial\varphi}{\partial c'}\,v_i'\;p_{c'}\;dc'. \tag{1.193}$$

Similarly, we can also present the term $\overline{(\partial^2\psi/\partial c'^2)D(\partial c'/\partial x_i)^2}$ bearing in mind that $D(\partial c'/\delta x_i)^2$ is nothing else than the 'dissipation' of the contaminant-concentration fluctuations ε_c (or the rate of leveling off the contaminant-concentration inhomogeneities). We will then arrive at:

$$\overline{\frac{\partial^2\varphi}{\partial c'^2}\,\varepsilon_c} = \iint \frac{\partial^2\varphi}{\partial c'^2}\,(\bar{\varepsilon}_c + \varepsilon_c')\;p_{\varepsilon_c',c'}\;d\varepsilon_c'\;dc', \tag{1.194}$$

where $p_{\varepsilon_c',c'}$ is the joint probability density distribution for the fluctuations ε_c' and c'.

Let us, similarly to v_i', introduce a function f defined by the expression

$$f = \frac{1}{\bar{\varepsilon}_c} \int \varepsilon_c' \, p_{\varepsilon_c'/c'} \, d\varepsilon_c', \tag{1.195}$$

where $p_{\varepsilon_c'/c'}$ is the conditional probability density distribution function for ε_c' at fixed concentration fluctuations c'. Using the expression for the function f, the formula (1.194) can be rewritten as follows:

$$\overline{\frac{\partial^2 \varphi}{\partial c'^2} \varepsilon_c} = \bar{\varepsilon}_c \int \frac{\partial^2 \varphi}{\partial c'^2} (1 + f) \, p_{c'} \, dc'. \tag{1.196}$$

Now, substituting the expressions (1.192), (1.193) and (1.196) into equation (1.188) we obtain

$$\begin{aligned} &\int \varphi(c') \left\{ \frac{\partial p_{c'}}{\partial t} + \frac{\partial}{\partial x} [(\bar{u}_i + v_i') p_{c'}] - D \nabla^2 p_{c'} \right\} dc' + \\ &+ \bar{\varepsilon}_c \int \frac{\partial^2 \varphi}{\partial c'^2} (1 + f) \, p_{c'} \, dc' - \frac{\partial}{\partial x_i} \overline{(c' u_i')} \int \frac{\partial \varphi}{\partial c'} p_{c'} \, dc' + \\ &+ \frac{\partial \bar{c}}{\partial x_i} \int \frac{\partial \varphi}{\partial c'} v_i p_{c'} \, dc' = 0. \end{aligned} \tag{1.197}$$

The following relation is valid for the function v_i' as well as for f:

$$\int \frac{\partial \varphi}{\partial c'} v_i' p_{c'} \, dc' = \int v_i' p_{c'} \, d\varphi - v_i' p_{c'} \, \varphi \Big|_{-\infty}^{\infty} - \int \varphi \frac{\partial}{\partial c'} (v_i' p_{c'}) \, dc'. \tag{1.198}$$

But because of the rapid decrease of the function $p_{c'}$ at $\pm\infty$ we may let the first term on the right-hand side of the relation (1.198) equal zero. Then the following equalities will be held:

$$\int \frac{\partial^2 \varphi}{\partial c'^2} (1 + f) \, p_{c'} \, dc' = \int \varphi \frac{\partial^2 p_{c'} (1+f)}{\partial c'^2} \, dc'; \tag{1.199}$$

$$\int \frac{\partial \varphi}{\partial c'} p_{c'} \, dc' = -\int \varphi \frac{\partial p_{c'}}{\partial c'} \, dc'; \tag{1.200}$$

$$\int \frac{\partial \varphi}{\partial c'} v_i' p_{c'} \, dc' = -\int \varphi \frac{\partial p v_i'}{\partial c'} \, dc'. \tag{1.201}$$

Now, if we substitute the expressions (1.199) – (1.201) into the formula (1.197), we will obtain

$$\int \varphi(c') \left\{ \frac{\partial p_{c'}}{\partial t} + \frac{\partial}{\partial x_i} [(\bar{u}_i + v_i')p_{c'}] - D\nabla^2 p_{c'} + \bar{\varepsilon}_c \frac{\partial^2 p_{c'}(1+f)}{\partial c'^2} - \right.$$

$$\left. - \frac{\partial p_{c'} v'}{\partial c'} \frac{\partial \bar{c}}{\partial x_i} + \frac{\partial p_{c'}}{\partial c'} \frac{\partial}{\partial x_i} (\overline{c'u_i'}) \right\} dc' = 0. \tag{1.202}$$

But since the function $\varphi(c')$ was assumed an arbitrary one the integrand in the figure brackets must become zero. Thus, we finally arrive at an equation for the function $p_{c'}$:

$$\frac{\partial p_{c'}}{\partial t} + \frac{\partial (\bar{u}_i + v_i') p_{c'}}{\partial x_i} = D\nabla^2 p_{c'} - \bar{\varepsilon}_c \frac{\partial^2 p_{c'}(1+f)}{\partial c'^2} -$$

$$- \frac{\partial p_{c'}}{\partial c'} \frac{\partial (\overline{c'u_i'})}{\partial x_i} + \frac{\partial p_{c'} v_i'}{\partial c'} \frac{\partial \bar{c}}{\partial x_i} . \tag{1.203}$$

Interestingly, changes of the function $p_{c'}$ in time are related to the transport $p_{c'}$ not only by the mean velocity $\bar{u}_i$ but by the conditional mean velocity v_i' (the second term of the left-hand side of the equation). Transformation of the function $p_{c'}$ also occurs under the influence of molecular diffusion and dissipation (the first and second terms on the right) and under the influence of the gradient of contaminant concentration flux and that of mean concentration (the third and fourth terms on the right). However, even assuming that the mean fluctuation current velocities $\bar{u}_i$ and u_i' as well as the mean contaminant concentration $\bar{c}$, the dissipation ε_c and the contaminant fluxes $\overline{c'u_i'}$ are known, equation (1.203) still remains unclosed since the functions v_i' and f are dependent on the unknown conditional density distribution functions $p_{\mathbf{u}'/c'}$ and $p_{\varepsilon_c'/c'}$. Therefore, the problem of closure of equation (1.203) appears again. A number of authors suggested various forms of closure. For example, it is proposed in [56] that we assume that $v_i' = (\overline{c'u_i'/c'^2})$, so under the assumption of statistical independence of ε_c and c we arrive at $f = 0$. Under such conditions, equation (1.203) is closed and, thus, can be solved in this way or another. For the case of stationary and horizontally homogeneous turbulence, a solution in the self-similarity form was obtained in [59].

Now we would like to finish our brief consideration of some statements and results of the contaminant-diffusion theory. Extensive information is available on this problem, and the list of relevant works is increasing from year to year. Unfortunately, in the present review, we did not have enough opportunity to touch upon many other important aspects of the problems. There are, for example, interesting publications on the uses of differential hyperbolic equations for the description of diffusion in turbulent media which give finite propagation velocities of the 'front' of the diffusing substance. Many important works have been written on the use of similarity theory for determining contaminant-concentration fields, on space-time 'memory' of diffusion and on other topics. Some of them will be mentioned in the following sections of the book, while for a more detailed understanding of the problem we would like to refer the reader to the fundamental monograph by A.S. Monin and A.M. Yaglom, "Statistical Fluid Dynamics" [68], which devotes several chapters to diffusion. Some recent results of and trends in the development of the turbulent-diffusion theory are given in book [63].

CHAPTER 2

Diffusion of Contaminants from Localized Sources in the Ocean

2.1. Experimental Methods of Studying the Diffusion of Contaminants in the Ocean

Estimations of transport coefficients for various substances using data on the natural fields of these substances enables us, as was shown above, to comprehend the orders of these coefficients. However, in order to make these estimations reliable it is necessary to conduct a detailed analysis of the existing conditions in the particular region of the ocean being considered. Such an analysis must result in revealing the main mechanisms of formation of a substance field as well as in the ways of neglecting the secondary factors. For example, when estimating transport coefficients using equations for the transport of a substance (the so called indirect coefficient determination methods), the contribution of each term in the balance equation needs to be evaluated so that only the governing terms are left in the final computation formulas. For instance, when the thermal wave method is used, it is necessary to show the possibilities of neglecting the horizontal processes, heat sources in the ocean body, heat fluxes across the seabed, etc. Only in that case are we justified in using for the estimation of the vertical-transport coefficient K_z the thermal-conductivity equation with a non-stationary term and a vertical heat diffusivity term, the solution of which makes it possible to estimate K_z using the formulas (1.33) – (1.36). Without a preliminary analysis of the factors influencing the process, indirect determination of transport coefficients may lead to the wrong results because of unaccounted factors rather than the mixing process itself.

When evaluating the transport parameters using observation data on the natural fields of substances, it is usually very difficult to set correctly the initial conditions for a problem. In the case of the thermal wave method, for example, it has to be assumed that there is a strictly periodical regime for heat and thermal fluxes on the ocean surface, with this periodical regime lasting so long that the peculiarities of the initial temperature distribution for the ocean layer in question may be totally neglected. Such an overidealized picture of the process may certainly lead to essential errors in determining the transport parameters.

For calculating transport coefficients with the use of transport equation, it is necessary to have rather complete information about the field of a substance being transported as well as about the current velocity field in the observation area. This information has to be so detailed as to allow a reliable estimation of the first and second space derivatives and the first time derivative of the studied field. When estimating vertical derivatives, it is desirable to operate with the continuous substance distribution profiles since measurements at discrete points may generate errors in the derivative estimations because of the fine structure peculiarities of the field of that are lost in the case of discrete measurements. At the same time, before calculating the derivatives the continuous curve should be smoothed in a certain way to conform with the scale of averaging used when deriving the averaged equation for the transport of a substance in the ocean. The choice of such a scale of averaging is a rather difficult question in itself, requiring adequate information on the peculiarities of the studied process, type of source of the substance, boundary conditions and their space-time variability, and on the accuracy requirements for a particular problem to be solved. After determining the scale of averaging, it is necessary to choose the optimal way of smoothing the original data so as to preserve in the averaged results the most valuable information as regards the given problem.

Usually, the most scarce information is on the variability of the field in question in horizontal directions at large depths. Measurements available from individual hydrological stations, separated from one another by large distances, only provide discrete and often unsynchronic data which are rather difficult to use for estimating space-time derivatives. The finite differences for the field of a substance in question, calculated with the use of such data, may significantly differ from the actual derivatives.

All the above difficulties account for the fact that indirect methods of estimating diffusion parameters can only give very approximate results. And sometimes the commonplace massive computations of transport coefficients with the use of standard hydrological observations for large ocean areas can hardly be welcomed.

More convenient substances for studying the pecularities of diffusion processes in the ocean turned out to be those of anthropogenic origin, e.g. the radioactive isotope ^{90}Sr. This isotope, formerly unobserved in nature, appeared on Earth as a result of nuclear testing. This has furnished an opportunity to define, with a certain degree of accuracy, the initial conditions for diffusion problems on the spread of isotopes in the ocean waters. Artificial radionuclides reached ocean regions, sufficiently remote from the nuclear testing sites, primarily from the so-called stratospheric reservoir formed as a result of shooting explosion products high into the upper layers of the atmosphere. On its information, this reservoir provided a global and rather stable (on the average, of course) supply of isotopes onto the ocean surface. From the standpoint of diffusion studies, this also meant the possibility of correct definition of and boundary conditions for isotope-diffusion problems for the ocean waters. These made it possible to evaluate, with a sufficient degree of confidence, the parameter of transport in the ocean. Solutions of these problems were a subject of a large number of papers (see Chapter 3).

Since the 1940's, direct methods of determining the transport parameters came into use in oceanology. Unlike the indirect ones, based on the analyses of averaged contaminant and current velocity fields, the direct methods analyze data on fluctuation characteristics of these fields. If the researcher possesses the corresponding data he can make a direct calculation for the contaminant fluxes using the left-hand sides of the formulas (1.25), without referring to the semi-empirical hypothesis on the proportionality of the fluxes of a contaminant to the gradients of its averaged field. At the same time, if, besides the data on the fluctuation components of the fields, there is also information available on the mean field of concentrations then, using the same formulas, the coefficients K_z and K_l can be estimated. Earlier, direct methods were used for estimating turbulent viscosity coefficients (see ref. [81]) since special multi-channel instruments were needed for the gathering of synchronic information on the fluctuations of velocity components and those of the concentrations of a contaminant being transported. In the U.S.S.R. such instruments were for the first time created by Kolesnikov [45], who attempted to determine the vertical turbulent coefficient for thermal diffusivity in the ocean K_z^T according to the formula

$$K_z^T = \frac{\overline{w'}\,\overline{T'}}{\partial \overline{T}/\partial z} \tag{2.1}$$

The values for the coefficient K_z^T were of the order of 1 cm^2/s, with K_z^T decreasing considerably with the growing stability of density stratification of water layers in the ocean. Later, similar measurements and calculations were made by Williams and Gibson [165] Lozovatsky and Ozmidov [58]. The first authors estimated the vertical heat fluxes in the region of the equatiorial Counter-Current in the Pacific at $5.9 \cdot 10^{-3} - 11.7 \cdot 10^{-2}$ J/(cm^2/s) and K_z^T in the range of $0.52 - 27$ cm^2/s. For the frontal zone of the Kuroshio in the layer from 80 to 710 m, the latter authors obtained figures ranging from 1.7 to 74.4 J/(cm^2/s) respectively and from 0.13 to 12.71 cm^2/s for the coefficient K_z^T. The above authors took measurements using low-inertia instruments, with velocity and temperature sensors that are able to register fluctuations with frequencies up to hundreds of hertz. The duration of synchronic records of velocity and temperature fluctuations subject to processing usually did not exceed 10 minutes. Direct methods of determining heat and salt fluxes and transport coefficients as well can, certainly, also rely on synchronic measurements of temperature, salinity and current velocities by standard hydrological instruments. Then the values of the parameters to be determined will, naturally, refer to much larger-scale processes as compared to those obtained under measurements by low-inertia instruments.

A rather simple method of studying the diffusion of contaminants in the ocean and, at the same time, producing interesting results is the one based on the observations of the motions of discrete indicators in the ocean. This method was first used by Richardson and Stommel [153] for studying the dependence of the horizontal transport coefficient in the ocean on the scale of the phenomenon. For this reason, Richardson and Stommel used small floats which were being observed from the pier for about 22 minutes with the discreteness of 30 seconds. As a result, some data on the relative distances between the pairs of indicators were obtained, and these were processed with the use of formulas from the relative diffusion theory. Floating indicator observations were well repeated afterward by Stommel [156] off the Atlantic coast of America and in the region of Bermuda. Immense observations, based on the same methodology, were undertaken in subsequent years by Ozmidov [81] in artificial basins, in the Caspian Sea and in the Pacific. These investigations made it possible to study the dependence of diffusion parameters on the phenomenon scale and other factors within the wide range of their variability. Later, diffusion studies with the use of discrete indicators were carried out in the Black Sea (see ref. [35]) and in other ocean areas.

Interpretation of experimental data on the diffusion of discrete particles within the framework of the relative diffusion theory makes it possible to obtain a lot of interesting data on diffusion; however, it is rather difficult, as was shown above, to derive from them any data on contaminant concentration fields. In this regard, experiments in which soluble contaminants are artificially introduces into the ocean and their diffusion observed in this way or another became widespread in oceanology. Such a research method enables the conducting of strictly controllable diffusion experiments under which both the initial and boundary conditions of a problem are known. In such experiments, radioactive isotopes or fluorescent dyes are most often used as contaminants. Using these substances makes it possible to determine their very small concentrations, if adequate instruments are used. Various radiation detectors serve as necessary instruments for measuring radioactive isotopes. Water may be samples and then analyzed on board a ship or on shore, with preventive concentrating of radioisotopes in the samples with the use of radiochemical or sorption techniques often taking place. In order to determine the activity of the samples collected gammaspectrometers can be used apart from standard radiometrical equipment. Naturally, it is quicker and more convenient to use either submerged or towed gammaspectrometers in the ocean [69].

The choice of these radioactive or other non-radioactive substances as indicators (tracers) is determined by the following requirements:

(a) the possibility of determining small indicator concentrations in the ocean water;
(b) harmlessness to the ocean flora and fauna as well as to man;
(c) low dependence of a tracing substance on the factors of the ocean medium (temperature, salinity, pressure, etc.) and its high persistence under the real conditions;
(d) low cost and availability.

Luminescent dyes, to a large extent, meet all these requirements (rodamin, uranin – the sodium salt of fluorescein, etc.); that is why they are widely used in diffusion experiments, not by water sampling with further analyses but by towing a luminescent indicator concentration sensor (fluorimeter). Instruments equiped with such sensors provide sufficient rates of concentration field surveying with high sensitivity and space-time resolution.

The design of a towed fluorimeter should, naturally, take into account the contaminant diffusion pecularities in the ocean as well the regularities of the luminescence of dye and the specifics of the sea experiments. The first serviceable make of towed fluorimeter was created in the U.S.S.R. by Karabashev [43]. The sensor design incorporated one of the traditional optical schemes, with a beam of light, which causes the excitation of fluorescent, directed at a angle of 90° to the axis of the luminescent light receiver. A normal incandescent lamp was uses as a light source. The light beam from the lamp passed, via a condenser and turning lens, through a interference lightfilter and port-light. The luminescent light, excited by lamplight, passing through the port-light and lightfilter came upon the photocathode of a photoelectronic multiplier. The electronic scheme of the device was a modified logarithmic photometer which made it possible to register dye concentrations varying from 10^{-2} to 10^{-10} gr/cm^3.

In the visible range of the spectrum fluorescein and rodamin have bright luminescent bands that are excited by electromagnetic oscillations within the optical frequency range. The spectral curves absorption maximum for fluorescein is at the wavelength of 480 nm and that for rodamin at 550 nm. The maximums of luminescent bands are, correspondingly, at the wavelengths of 530 and 590 nm. Glowing dye is isotropic and, particularly, non-inertial (the excited state of molecules is about 10^{-8} s). Lightfilters in the instrument were chosen based on these characteristics.

Calibration of the fluorimeter was done using a set of dye solutions with precisely determined concentrations. The calibration curve was practically straight for most of the scale, with slight deviations from linear dependence only observed at extreme concentration.

Operation of the first model of the fluorimeter in the Black Sea showed its efficacy for diffusion studies. The main design features were also proven correct. At the same time, a number of shortcomings in the instruments needed to be overcome. First and foremost was the exposure of the photodetector to bright daylight, which hampered working in daytime. The solution found was the modulation of the exciting radiotion with the help of a rotating disc with holes. Such a device in the electronic scheme made it possible to further filterout the variable component of the signal from the constant background daylight exposure. Protective screens of the port-lights and corrugated plates serving as light traps were used for reducing the level of exposure. The optical scheme of the fluorimeter is shown in Figure 2.1.

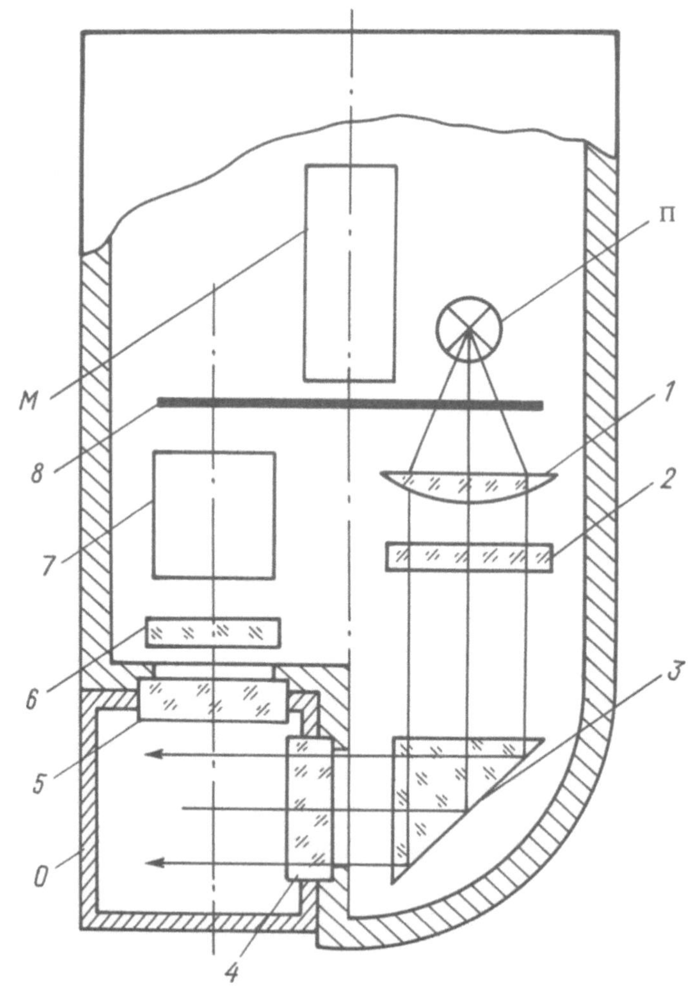

Fig. 2.1. Optical train of fluorimeter. П – light source; 1 – condensor; 2 – primary light filter; 3 – deflecting prism; 4, 5 – illuminators; 6 – secondary light filter; 7 – photomultiplier; 8 – disc with holes; *M* – electric motor; *O* – protection screen.

Electricity was supplied to the part of the instruments that was submerged into the sea by cable from batteries or the supply line of the ship. The same cable was used to receive signals on board the ship from the fluorimeter's sensor as well from the sensor of submergence depth. Registration of the signals can be done by pen recorders or onto magnetic media. The timelag of the photometric scheme was close to 0.2 seconds, and the water volume whose luminescence was registered by the instrument was close to 100 cm^3. The consumption of power was about 80 W, the duration of continuous operation for zero point departure of $\pm$ 4% of the entire scale was not less than 7 hours and the range of depth was from 0 to 100 m for any levels of natural lightning of the sea surface by the Sun.

The above model of fluorimeter was successfully used in diffusion experiments in water with various optical characteristics – from clear Black Sea water in surf areas of shallow sandy shores. In one of the experiments, synchronic registration of the indicator content was carried out for the same value of water by two different instruments – the towed fluorimeter and the laboratory running-water fluorimeter made by the Aminco company (with linear scale), with the outside water being supplied to the basin of the latter instrument by hose. The comparison of the thus-obtained records of the indicator distribution revealed

(1) noticeable delay in registering by the running-water fluorimeter as compared with the towed one;
(2) distortion of the form of the indicator distribution by the running-water fluorimeter at large concentration gradients;
(3) no fine structure of distribution when registered by the running-water fluorimeter, which is explained by spacial averaging of the concentration profile by the pumping channel several meters long.

These shortcomings of the running-water registering instrument are especially important at earlier stages of the diffusion process when the inhomogeneity of the indicator spot is very high. The latter circumstances also makes linear scales inconvenient for registration. However, after 3–4 hours from the moment the indicator was placed into the sea both instruments provided almost identical records. Thus, that experiment clearly showed that more complete information on the diffusion of contaminants in the ocean under the field simulation of the process with the help of luminescent indicators can be obtained with the use of towed equipment, with the registration being provide in both logarithmic and linear scales. For that matter, the linear instrument ought not to be used at an early stage. The crucial requirement of such an instrument is high sensitivity. In surface sea layer experiments, the substitution of the running-water on-board instrument for the towe linear fluorimeter for registration of the indicator distributions may only be done after several hours from the time when the indicator was placed into the sea.

The high operational qualities of the instrument made it possible to work out sound experimental techniques that ensure a body of initial data sufficient for confident determination of the mean characteristics of the stochastic diffusion process within a wide range of space-time scales. Here, the main attention was devoted to refining the techniques for experiments with stationary point and instantaneous indicator sources that simulate two cases of the scattering of contaminants in the ocean and are of the utmost practical and theoretical interests.

Under a continuous and uniform outflow of an indicator from a point source at rest, the 'plume'-shaped indicator body, which stretches along the current and widens due to diffusion with the distance from the source, appear in the water flow. Owing to the statistically stationary character of the process, instantaneous indicator concentrations in any point of the 'plume' fluctuate around the most probable or mean value; however, the latter must remainconstant after the initial transition period, provided the mean current velocity and other medium parameters do not change with time. Under these conditions the mean indicator-concentration field can be determined by multiple registration of the indicator distribution in the cross-section of the 'plume' at several fixed distances from the source and at different moments of time. The latter circumstance greatly facilitates the experiment. If the number of measurements is large enough the mean indicator concentration distribution at a given distance from the source can be found with a accuracy determined by the errors in measuring the coordinates and by the instrumental errors in estimating the indicator concentration. It is the main advantage of stationary sources that mean indicator distributions can be obtained without serious difficulties and are directly comparable with the theoretical expressions describing the distributions of averaged contaminant concentrations.

In order to simulate a stationary source a special installation is used where the indicator solution is forced by compressed air out in turn of two containers through a hose to a given depth (see Figure 2.2). The installation is completely autonomous; it is suitable for all ships and provides a uniform supply of indicator over a long period of time within a range of depths. In some cases, the use of a simpler stationary source – an open tank with a hose through which a dye solution flows out into the sea – does not turn out to be less successful. The intensity of such a source is regulated by the outer hose diameter and the height of the tank above sea level.

The determination of the indicator concentration field from a stationary source is provided in the following way (see Figure 2.3). The device for simulating a stationary source should be installed on board the ship C_1. This ship is brought to anchor in the area of the experiment. After the device is assembled and both its containers are filled with indicator solution the free end of the hose is submerged to the required horizon. Given excessive pressure in

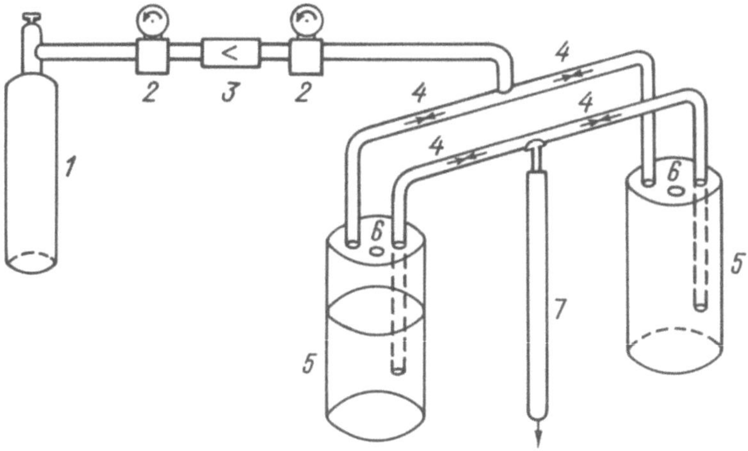

Fig. 2.2. A device for continuous point source of indicator. 1 – pressurized air tank; 2 – manometers; 3, 4 – valves; 5 – dye-solution tanks; 6 – locked outlets for filling the solution; 7 – hose for dye releasing into the sea.

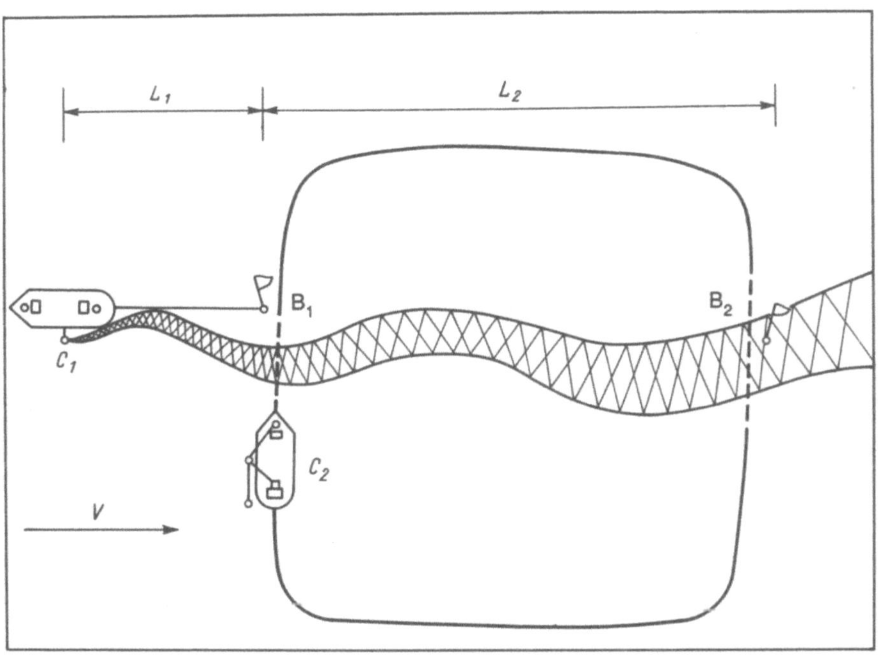

Fig 2.3. Experimental scheme for stationary point source of indicator (see text for details).

one of the containers, the solution begins to flow out and the indicator body to form. Actually, the measurements take place some time after the process is started with the help of the a ship C_2 equipped with measuring instruments. With an outboard sensor the ship C_2 makes, at different distances from the source $10-20$ measurements for each L. The distance from the source to the point of measurement can be determined by radar, rangefinder of simply from the length of the capron halyard which connects a drifting buoy B_1, equipped with an underwater sail, with the ship C_1. If the experimental conditions are favorable, it is desired to fix an anchored buoy B_2 at a distance L_2 down stream from the buoy B_1. Then the ship C_2 can make circulation paths, as shown in Figure 2.3, making two measurements per circulation path at distances L_1 and $L_1 + L_2$ from the source. The value of L_2 is determined either by instruments from the ship C_1 or from the current velocity and time of drifting of a float from B_1 to B_2.

Such a experimental scheme with a stationary source allows the determination of the coordinates of the measurement points by available and unsophisticated techniques from vessels not equipped with up-to-date navigational instruments. For instance, the work in the Black Sea was carried out from trawlers with displacement of as little as 78 tons. If, as is done under theoretical analyses, the origin of the coordinates is placed in the indicator source, the axis X is directed along the mean vector of current velocity V and the axis Z is oriented downward, then under such an experimental scheme the coordinate X is determined by L_1 and L_2; the coordinate y can be found from the length and scale of the record taken by the strip-chart recorder; and the coordinate Z by the depth-sensor indications.

In the case of a stationary source, the size of the indicator body depends both on the intensity of the source (the amount of indicator being brought into the sea per unit of time) and the kinematic structure of the flow. The above derive for simulating a stationary source made it possible to indicator 'plumes' up to $1-1.5$ km long and $40-60$ m wide (within the sensitivity range of the sensor) at current velocities of $10-40$ cm/s and source intensities of $300-600$ g/h. Variation of the source intensity was achieved by regulating excessive pressure in the containers while the concentration of the initial solution in all the experiments was the same, $1 \cdot 10^{-3}$ g/ml.

It is much more difficult to obtain experimental data convenient for theoretical comparisons in experiments with instantaneous indicator sources. In such a case, it is necessary to conduct several identical experiments with 'instant' surveys of the indicator spots, made at the same diffusion times, in order to obtain averaged pictures of the diffusion process. Here, the mean hydrometeorological conditions must remain unchanged. The complexity of such a procedure, even if reliable equipment is available, is rather high, therefore, diffusion from instantaneous sources is usually studied by periodical quasi-synchronic surveys of a sole indicator spot over a time from several hours to several days from the moment the indicator was released.

A device designed to simulate an instantaneous contaminant source, besides meeting the usual requirements such as compactness, transportability, reliability, etc., must also allow a sufficiently quick (over about one second) release of large amounts of dye (from 1 to 100 kg) into the given depth. Even now there are no standard devices, apparently, that totally satisfy these requirements. Elaboration of such devices is mainly retrained by the fact that, because of the relatively low indicator solubilities, the mass of the initial solution should exceed, at least several times, that of the indicator itself. Since the spot size depends directly on the released indicator amount it is desirable, in order to ensure sufficient duration and scales of the processes, to release into the sea as much indicator as possible in each experiment.

During the Black Sea experiments, instant releases of the indicator onto the sea surface was done from a 200 liter metal tank. Below its center of gravity a horizontal axle was welded in, with the ears of the frame used for fastening the tank to the winch cord put onto its ends. A simple locking device on the frame prevented the rotation of the tank around its axle. After filling the tank with initial solution and using a cargo boom to place it overboard the locking device is unfastened from the tank, and when the latter is turned over the solution pours out onto the sea surface. A special 10-liter container was designed for introducing smaller dye quantities (up to 1 – 2kg). The container is a rigid PVC plastic cylinder with lids on its ends. Filled with initial material, it is fastened by clamps to the whinch cord and, unlike a conventional water sampler, is lowered in its closed position to a given horizon. A dispatch load triggers the mechanism controlling the lids, with the latter taking a vertical position. When the container is lifted thereafter its contents remain as they were, practically preserving the container shape.

Observation of the diffusion from an instantaneous source is conducted, in the case of quasi-synchronic surveys, in the following way. Some time after the release of the indicator, a ship with outboard sensors makes a series of measurements across the spot for as little time as possible. The following surveys are made in fixed periods of time until the spot disappears completely. Normally, the directions of the measurements and intervals between surveys are chosen in the course of experiments since it is difficult to predict the forms, orientations and velocities of moving spots.

If such experiments should result in a spacial picture of the indicator distribution relative to a fixed system of coordinates, special importance should be attached to the navigational support of the experiments. However, the accuracy of navigational referencing, even of that provided by the DECCA system, is insufficient for surveying spots with sizes from 1 to 10 km. For that reason, the "to-water-navigation" principle may be used. It is based on measuring with the help of an onboard radar the distances of the ship from an azimuth relative to a buoy equipped with a reflector and underwater sail, with the mutual orientation of the buoy and maximum orientation area in the spot considered being unchanged throughout the entire period of measurements. However, experience says that such an orientation is maintained in the course of the experiments over a sufficiently long period of time only under favorable weather conditions. But in a wind of even moderate strength, the drifting buoys usually moved outside the spots two or three hours after the experiments had begun. Thus, the "to-water-navigation" does not meet all-weather requirements. In coastal sea areas, sufficient referencing accuracy can be provided by theolodite intersection of a vessel by on-shore observers, by radionavigational means and radars as well as by aerial photographic surveys. For interpreting the results of experiments on diffusion within the framework of the relative diffusion theory the "relative-to-the-center-of-gravity-of-the-spot navigation" is quite enough. Such a navigation is, apparently, very simple as it only requires data on the speed of a ship and charting its tack each time across the central part of the spot, which usually can be determined accurately enough visually.

The three-dimensional distribution of an indicator can be evaluated by two methods: either by vertical sounding with the help of a sensor from a drifting vessel or by towing the sensor across the indicator body at different depths. The first method requires a large number of drifting stations within the visible margins of the indicator body or within a sea area designated beforehand. This procedure is not very appropriate for studies on diffusion from an instantaneous source because of the additional time needed to halt the ship for sounding, thus violating the 'quasi-synchronic' measurement requirements. The time factor does not play an important role in studying stationary distributions of an indicator; however, even in that case it is difficult to obtain a three-dimensional picture by vertical sounding, since the cross-section sizes of indicator bodies are generally smaller than or comparable with those of expedition vessels, which makes it possible to determine the location of vessel against indicator 'plume'. Obviously, vertical sounding is only justified in some specific cases as an auxiliary method used for an approximate estimation of the spreading of a contaminant in the vertical direction.

A most successful method is measuring the indicator concentration from a moving vessel when the vertical distribution of the indicator can be obtained if varying the depth of submergence of the towed sensor en route. This possibility was realized, for example, during the Black Sea experiments. The way of towing the fluorimeter is shown in Figure 2.4. In order to obtain a three-dimensional picture of indicator distribution in the 'plume' from a stationary source at each distance from the latter it is necessary to make a series of measurements, each of them corresponding to a certain depth of sensor location. Similarly, multi-depth sounding of the indicator spot from an instantaneous source can also

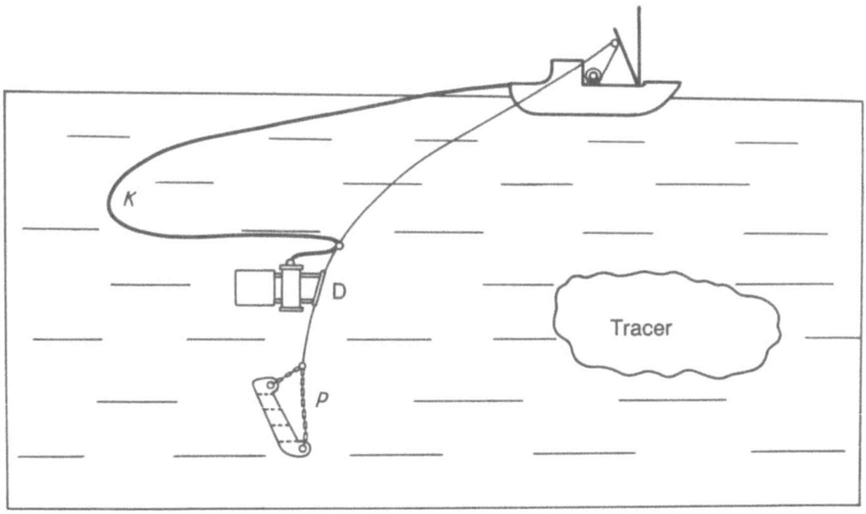

Fig. 2.4. Scheme for measuring indicator distribution at deep horizons with the help of towed fluorimeter *D*. *P* – sinking device; *K* – cable; the length of support wire may be altered in the course of experiment by winch. The cable and support wire may be substituted by one single cable.

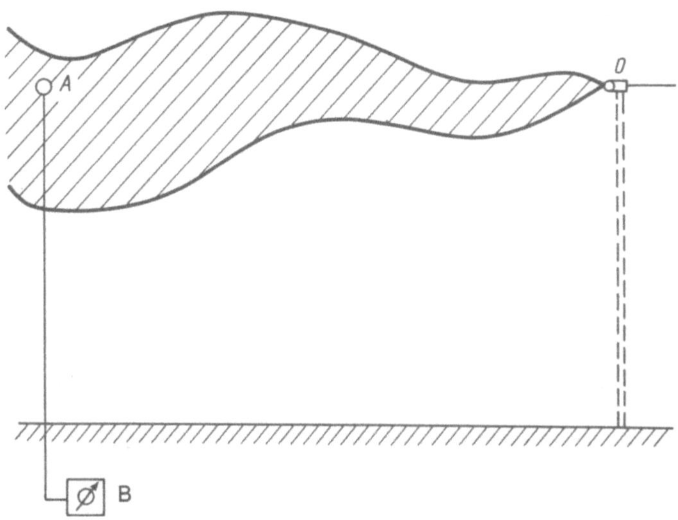

Fig. 2.5. Experimental scheme for stationary indicator sources (*O*) and fluorimeter (*A*) in a coastal sea area. *B* – registering instrument at a shore-based facility.

be done; in that case, however, non-synchronic sounding of different horizons may create additional errors. The best decision it to use several sensors towed at different depths by one ship and on one cord.

In shallow coastal sea areas, methods of conducting diffusion experiments that enable the studying of time variations of the indicator concentration at a fixed point of the 'plume' from a stationary source can be used (see Figure 2.5). For this purpose, a tripod with attached fluorimeter is placed on the seabed, and a stationary indicator source is positioned upstream. With the use of several sensors this experimental methods proves to be exceptionally effective for solving practical problems, for example when modeling a sewer for waste discharge into the sea in conditions in which towing turns out to be unfeasible (small depths, complicated seabed topography, severe weather conditions, etc.).

By now a large number of fluorimeter modifications have been made. For example, the P.P. Shirshov Institute of Oceanology of the Academy of Sciences of the U.S.S.R. uses a device which, apart from the fluorimeter itself, has a water-temperature sensor that enables the determination of the thermal structure of a water body along with the contaminant-concentration field. Many foreign companies have put fluorimeters into commercial production.

Very promising methods for studying contaminant diffusion in the ocean are remote-sensing and aerospace methods. Undoubtedly these methods have a number of advantages over contact measurements. Firstly, it is a possible to obtain series of synchronic measurements, with the time intervals set beforehand, which enables the studying of the variability in time of a phenomenon within a wide range of scales. Another advantage is the possibility of investigating processes without disturbances caused by instruments or the vessel, which is especially important for processes with scales from several to dozens of meters. Naturally, remote-sensing methods do have some shortcomings as compared with contact measurements. For example, using aerial photography it is not always feasible to determine the true dye concentrations in a diffusing spot or jet. Besides, the photograph presents some integral concentration value along the line of sight, which is why three-dimensional pictures of diffusion are hard to obtain, at any rate. Therefore, in diffusion experiments the best way apparently is to combine contact and remote-sensing methods that are complementary to each other.

When processing aerial or space photographs we should, as a matter of fact, take into account the absorbing properties of the atmosphere, which do change, depending on its state. Multi-zonal surveys help correctly interpret the obtained aerospace pictures. In order to determine the density level of the negatives it is desirable to use microphotometers and densitometers. In that case instruments automatically translating pictorial contrasts into digital information for further computerization are most suitable. Calibration of the obtained photographs should be done with the help of simultaneous contact measurements of dye concentrations by a fluorimeter. Here, not only the towing of instruments but also sounding is desirable, both of which ensure a better comparability between the real and computed pictures of light-beam penetration through dye toward the photographic camera.

2.2. Diffusion from Instantaneous Point Sources

Fluorescent dyes were, apparently, uses for studying diffusion processes in the ocean for the first time by Hunnerson in 1956 and the Japanese scientists in the following year (see ref. [81]). The Japanese oceanologists carried out an experiment with the use of 4 kg of dye, which made it possible to observe the diffusion for about 4 hours, until the radius of the diffusion indicator spot reached 400 – 500 m. In 1960, Hela and Voipio conducted experiments in the Golf of Bothia with the release of 1 – 2 kg of rodamin [126], with the maximum horizontal sizes of dye spots reaching 500 m while the indicator did not penetrate deeper than 10 m. The authors managed to obtain a distribution picture for dye concentrations in the spot using the sampling method. They found that the concentration in the center of the spot decreased in inverse proportion to the third power of the diffusion time. In the following experiments conducted by other authors, with the release of smaller indicator amounts, some interesting diffusion pecularities were noted, e.g., the stretching of spots and their fragmentation. Experiments in the North Sea with the release of larger quantities of rodamin were carried out in 1960 – 1962 by Joseph, Sendner and Weideman [130]. In the three experiments, 175, 350 and 450 kg of dyes were used, respectively, which made it possible to observe the diffusion spots for 3 – 5 days, until the mean radii of the spot reached several kilometers. The experimental results were compared with the solutions of the diffusion equation with a linear dependence of the horizontal-diffusion coefficient on the radius vector of the observation point [129]. They managed to determine the values for the coefficient of proportionality in this dependence – the so-called diffusion velocity U_1 (see Part 1.5). These values appeared to be close to 0.2 – 0.4 cm/s. The depths of the penetration of dyes in these large-scale experiments reached 30 – 50 m.

In the U.S.S.R. proposeful experiments on diffusion with the use of fluorescent indicators were started in 1964 in the Black Sea. The experiments were being carried out from small research vessels with the help of towed fluorimeters. The amounts of dye released in each experiment reached several kilograms. After 1 – 2 hours from the moment of dye release, the spots reached sizes up to 500 m and were quite visible from the surface of the sea. Usually the spots did not diffuse in a strictly symmetrical manner but became deformed and acquired irregular shapes. The contaminant concentration fields in the spots were rather inhomogeneous; the dyes formed separate ‘clouds’ and ‘plumes’ of various forms and orientations in the spots. Figure 2.6 presents records of dye concentrations along the number of tacks crossing the spot. The curves in the figure can be regarded as ‘instantaneous’ contaminant-concentration distributions along different directions. However, as was said above, it is difficult to obtain concentration-distribution profiles averaged over several realizations using these data, since every new crossing of the spot by instruments takes place at a dif-

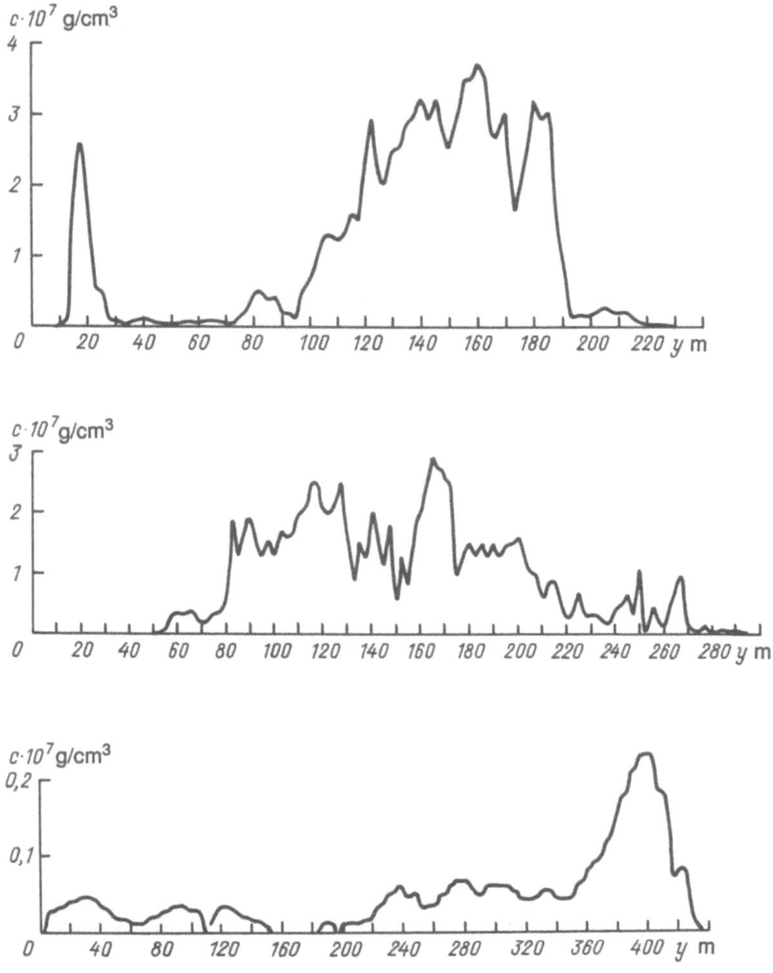

Fig. 2.6. Examples of contaminant-concentration distribution in diffusing rodamin spots in horizontal sections (based on the 1965 experiments in the Black Sea). y-axis; $c \cdot 10^7$ g/cm^3; x-axis; y m.

ferent time of the non-stationary diffusion of the spot. Here, the only feasible way of smoothing (filtration) every individual curve is with the help of some smoothing filter.

Analytically, the filtration of a function of one variable $u(t)$ can be presented by the following expression:

$$\bar{u}(t) = \int_{-\infty}^{\infty} u(\theta)\, R(t - \theta)\, \mathrm{d}\theta, \tag{2.2}$$

where $\bar{u}(t)$ is a smoothed function $u(t)$; $R(\theta)$ is the smoothing kernel; and θ is the integration variable. In an ideal case, low-frequency fluctuations of $u(t)$ up to the chosen scale T_0 will 'pass through' the filter (2.2) without distortion while short-period components will be completely 'damped'. The choice of T_0 (scale of averaging) must be governed by the physical pecularities of the process studied, preliminary analysis of the function $u(t)$ or some other considerations. The frequency (spectral) characteristic of an ideal filter must apparently have the form of a rectangle with the value of 1 from zero frequency to the frequency $\omega_0 = 2\pi/T_0$ (see Figure 2.7). The function $R(\theta)$ with such a characteristic has the form

$$R(\theta) = \sin(\omega_0\theta)/(\pi\theta). \tag{2.3}$$

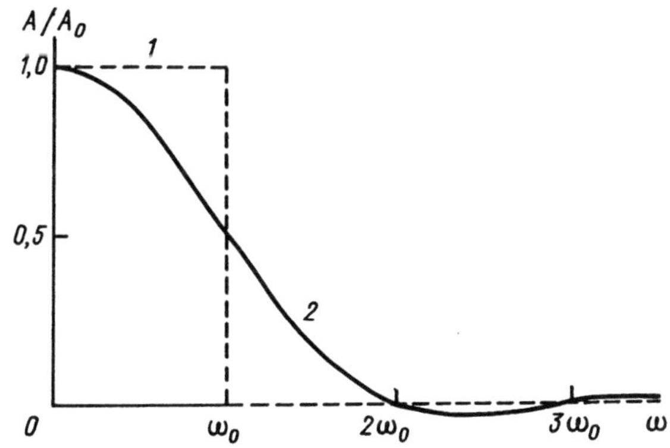

Fig. 2.7. Spectral characteristics of ideal smoothing kernel (1) and cosinus-filter (2)

Actually, the integral (2.2) with such a form of function $R(\theta)$ and at $u(t) = \sin(\omega t)$ equals zero for frequencies $\omega > \omega_0$, while at $\omega < \omega_0$ sinusoidal oscillation remains unchanged after it has been influenced by the smoothing procedure. At $\omega = \omega_0$ the value of the integral for the case in question is equal to 1/2.

Practical use of the ideal filter (2.3) is impossible, however, because integration in the infinite limits is impossible for experimentally obtained functions (or observations series). Therefore, for practical purposes, kernels of averaging, determined at finite segments of the axis θ, are used. Natural, the filter characteristic get changed – it becomes 'smeared', in this way or another, along the frequency axis (wave numbers). The more localized the segment of the axis θ, where the function $R(\theta)$ is determined, the wider is the 'smearing' of the frequency (spectral) characteristic of the filter. In oceanology, the most widely used filter is the so-called 'cosinus-filter', for which the function $R(\theta)$ has the form

$$R(\theta) = \begin{cases} [1 + \cos(\omega_0\theta)]/2, & -\pi/\omega_0 \leqslant \theta \leqslant \pi/\omega; \\ 0, & |\theta| > \pi/\omega_0, \end{cases} \tag{2.4}$$

where the frequency ω_0 (or wave number k_0 for spacial averaging, correspondingly) determines the scale of the division of harmonic components of the function $u(t)$ into the mean and fluctuation ones. The frequency characteristic of the filter (the ratio of the amplitude of the harmonic oscillations entering the filter so that of the oscillations of the same frequency ω leaving the filter) is given in Figure 2.7. As is seen, it is significantly differs from the ideal one. The cosinus-filter partly passes the components of a function to be smoothed with frequencies smaller than ω_0 (and correspondingly, for spacial variables, disturbances with scales larger than $L_0 = 1/k_0$) and at the same time it damps, to a certain extent, components with frequencies smaller than ω_0.

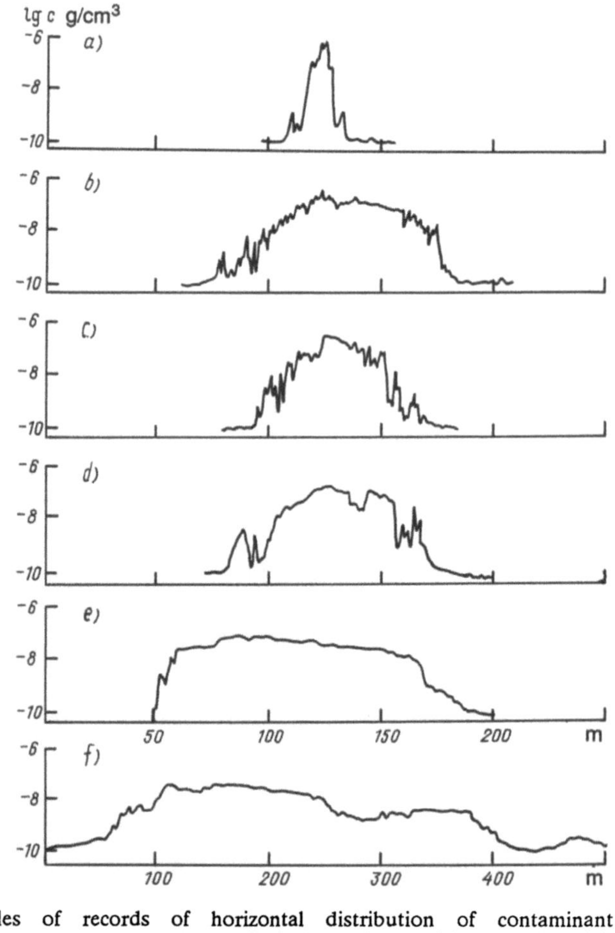

Fig. 2.8. Examples of records of horizontal distribution of contaminant concentration in diffusing spot at consequent moments after dye release: (a) 44 min; (b) 93 min; (c) 108 min; (d) 128 min; (e) 172 min; (f) 258 min. The 1965 experiment in the Black Sea. y-axis: lg c g/cm^3; c-axis: m.

Dye concentration distributions in horizontal cross-sections of the spot in different moments of time gives us perceptions of the growth rates for the contaminant diffusion spot. These were obtained in the 1965 experiments in 44, 108, 128, 172, and 258 minutes after releasing a rodamin 'point' source (curves a – f in Figure 2.8). Naturally, the spot increases averagely in size during diffusion while the contaminant concentration decreases. Again, as is seen from the figure, individual realizations of the function $c(x, y, z, t)$ are irregular, which is accounted for by the complexity of the turbulent structure of the motions of seawater masses. A non-monotic growth of the spot along the sections given is explained by its irregular form so that individual sections may cut the spot either across or along its elongation.

The Black Sea experiments with instantaneous point sources of contaminants also surely reflected the clear cut anisotrophy of the diffusion process in horizontal and vertical directions. Figure 2.9 presents a example of a record of a vertical dye-concentration distribution in the central part of a spot, with its size in the horizontal direction being about 500 m. The record was obtained from an immovable ship through sounding the water body by fluorimeter. As is seen

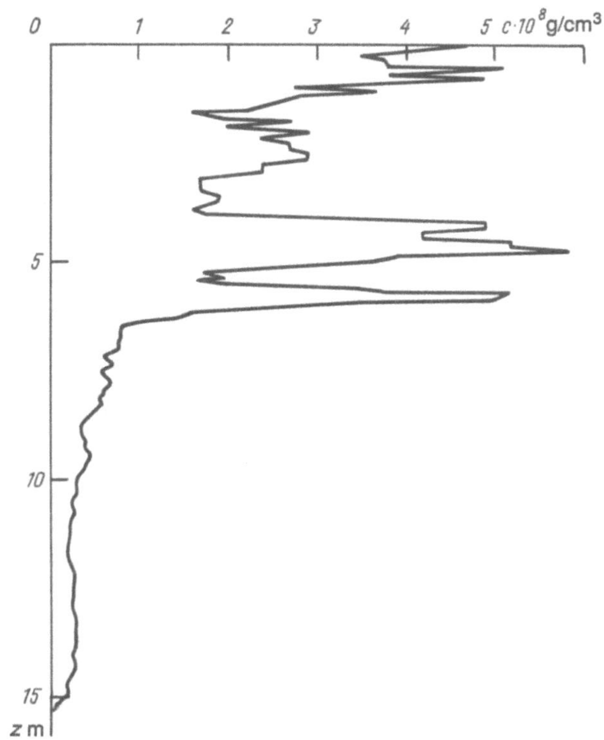

Fig. 2.9. An example of vertical distribution of contaminant concentration in the central part of the diffusing spot. y-axis: $c \cdot 10^8$ g/cm^3; x-axis: z m.

from the figure, the penetration depth for the indicator did not exceed 15 m in the case being considered, with the record segment, approximately from 7 to 15 m, having extremely small concentration values. In other words, a pronounced penetration of the contaminant into the depths at the moment of sounding did not exceed 7 m and, consequently, the ratio of the horizontal to vertical axes of the diffusing 'body' reached 70. This example, as well as the results obtained by other authors, apparently can serve as an explicit illustration of the considerations postulated in Part 1.4 on the influence of the density stratification of sea waters on the process of diffusion.

The 1965 experiments in the Black Sea staged two releases of dye. In one case 6 kg of rodamin were used and in the other 10 kg. The survey of the first spot was conducted for about 6 hours, during which time 26 crossings of the spot by towed fluorimeter were made. The second spot was crossed 15 times.

A convenient characteristic for comparing theoretical predictions with experimental data may be, as we have already seen, the contaminant concentration in the center of the diffusing spot. It is not very easy, however, to evaluate this characteristic from individual records made when crossing the spot. First, the obtained concentration data need to be averaged somehow. Second, we should not disregard the fact that individual crossings of the spot may have a chordal rather that a diametrical direction since it is rather difficult to identify the center of a spot visually. For overcoming thes difficulties, the following techniques was suggested [88]. Maximum dye concentration in each measurements are plotted in the logarithmic scale, with the time from the moment of dye release being plotted on the abscissa axis. These points are usually located in the graph in a chaotic manner. But if we consider that the maximum contaminant concentration in the spot is monotonically decreasing function of time then, in order to identify the form of this function, it is only natural to use on only the points the graph that are farthest away from the abscissa axis (as compared with the neighboring points). The other points should naturally be referred to the 'chordal' cuttings across the spot.

The above technique was use when analyzing 26 concentration records made while crossing the spot that was formed after releasing 6 kg of dye. It has turned out that, in relation to the neighboring points, extreme values were the maximum concentrations for the third, fourth, fifth and 22nd measurements. These points were plotted on the graph of Figure 2.10 where the approximating straight line saw also plotted with the use of the least-squares-method. Its slope appeared to be close to 2 (2.03), which corresponds to the inverse-square dependence of the mean maximum concentration in the spot on time. Such a dependence follows, as was shown above in Part 1.5, from certain theoretical models for the phenomenon with a dependence of the horizontal-diffusion coefficient on the scale of the phenomenon and time.

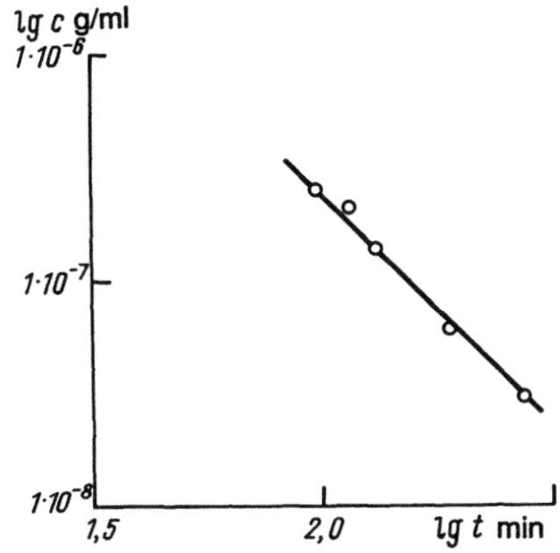

Fig. 2.10. Changing contaminant concentration in the center of diffusing spot with time (based on the 1965 experiment in the Black Sea). y-axis: lg c g/ml; x-axis: lg t min.

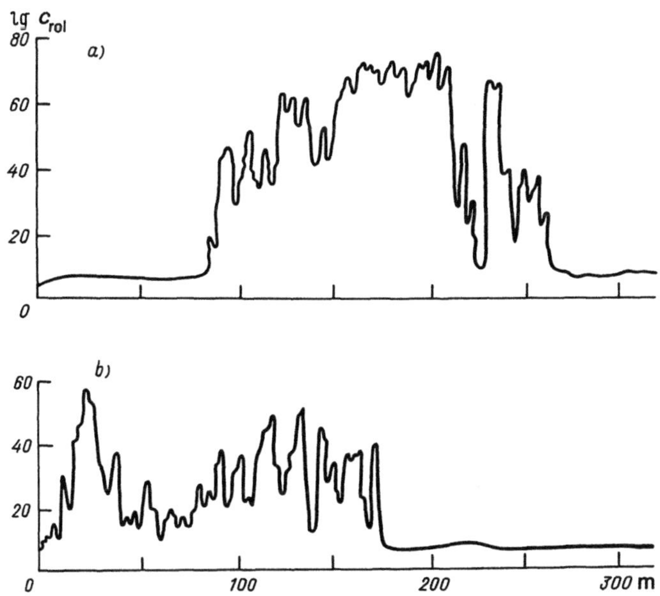

Fig. 2.11. Records of horizontal distribution of contaminant concentration (in relative units) at the depths of 5 m (a) and 10 m (b) (based on the 1966 experiment in the Black Sea). y-axis: lg c_{rel}; x-axis: m.

In the Black Sea experiments in 1966, measurements of dye-concentration distributions in diffusing spots at various depths were made. Here, the interesting fact was found when the indicator layers shifted relatively to each other at different depths. Figure 2.11 gives the distributions of dye concentrations at the depths of 5 and 10 m. As is seen, the shift of the dyed layers relative to each other reached to 80 – 90 m. This phenomenon might, apparently, be caused by the changing with depth of the current-velocity vector which occurs, for example, in drift currents (the Ekman current). The velocity shears in drift currents are hard to observe with conventional current meters since they are not large, as a rule, and can be obscured by other processes. In the case being considered, the relative shift of water layers 5 m away from each other reached 80 – 90 m after approximately 2.5 hours since the diffusion has begun. Therefore, the mean velocity difference in these layers was always 2 cm/s, which is difficult to detect by conventional current meters. A similar shift in rodamin-dyed layers in the upper sea, clear from an aerial photograph, was reported by the authors of paper [131], who suggested a 'comet-like' model for the diffusing spot, with a higher dye concentration in the forward (in the current direction) part of the spot as compared to that in the rear, deepened.

The current-velocity shear (along with the anisotropy of turbulence) may lead, as was shown in Part 1.6, to the stretching and rotation of diffusion dye spots. Experimental evidence of such phenomena were obtained during the experiments in the Black Sea in 1967 [88]. A release of two spots with the use of 5 and 13 kg of rodamin was performed. Observation of the first spot lasted for about 6 hours and of the second for more than 16 hours. We must also note that dye-spot measurements with simultaneous determination of the location of the ship relative to the on-shore referencing points (or by other observation means) enables us to quite accurately determine mean velocities in the sea. This method of measuring velocities is free of the errors inherent in the method of drifting floats, which are always, to this or that extent, subject to the influence of wind.

The mean current velocities in the 1967 experiments reached as much as 63 cm/s. Such fast currents and shear effects related to them led to charge elongations of the spots in the course of diffusion. For example, the first spot, 2 hours after the moment of release, was 162 m long and 80 m wide (in the threshold instrument sensitivity). Five hours and 45 minutes from the release the length and wide were 275 and 87 m, correspondingly. Even greater elongation was observed in the second experiment. Three hours from the release moment the spot was 725 m long and 162 m wide, and after 4 hours and 45 minutes the length of the spot reached 1137 m while its wide had reduced to 80 m. Figure 2.12 presents schematic pictures of the spot at the moments of time mentioned. It also give mean current directions and those for the wind and coastline in the area of observation. During the first experiment, the wind velocity was 5 – 6.5 m/s, and during the second one, it varied from 10 – 6 m/s, which its direction changing from north-west to north-north-east. As is seen from the figure, the main axes of the spots generally do not coincide with either of these directions.

A reason for the elongation of spots may be the anisotropy of turbulent transport in the coastal sea area. Some authors have mentioned the possibility of such an anisotropy, e.g. Gezentsvei [20] for the Black Sea coastal zone. These conclusions, however, were obtained for processes with time scales of several

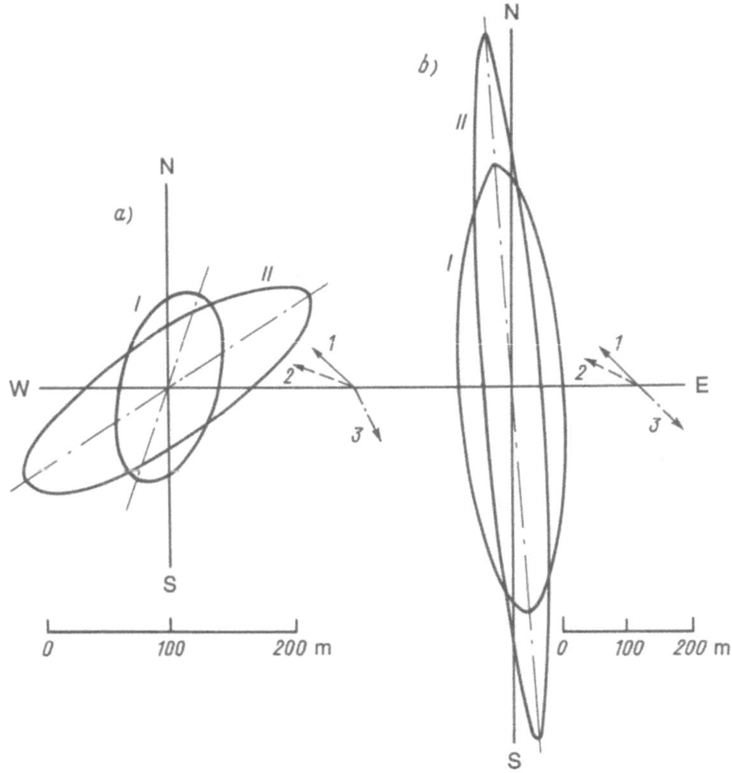

Fig. 2.12. Schemes of two diffusing contaminant spots (according to the 1967 experiment in the Black Sea). Spot a – 120 min (I) and 345 min (II) after release moment. Spot b – 180 min (I) and 255 min (II) after release moment. Figures 1, 2 and 3 refer to the directions of current, wind and coastline in observation area.

weeks. But in the experiments being described this scale was much smaller, therefore the supposition of the influence of the coastline on the process does not seem well grounded. Mainly, the elongation of spots was obviously influenced by the combined effects of vertical diffusion and vertical and horizontal gradients of current velocity. But if to estimate these effects we take, for example, the mean velocity differences from the shore to the observation point and from the sea surface to the sea bed, according to formula (1.117), then these velocity gradients will appear to be inadequate to explain such intensive deformations of the spots. Apparently the essential, for the phenomenon being considered, current-velocity gradients can be related to the meso-scale inhomogeneities of the velocity field, in particular, with the so-called Langmuir circulations which often appear in the sea lines (along which the so-called 'streaks' or 'windows' are formed on the sea surface) form current convergence where floating objects and foam accumulated. Velocity gradients in such circulations of waters can reach rather noticeable values because the distance between the neighboring convergence lines usually do not exceed several hundred meters. In the 1967 diffusion experiments, apparently, the dye-spot elongation observed was accounted for by such wind circulations combined with the strong fringing current. Similar phenomena were noticed visually by Ichiye [128], Krauss [133] and other authors in experiments with rodamin spots. In their experiments the rodamin spots were often split, under sufficiently strong winds, into a number of streaks parallel to the wind direction.

In approximately the same years, diffusion experiments were carried out in various ocean areas. For example, several of them were conducted off the coast of the Netherlands with the use of instantaneous point sources of dye [10]. Indicator concentrations were measured in the samples as well by aerial photographic surveys. Attempts were also made to compare the observed concentration distributions with the available theoretical expressions; however, they failed to give preference to either of them. This obviously, is explained by the complicated shape of the shoreline, variable seabed topography and other factors influencing the process in the observation area but neglected in theoretical models. Ichiye [128] managed to approximate the dependence of the dispersion of dye spots on the time of diffusion by square and cubic laws. According to his observations, the elongation of spots occurs more frequently in the direction of the wind or current, and the distortion of the underwater parts of the spot occurs more often in the clock-wise direction (following the Ekman spiral).

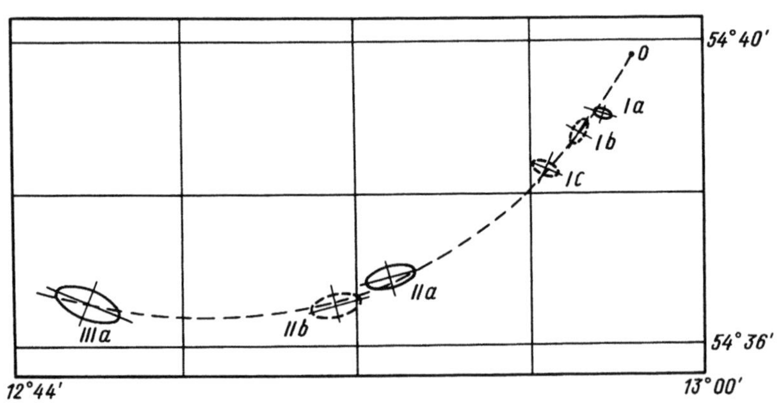

Fig. 2.13. Trajectories of dye spot during the August 23, 1968, experiment in the Baltic Sea. Spot location during surveys I, II and III at horizons 1 (a), 4 (b) and 6 (c) m. 0 is the dye release point. The intervals from dye release to survey Ia – 92 min; to survey IIa – 340 min; to survey IIb – 390 min; to survey IIIa – 516 min.

A larger series of diffusion experiments was carried out in 1968 in the Baltic Sea in a joint expedition of COMECON oceanologists [82]. The work was done from the research vessel *Professor A. Penk*, belonging to the GDR Academy of Sciences. Towing the fluorimeter made it possible to obtain sufficiently detailed three-dimensional pictures of diffusing dye spots, with accurate referencing of spot locations to the system of coordinates [115]. Figure 2.13 gives an example of a contaminant spot trajectory, according to the measurements taken in the 1, 4 and 6 m horizons during three surveys. The spot is contoured by the isoline of dye concentration of 10^{-9} g/cm^3. Individual records of dye concentrations in the cross-sections of the spot underwent smoothing by a cosinus-filter. Figure 2.14 presents an example of such smoothing with the scales of averaging 15, 35, 70 and 140 m. Pictures of dye-concentration isolines in the diffusing spots for the moments of time and horizons, corresponding to those given in Figure 2.3, are shown in Figure 2.15. These figures also present the trajectories of the measurement tacks and the north-south direction. The spots are bounded in the pictures by the isolines of dye concentration of 10^{-9} g/cm^3. Similar pictures of dye-concentration isolines in the experiment with a second spot are given in Figure 2.16. Using the obtained pictures of concentration distributions for four

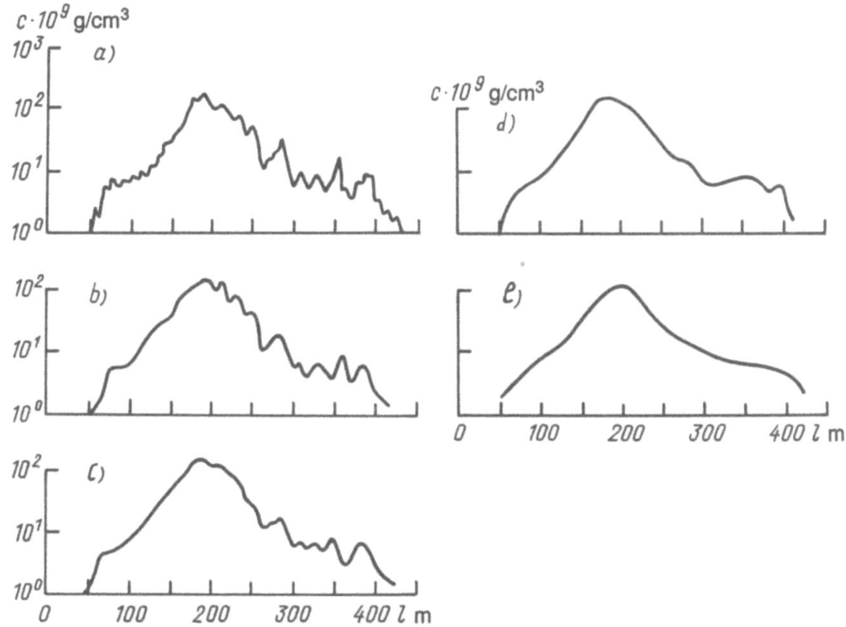

Fig. 2.14. An example of smoothing contaminant concentration records by cosinus-filter with averaging scales 15 (b), 35 (c), 70 (d) and 140 (e) m. Curve *a* is given without smoothing. *y*-axis: $c \cdot 10^9$ g/cm^3; *x*-axis: *l* m.

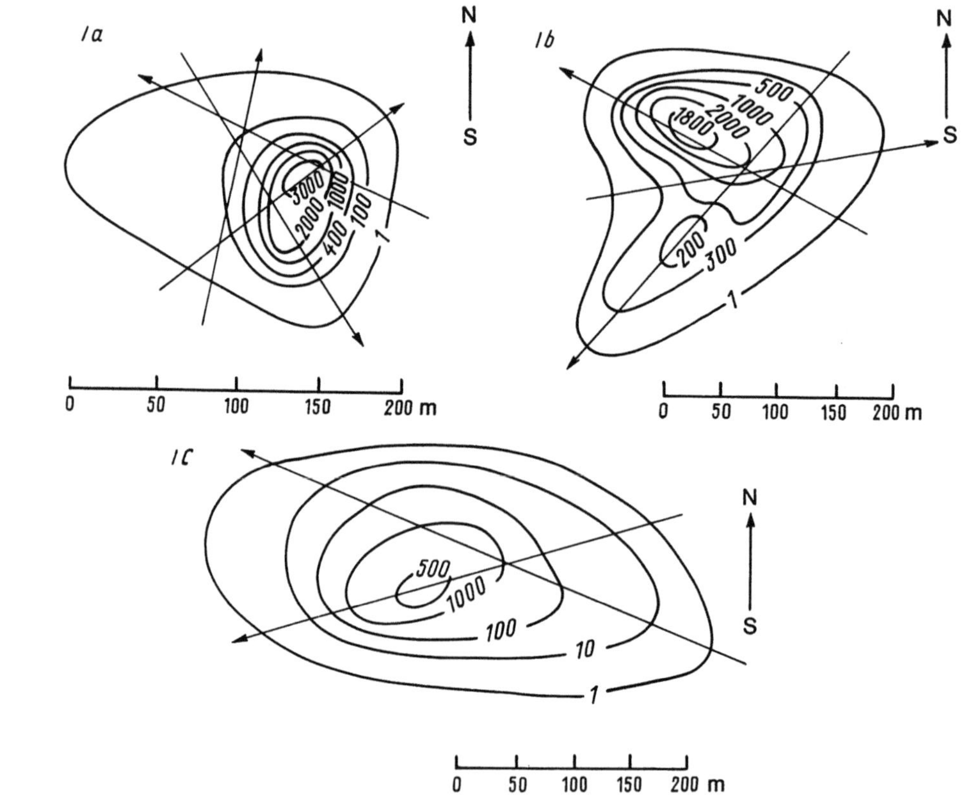

dye releases, dependence of maximum dye concentrations in the centers of the spots on the time of diffusion was plotted. these dependences for different horizons of observation are given in Figure 2.17. As is seen from the figure, all the data obtained during the Black Sea experiment satisfy very well the inverse cubic dependence of dye concentrations in the center of the spots on the time of diffusion. The approximation of dye-concentration-distribution pictures by the solutions of diffusion equations without advective terms but with different diffusion coefficients for the three coordinate axes (K_x, K_y, K_z) made it possible to evaluate these coefficients. The data obtained from two experiments on the horizontal diffusion coefficients are given in Table II.I. In those experiments the vertical-diffusion coefficients K_z appeared to be equal to 4 and 5 cm^2/s, i.e., they were, as had been expected, many times smaller than the values of K_x and K_y which, as it is easy to see from the table, tend to increase with the growth of diffusion time. The fact that the coefficient K_x (in the direction of the mean current) significantly exceeds the coefficient K_y is accounted for by the actual inclusion into K_y of the effects of current-velocity shears, neglected when approximating experimental distributions by model curves. Thus in this case, the coefficient K_l^{eff}, which is determined, for example by formula (1.112), was in fact estimated.

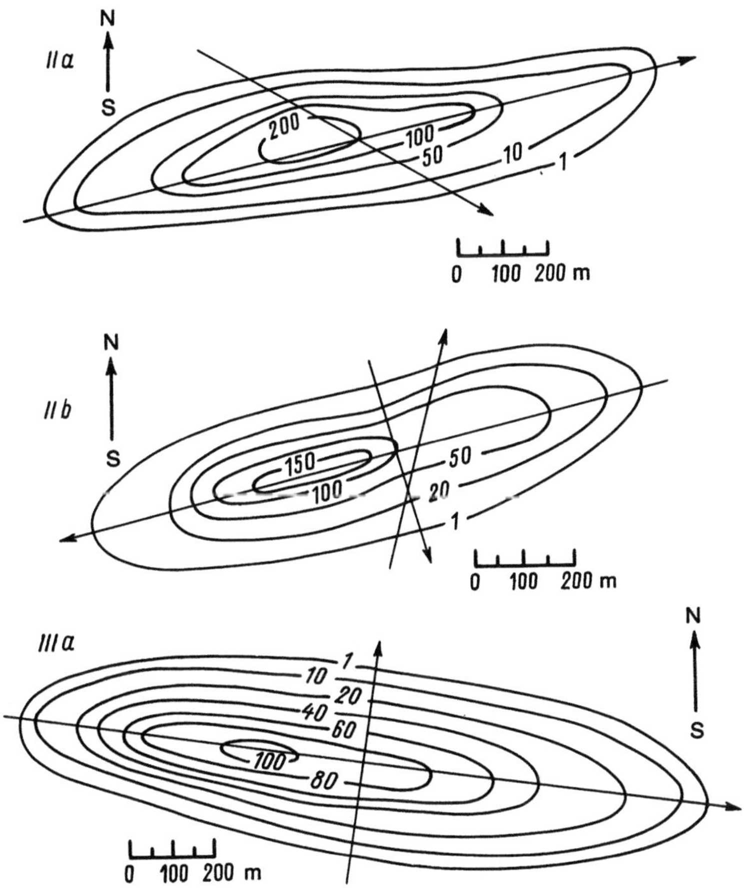

Fig. 2.15. Pictures of dye-concentration distribution in diffusing spot in the August 23, 1968, experiment at the moments of time and horizons corresponding to those in Figure 2.13.

TABLE II.I
Horizontal diffusion coefficients (according to the Baltic Sea experiments)

Experiment date	Observation horizon, m	Time elapsed from dye release, min.	K_x cm^2/s	K_y cm^2/s
08.23.68	0	92	420	70
	0	340	2630	107
	4	390	6740	382
	0	516	26800	1060
08.30.68	0	142	2330	642
	5	180	5130	733
	5	493	2380	210
	7	523	1950	103

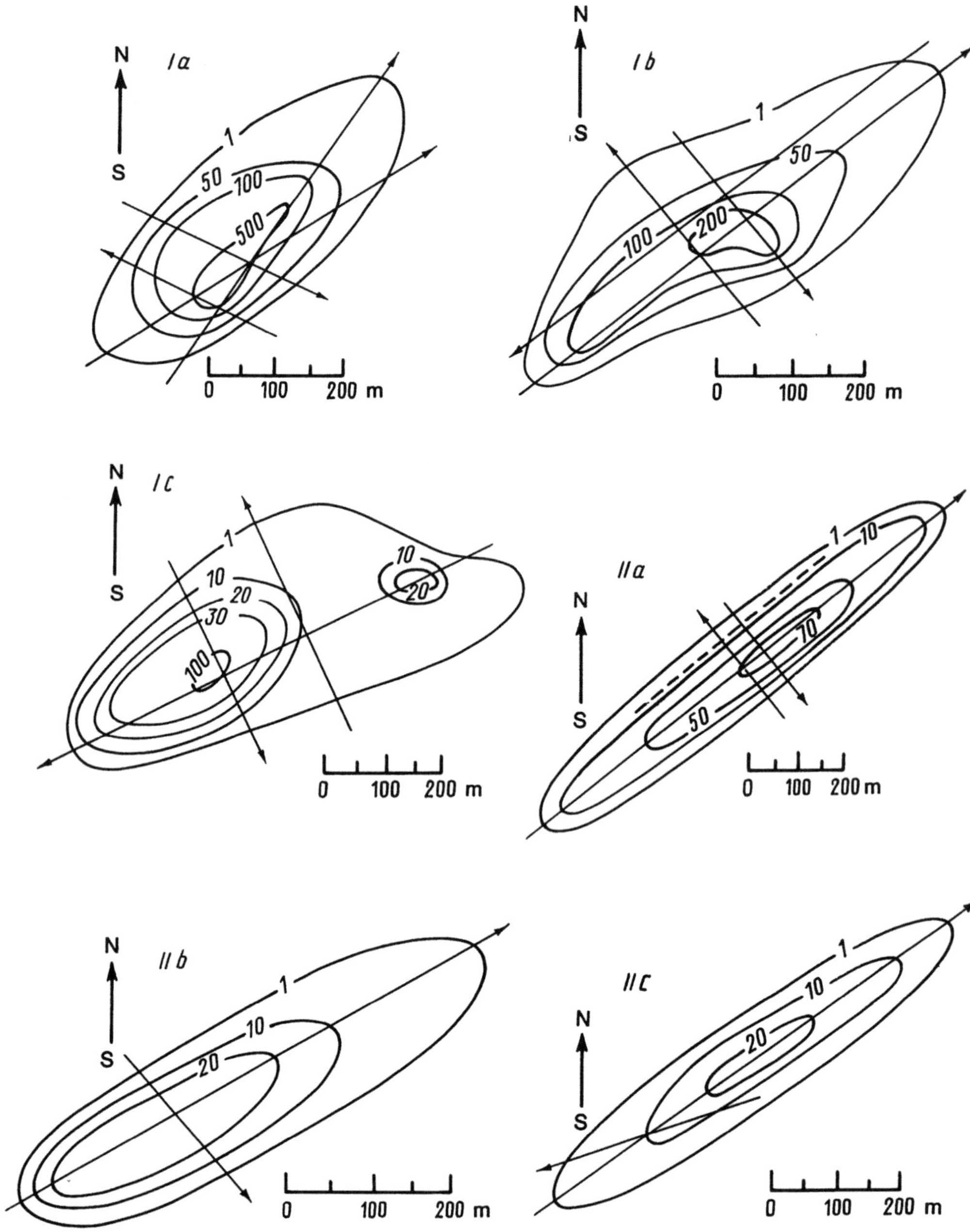

Fig. 2.16. Pictures of dye-concentration distribution in diffusing spot in the August 30, 1968, experiment at horizons I (a), 5 (b) and 7 (c) m in the course of two surveys (I and II). Time intervals from dye release to survey Ia – 142 min; to survey Ib – 180 min; to survey IIb – 493 min; to survey IIc – 523 min.

A detailed study of the influence of current-velocity gradients on the contaminant-spot diffusion was conducted during the 1968 experiments in the Black Sea

[88]. Twenty-five kilograms of rodamin were released to form the spot. Two buoy stations with current-velocity meters were positioned in the area of experiment. Three surveys were made in the course of the experiment, with one crossing along and several crossings across the elongated spot being made in each survey. Contaminant-concentration distributions in spots for each survey are given in Figure 2.18. As in the 1967 experiments, elongation of dye spots in the diffusion process was observed. For example, 2 hours and 16 minutes from the release mo-

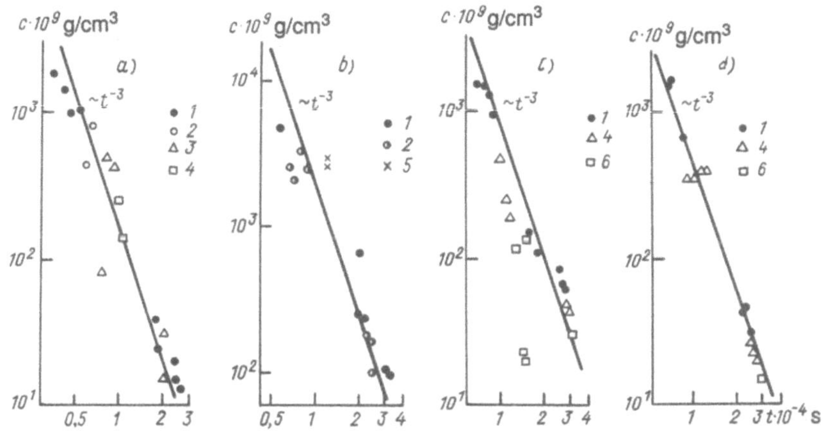

Fig. 2.17. Dependence of maximum dye concentration in the center of spots in the Baltic Sea experiments of August 21 (a), August 23 (b), August 30 (c) and September 1 (d), 1968, at horizons 1 (1), 3 (2), 4 (3), 5 (4), 6 (5) and 7 (d) m. *y*-axis: $c \cdot 10^9$ g/cm^3; *x*-axis: $t \cdot 10^{-4}$ s.

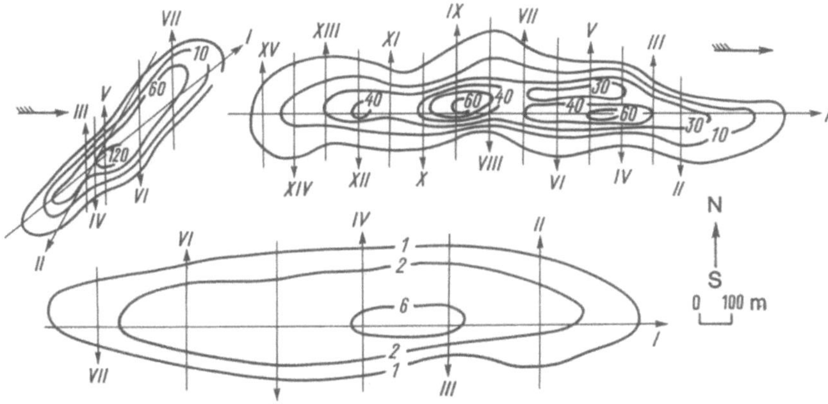

Fig. 2.18. Dye-concentration distribution in diffusing spot during three surveys in the 1968 experiment in the Black Sea (in units of 10^{-9} g/cm^3). Longer arrows correspond to ship tacks during measurements; shorter arrows are for wind direction. Roman figures denote numbers of tacks in each survey.

ment (the middle of the survey period was taken for the moment of survey) the spot was 750 m long and 160 m wide (within the isoline 10^{-9} g/cm^3). After 4 hours and 33 minutes the length of the spot reached 1650 m and its width 300 m. Seven hours and 46 minutes from the release moment, the length of the spot exceeded 2200 m while the width was 400 m. Thus, the length-to-width ratio was equal to 4.7, 5.5 and 5.5 for each survey, respectively. The inclination angle of the longtitudinal axis of the spot to the axis X, directed along a parallel of latitude, also varied with time. This angle appeared to be 48° in the first survey and about 0° in the two following ones. The observed elongation of the spot and the changing orientation of its longtitudinal axis in the time period from the release moment to the first survey was explained by the joint effect of the vertical turbulent transport and vertical current-velocity gradient. The experimental result interpretation was kept within the framework of the diffusion equation (1.126). In that case, critical time, the inclination angle of the ellipse axis to axis X and the ellipse half-axes ratio were described, respectively, by the expressions (1.127) – (1.129). For the phenomenon scale in question, the ratio K_L/K_z, according to the observations in the Black Sea and Baltic Sea, turned out to be close to 20. Then transforming (1.129) to the form

$$\sqrt{a_1^2 + a_2^2} = \frac{1}{t}\sqrt{12(\lambda^2 - 1)\frac{K_L}{K_z}} \tag{2.5}$$

and using the schematic pictures of the diffusing spots in the form of ellipse instead of the complicated curves shown in Figure 2.18, for the vertical-velocity gradient we obtain $\sqrt{a_1^2 + a_2^2} = 58.5\cdot10^{-4}\ \text{s}^{-1}$. A direct calculation of the vertical-velocity gradient with the use of the current-meter data from the buoy stations in the area of the experiment resulted in $59.9\cdot10^{-4}\ \text{s}^{-1}$ (in the case of a 6 hour averaging period for velocity values). There is also good agreement for the angle of inclination of the larger axis of the ellipse to axis X : 48° according to the experimental data and 54° according to formula (1.128). The critical time, as calculated with the use of formula (1.127), turned out to be 43 minutes; i.e., the first survey took place at a time when it is already necessary to take into account the effect of vertical shear of current velocity.

In order to explain the dye-spot behavior in the period from the first to the third surveys it appeared necessary to take into account the influence of horizontal current gradients on the process of diffusion. The following dependence of the velocity components on the horizontal coordinates was used:

$$u = u_0 + ax + by; \qquad v = v_0 - ax - by, \tag{2.6}$$

where the constants a and b were estimated based on the current-velocity measurements at the buoy stations. The solution of the two-dimensional diffusion equation with velocity components of the form (2.6) can be derived from a more general solution (1.117) in a coordinate system moving along the axes x and y with the velocities U_0 and V_0, respectively. The solution thus found was used to plot ellipses of concentration isolines. When compared with the diffusion pictures observed they showed good agreement. Therefore, for the first stage of diffusion the main determining factor was the effect of vertical current-velocity gradients. But when the diffusing spot grew in size the influence of horizontal velocity gradients was becoming dominant.

Based on the data of the Baltic Sea diffusion experiments, the important characteristics of dye spots – dispersions of dye-concentration distributions along the horizontal axes x and y – were calculated. Since concentration is a function of the coordinates and time dispersions σ_x^2 and σ_y^2 were determined according to the formulas

$$\sigma_x^2(t) = \frac{\iiint_{-\infty}^{\infty} x^2 c(x, y, z, t)\mathrm{d}x\mathrm{d}y\mathrm{d}z}{\iiint_{-\infty}^{\infty} c(x, y, z, t)\mathrm{d}x\mathrm{d}y\mathrm{d}z}; \tag{2.7}$$

$$\sigma_y^2(t) = \frac{\iiint_{-\infty}^{\infty} y^2 c(x, y, z, t)\mathrm{d}x\mathrm{d}y\mathrm{d}z}{\iiint_{-\infty}^{\infty} c(x, y, z, t)\mathrm{d}x\mathrm{d}y\mathrm{d}z}. \tag{2.8}$$

The dependence of the dispersions σ_x^2 and σ_y^2 on the diffusion time t appeared to vary considerably from one experiment to another. When approximating this dependence by power functions of t the exponents varied from 1 to 3, with the diffusion times lasting for about several hours (see ref.[55]). We ought to remind the reader that such dispersion dependences on time have resulted from horizontally symmetrical diffusion models with different forms of setting the dependence of horizontal diffusion coefficients on the phenomenon scale.

Interesting results were also obtained in the Baltic Sea on the contaminant-concentration decrease rates in the center of the spot. Under power approximation, these dependences resulted in exponents mainly in the range from −2 to −3. Measuring mean hydrometeorological characteristics during experiments on diffusion made it possible to study the relation between the exponent and the determining factors. For example, in one of the experiments in the Baltic Sea was noted that the exponent −3 was characteristic in those cases in which the current-velocity vector rotated with depth, while in the conditions of a uniform flow the figure was closer to −2. Under rapid variations of hydrometeorological conditions in the course of an experiment, power approximation of the obtained dependences appeared to be difficult [9]. It was also found out that with increasing diffusion time the square law tends to transit into the cubic law. This transition was observed in a coastal sea area 2 hours after the beginning of diffusion, with this time-lapse regularly increasing with the growing distance from the experiment site to the shore [117]. The 1974 experiments also revealed a dependence of the exponent on the wind velocity observed in the course of experiment.

In more detail, the problem of relation between dye-concentration decrease rates in the center of the spot and hydrometeorological conditions was studied by Kremzer [54]. The vast amount of information from the experiments jointly conducted by COMECON oceanologists in 1968 – 1978 was analyzed. It turned out that in those experiments the exponents in the dependence $C_{max} \sim t^n$ changed in the range from −0.3 to 11. At the same time, according to the theoretical models with constant diffusion coefficients, the exponent n, as was seen from Chapter 1, could only change from −1.0 to −4.5. However, in order to explain a wider range of variability of n it should apparently be supposed that the transport coefficients themselves can be variable. In a general case (for constant transport coefficients) C_{max} can be considered proportional to $(K_{xx}\ K_{yy}\ K_{zz})^{1/2}\ t^{-n}$.

Fig. 2.19. Variability of maximum dye concentration in the center of diffusing spot during the August 3, 1968, experiment in the Baltic Sea (after [54]). Figures denote exponents n^* in the formula (2.9). y-axis: $C_{max} \cdot 10^9$ g/cm^3; x-axis: $5 \cdot 10^4$ t s.

But if the diffusion coefficients themselves can change with the time of diffusion according to power dependences, then for C_{max} we will obtain

$$C_{max} \sim t^{-n+(\lambda_1+\lambda_2+\lambda_3)/2} = t^{-n^*}, \tag{2.9}$$

where λ_1, λ_2 and λ_3 are the exponents in the dependence of transport coefficients on t. The dynamic activity of the water medium, caused primarily by the influence of wind, should apparently be considered the main factor intensifying turbulent transport in the sea. The dependence of the exponent n^* on the variability of the wind velocity V was traced with the help of the data from the diffusion experiment conducted on August 30, 1968, off the Rügen Island in the Baltic Sea. In the course of that experiment, extremely unstable winds were observed, and the exponent n^* appeared to be highly varying (see Figure 2.19). Figure 2.20 presents curves of the variations of the exponent n^*, observed during the experiment, and the time derivative of the wind velocity observed over the area of experiment. As is seen from the figure, the dependence of n^* on dV/dt is clearly evident.

The experiment carried out on March 7, 1977, in the western part of the Baltic Sea showed very large variations of the exponent n^*: from −10.6 to −1.4 [54]. This, apparently, was related to the severe storm (with winds up to 17.4 m/s)

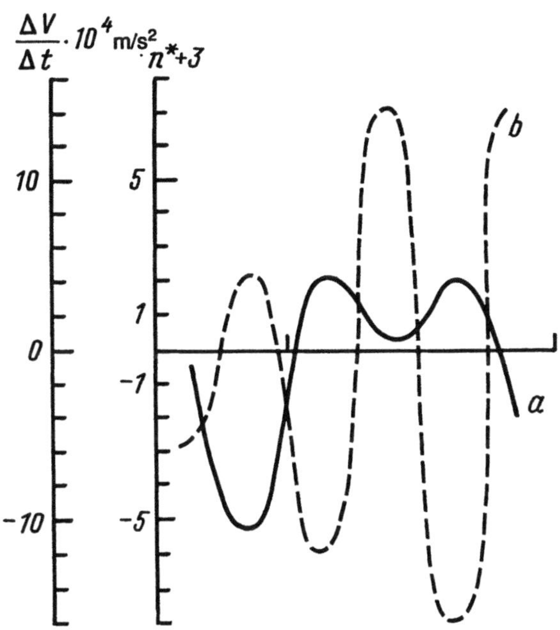

Fig. 2.20. Relation between exponent n^* (a) and wind-velocity variability dV/dt (b) in the August 30, 1968, experiment (after [54]). y-axis: $\Delta V/\Delta t$ 10^4 m/s^2.

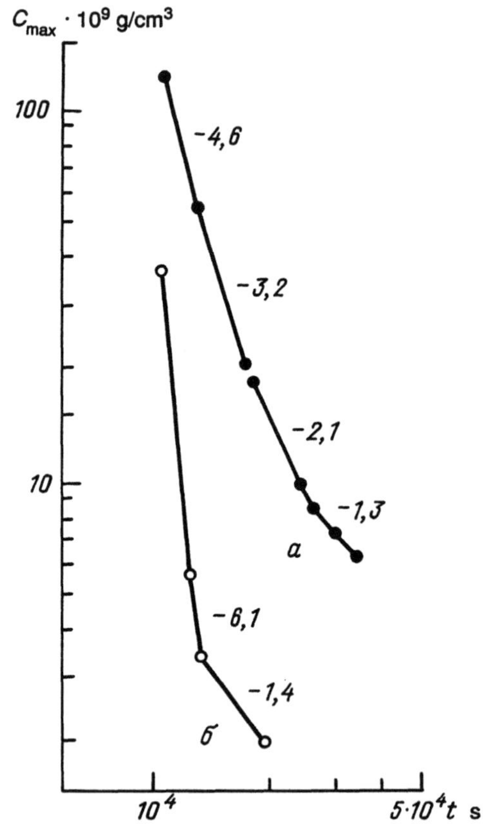

Fig. 2.21. Maximum dye-concentration variations in the centers of diffusing spots with time in the October 13, 1974 (a) and March 7, 1977 (b) experiments. Figures denote exponents n^* in the formula (2.9) (after [54]). y-axis: $C_{max} \cdot 10^9$ g/cm^3; x-axis: $5 \cdot 10^4$ t s.

that had passed over the area of field observations on March 5 but had completely subsided by March 7. Figure 2.21b shows that the exponent n^* decreased with time, which was evidence for the subsiding sea-water turbulence causes by the storm. During another experiment in the southern part of the Baltic Sea, held October 13, 1974, from the moment of dye release until $t = 2 \cdot 10^4$ s the wind velocity increased from 1.5 to 12.6 m/s then declined to 2.6 m/s. At the same time, the current velocity increased over the period $2 \cdot 10^3 < t < 1.9 \cdot 10^4$ s from 7 to 20 cm/s. Such a combination of determining factors led to a monotonic decrease (taken absolutely) in n^* from 4.6 to 1.3 (see Figure 2.21a). On the contrary, in some other experiments the module of the exponent displayed its increase with time (see Figure 2.22). This phenomenon, however, was not often easily related to the increasing velocities of wind or currents in the area of

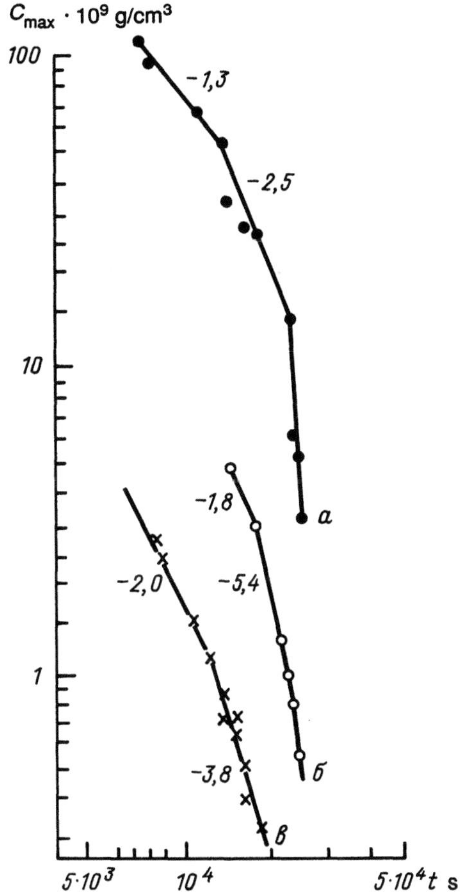

Fig. 2.22. The same as in Figure 2.21 for the October 9, 1974 (a), June 27, 1971 (b) and June 30, 1971 (c) experiments (after [54]). *y*-axis: $C_{max} \cdot 10^9$ g/cm^3; *x*-axis: $5 \cdot 10^4$ t s.

observation. Increasing turbulent transport in these cases can be accounted for by other reasons, for example, by the entry into the area of observation of a large turbulent eddy, which cannot be determined only by observing mean current and wind velocities. The increase in transport coefficient values under this or that form of sea-water turbulence can be estimated by the formula (2.9). For instance, according to the data of the experiment conducted on October 9, 1974, the product of the coefficient K_{xx} K_{yy} K_{zz} increased ninefold over about one hour, while, according to the data of the August 30, 1968, experiment, it decreased twice. The values of the diffusion coefficients themselves, according to the Baltic Sea experiments, turned out to be the following. On the average, vertical-diffusion coefficients for the upper sea layers were on the order of several square centimeters per second, with this figure, however, varying from one experiment to another from as little as hundredths to dozens of square centimeters per second. Horizontal-diffusion coefficients were growing substantially, as had been expected, with the growing scale of the phenomenon (diffusion time). For instance, 1 – 2 hours after the dye point source began to operate, when the spot sizes were about a hundred meters, the horizontal diffusion coefficients usually did not exceed several hundred square centimeters per second, while at diffusion times of about 10 hours (with spot sizes of about a kilometer) they increased up to thousands and even dozens of thousands of square centimeters per second.

Certainly, these figures varied considerably in individual experiments, which in some cases could be related to the hydrometeorological conditions in the course of the experiment. The characteristic diffusion velocity u_1 (determined e.g., by the formula $u_1 = K/l$, where K is the coefficient of diffusion in this or that direction and l is the scale of the phenomenon) also varied greatly in different experiments, usually amounting to hundredths to tenths of centimeters per second.

Massive diffusion studies, which enable the determination of basic diffusion parameters as well as their dependences on the determining factors, were carried out in the following years in the Black Sea by Zats *et al.* [34] and also during the joint ventures of the COMECON oceanologists off the Bulgarian coasts [11, 12]. For example, in 1977, during the international joint venture Kamchiya–77, experiments with instantaneous dye sources were staged near a trestle from where a complex of hydrological and meteorological observations was made. In those experiments, the diffusing dye spots often split into fragments elongated in the direction of wind or of the front of swell waves refracting near the shore. The decrease in maximum dye concentrations in the centers of the spots was again well described by power dependences with exponents close, on the average, to –2.5 (see Figure 2.23). Similar results were obtained in the coastal area of the Baltic Sea during the 1974 and 1976 joint experiments on diffusion conducted by COMECON oceanologists.

Immense amounts of information on dye-spot diffusion from instantaneous point sources was obtained through aerial photographic surveys. The aerial methods of studying the diffusion of contaminants in the sea were put into operation in the early 1960's. A description of the methods of using aerial photographs to follow diffusion dye tracer spots was made by Katz *et al.* [131] and Pritchard *et al.* [151]. Aerial photographic surveys of dye tracer spots in coastal sea areas were done by Japanese scientists (see ref. [34]). Diffusion studies in the Black Sea using aerial photography were started in 1968 and continued in the following years [150]. In 1969 – 1970, a series of experiments was conducted off the Crimean shores [34]. Spot releasing was performed from a vessel staying at a distance of 4 – 6 miles from the shore. Aerial photographic survey was done with the use of spectral-zonal film with the frame sizes of 18×18 and 30×30 cm. The intervals

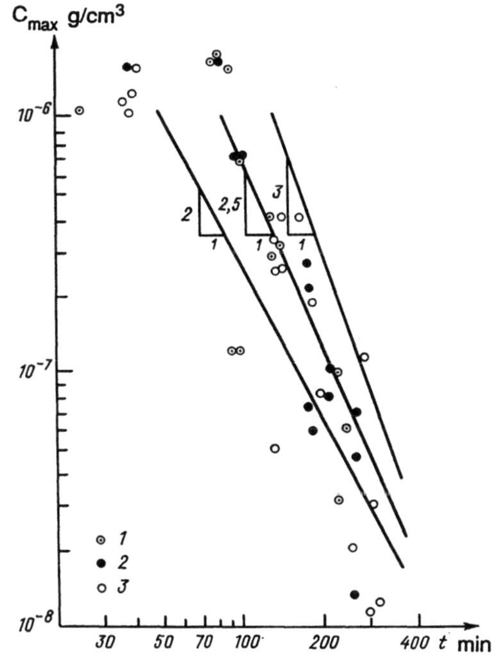

Fig. 2.23. Dependence of maximum dye concentrations in the centers of three spots in the October 7, 1977, experiment in the Black Sea off the Bulgarian coast (after Zats, Nemirovsky and others in [11]). *y*-axis: C_{max} g/cm^3; *x*-axis: *t* min.

between the sequent shots were 5 to 10 minutes during the initial stage of diffusion and in creased to 20 to 30 minutes at the end of the photographing. The overall duration of the experiment with one spot was 5 to 8 hours. In the 1969 experiments, the spots released contained from 17 to 30 kg of dye tracer whose solution was stored on board the ship in tanks capacities from 4 to 8 m^3. From the tanks, the solution was passed over onto the sea with the use of a powerful pump by hose whose free end was kept afloat by floats at a distance of 15 – 20 m from the ship. The duration of solution release by the pump was about 5 minutes, after which the spot area usually was about 50 – 200 m^2, while in the course of diffusion it grew up to 80 – 200 thousand m^2. In the 1970 experiments, the technique of contaminant-spot formation was altered. A dye solution was poured into a tank (having the form of a conical frustum) made of a rubber-proofed fabric, with the upper side of the cone being fixed to a metal ring. The ring, in turn, was fastened by special locks to a round rubber pontoon floating on the water's surface. When the tank, open on top, was filled with dye the locks released the ring, and it descended, forcing down the wall of the tank. In other words, the tank was somehow turned inside out, letting the dye solution freely diffuse in the upper sea layer. The original volume of that source was about 5 m^3 which can, with a fair approximation, be considered a 'point' one for experiments of such a scale. In the 1970 experiments, at the maximum stages of the development of spots from such dye sources, the spot areas reached a hundred thousand square meters. At the same time, the depth of dye penetration usually did not exceed 2–5 meters.

Based on the results of the 1969–1970 aerial photographic surveys of dye spots in the Black Sea, the areas covered by dye tracers and changing in the course of diffusion were determined. The spot areas were calculated based on the spot margins visible on the photographs. These margins referred to the isolines of a certain dye concentration corresponding to the threshold sensitivity of the film. Certainly, this assumption is not quite correct since the darkening of a negative is accounted for, by a number of factors, and not solely by contaminant concentration. However, under approximately similar hydrometeorological conditions and illumination intensity of the sea this assumption apparently does not extend far from reality. In particular, this was proven by using synchronic contact methods of dye-concentration measurements along with aerial photography. Based on the dye-spot areas S, obtained through planimetry, the radii of equivalent circles R were further calculated. The calculation was made according to the formula $R = \sqrt{S/\pi}$. In most experiments, observations were made at the stage of growth of the spot, i.e., from R being close to zero to R_{max}. On reaching the spot size R_{max} the spot begins to diminish (within the boundaries of an isoline corresponding to the film sensitivity threshold!) and, finally, disappears completely. At the stage of growth of the spot, the dependence of its area on the time of diffusion t can be, at a fair rate, approximated by a power expression of the form $S \sim t^n$. According to the data from the 1968–1969 experiments in the Black Sea, the exponent n varied from 0.63 to 1.92, with the spot radii varying from 20 to 350 m. Here, it is natural that $n \longrightarrow 0$ at $R \longrightarrow R_{max}$. Values for n, according to the Japanese scientists, were within the range of 0.93 to 1.96, with the spot radii varying from 40 to 378 m.

It would be more correct, naturally, to approximate the dependence $S(t)$ [or $R(t)$] not by a power function but by some function increasing as the spot grows and decreasing when the spot is beginning to disperse. Function $R(t)$ for horizontal two-dimensional diffusion with a constant diffusion coefficient K can be obtained from the fundamental solution (1.13) of a diffusion equation, if assuming that the concentration values c (z, y, t) are equal to a certain fixed threshold value c_δ. Simple computations will lead to an expression for $R(t)$ of the form

$$R(t) = \sqrt{-4K_L t \ln(c_0 4\pi K_L t)}. \tag{2.10}$$

if there is a mean current in the sea with a constant velocity U_0 (along the axis x) the diffusion time t in the formula (2.10) is substituted by x/u_0, where

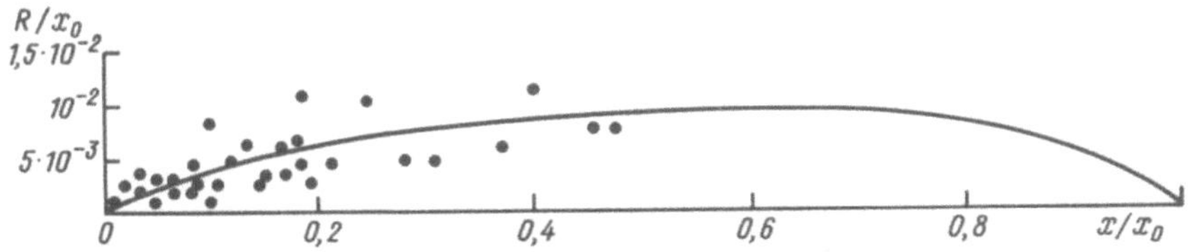

Fig. 2.24. Dependence of visible non-dimensional radius R/x_0 of diffusing spot on non-dimensional distance x/x_0.

x is the distance covered by the central point of the diffusing spot over the time t from the dye release point (the origin of coordinates).

A graph of the function (2.10) in non-dimensional variables R/x_0 and x/x_0, where x_0 is the coordinate at which the spot completely disappears, is given in Figure 2.24. Also plotted there are experimental points obtained by other authors. On the whole, the experimental points agree fairly with the theoretical curve, but only for the stage of growth of the spot since there were no data available for its diminishing stage.

Studies on dye-spot diffusion in the Black Sea with the use of aerial photographic surveys were continued in the following years [150]. The photographic cameras used in those experiments has resolutions of 40 lines per mm for the center of the frame and up to 20 lines per mm for the frame edge. The quality of pictures depending on the illumination intensity of the sea, the state of its surface, camera characteristics and the method of photographing were also studied. The deviations of the main axle of the camera from the vertical direction did not exceed $1-2°$ during the photographing, and the accuracy of determining the flight height by radiometer was $\pm$ 1.5 m. The scales of the pictures varied from 1 : 2500 to 1 : 7500. Microphotometers and densitometers, particularly systems translating videopictures into digital form suitable for computer processing, were used for determining the degree of darkening of the negatives, with the number of density discreteness levels for the darkening of the negative equalling 10. Calibration of photos was made with the help of simultaneous contact measurement data on dye concentrations obtained by towed fluorimeters.

Based on the aerial photographs of dye spots made in 1981–1982 off the shores of the Crimea and Sukhumi, diffusion of the spots within the diffusion time intervals from 10^2 to 10^5 s was investigated. The spot area when approximating the function $S(t)$ by a power dependence increased with time, with n being within the range of 1.11 to 1.60. In some cases, better approximation was achieved by using power expressions with different n for separate intervals of t. Variability of spot anisotropy with time was also observed. The ratio of the larger axis of the spot to its mean diameter α was a function of the diffusion time t. At earlier stages of diffusion (up to 30–60 min after release) the spots were usually sim-

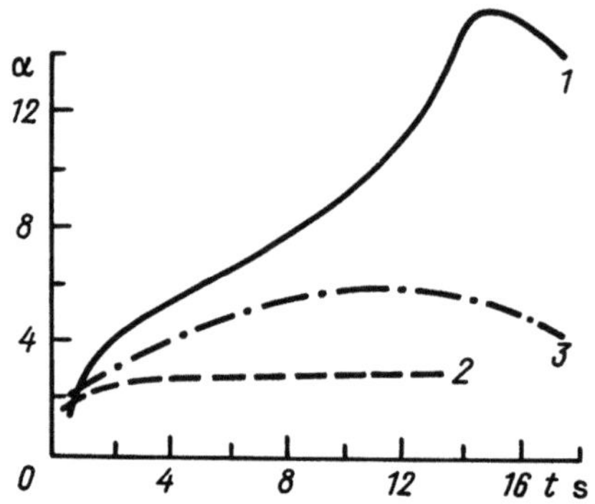

Fig. 2.25. Dependence of spot anisotropy parameter α on time, according to three experiments in the Black Sea (after [150]). *x*-axis: *t* s.

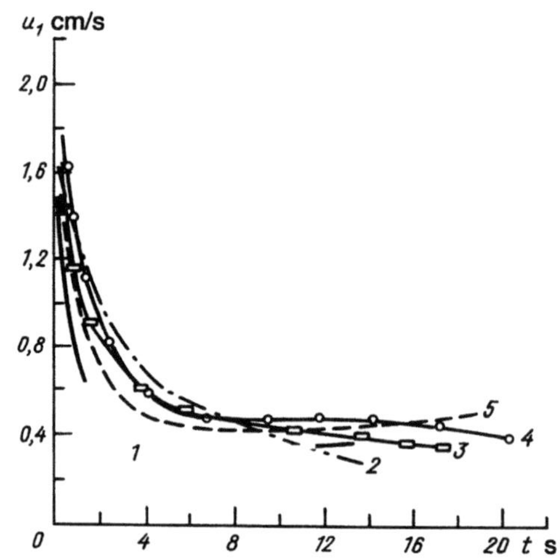

Fig. 2.26. Dependence of diffusion velocity u_1 on time, according to five diffusion experiments in the Black Sea (after [150]). *y*-axis: u_1 cm/s; *x*-axis: *t* s.

ilar in their shapes, and then they began to deform, with α increasing from 1 – 15 (see Figure 2.25). On the contrary, the diffusion velocity u_1 usually declined with the increase of *t*, staying within the range of 0.31 – 2.1 cm/s (see Figure 2.26).

Aerial photographs of dye spots allows us to form an opinion of the influence of the effects of velocity shears on the process of diffusion. Values for velocity shears were determined in the 1981 – 1982 experiments based on the current-velocity data obtained at buoy stations. Horizontal variations of the velocity components appeared, according to those data, to lie within $(0.1-14.5)\cdot10^{-5}$ s^{-1}, varying considerably depending on the size of the area where the velocities were measured (see Figure 2.27). For the Caucasus coastal area, this dependence appeared to be stronger, which could be related to the straight-line character of

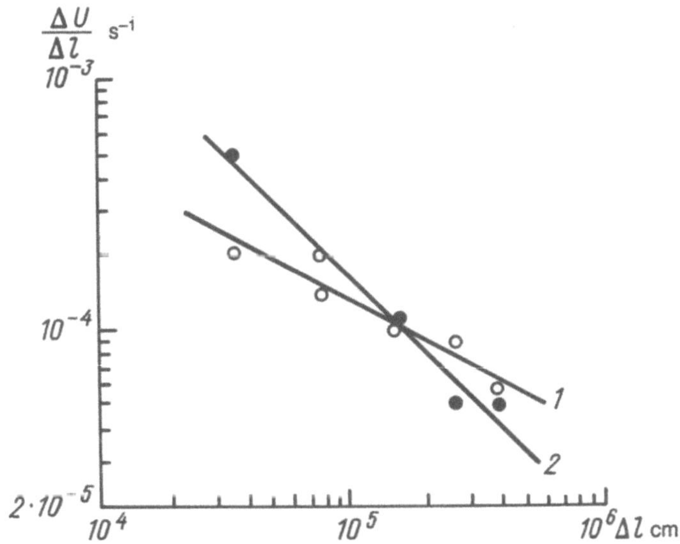

Fig. 2.27. Dependence of horizontal current velocity gradient on distance between buoy stations, based on the measurements off the Crimea (1) and the Caucasus (2) (after [150]). *y*-axis: $\Delta V/\Delta l$ s^{-1}; *x*-axis: Δl cm.

its shores, thus making the flow there closer in character to the turbulent boundary layer. Vertical-velocity shears in the upper 100 m layer, as measured by current meters at buoy stations, appeared to be, on the average, close to 10^{-2} s^{-1}; however, sometimes they reached 10^{-1} s^{-1}, which is a higher figure than those usually assumed for the open ocean. In the absence of wind, or if the wind was blowing along the current, dye spots usually elongated in the direction of the current. But if the wind direction did not coincide with that of the current, the axis of the spot, most commonly, was oriented differently (see Figure 2.28). An example of spot deformation due to vertical-velocity shears is given in Figure 2.29. At the spot scales 1 to 10 km, horizontal-velocity shears also began to cause the deformation of spots. The elongation and warping of dye spots could be related, in some cases, to the effects of drift currents and the Ekman spiral. One such case is illustrated by Figure 2.30. In that experiment, the measurements of the spot were made in a light wind and calm sea when the penetration depth of the drift current (the Ekman depth) apparently did not exceed 10 – 20 m, while the dye penetrated as deep as 15 m. In the case of stronger winds and rougher sea, the Ekman depth increased and pictures similar to the one in Figure 2.30 were not observed.

Aerial photographs of dye spots also present ‘streak’ structures of dye-concentration fields. Such a phenomenon can be related to the presence in the upper sea layer of organized water motions like the Langmuir circulations. Ac-

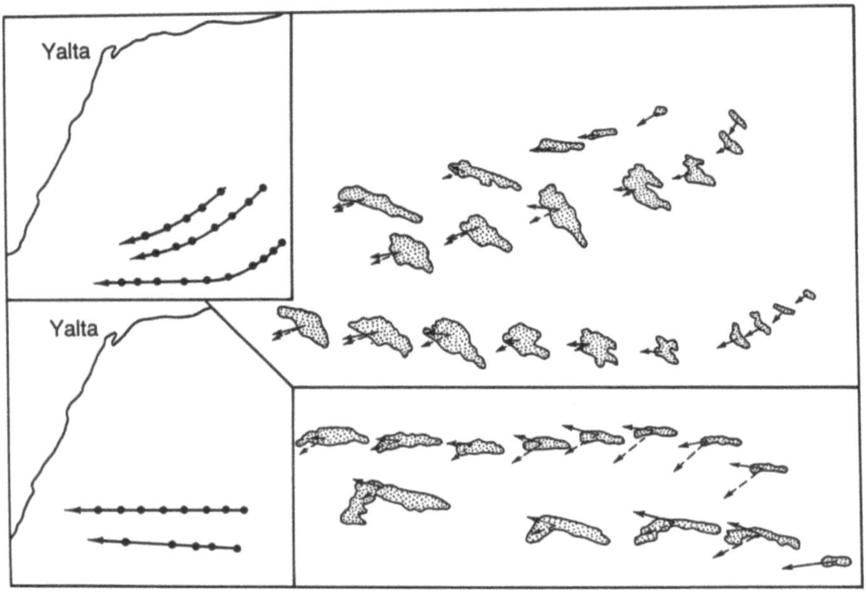

Fig. 2.28. Schemes of two drifting dye spots observed in a diffusion experiment off the Crimea. Arrows show wind and current directions (after [150]).

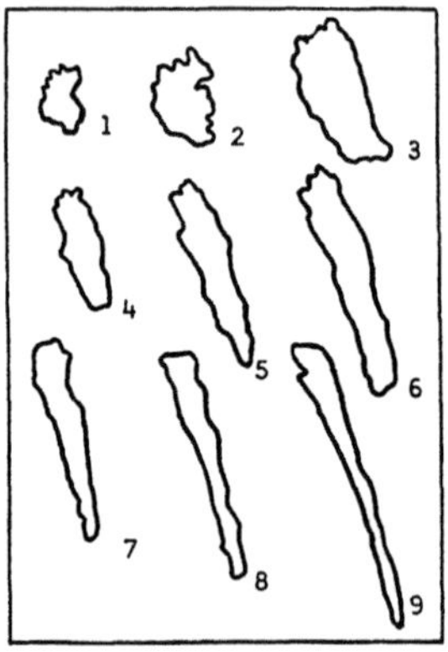

Fig. 2.29. Elongation of diffusing dye spots as a result of the influence of vertical velocity gradient (after [150]).

cording to the Black Sea experiments, the distances between individual circulations (streaks) varied from 5 – 10 to 20 – 30 m. These figures are much smaller than those given by Assaf *et al.* based on observations in the ocean [113]. Such a

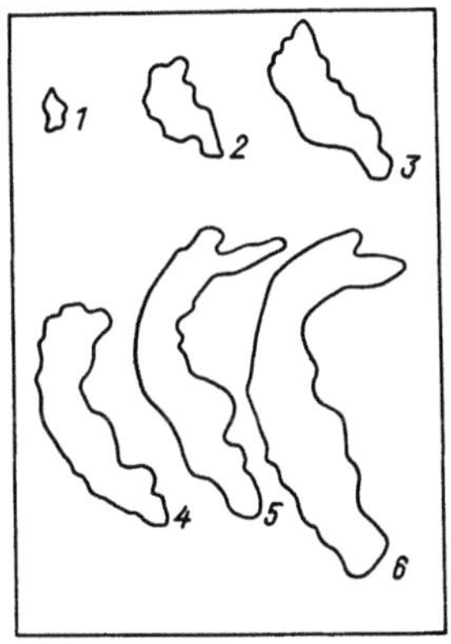

Fig. 2.30. Elongation and deformation of diffusing dye spots under the influence of drift current (the 'head' of spot is located in its lowest part) (after [150]).

difference is obviously related to the specific hydrological structure of the Black Sea, particularly with the shallow location of the pycnocline in the sea as compared to the ocean. And the depth of the mixed layer is known to determine, to a larger extent, the scale of Langmuir circulations developing in upper layers of natural water bodies.

Aerial photographic surveys of dye spots make it possible to determine, along with the diffusion parameters, many other interesting pecularities of the kinematics and dynamics of hydrophysical fields. For example, during the Black Sea experiments they managed to study the structure of the hydrological front in the coastal sea area. First, two dye spots were release 500 m away from the front on both its sides. The front line was noticeable owing to the different transparencies of coastal and seaward waters. The released spots began to quickly approach the front line, stretching in its direction. On reaching the frontline the spots abruptly disappeared, with their speed of approach reaching 5 – 10 cm/s. The quick disappearance of the spots on the front line was evidence of the fact that the values of vertical velocities of water motions at the front line were of the same order. The next dye spots were released farther out to sea at the distances of 1.5 and 3 miles away from the fronts. For these spots, the velocities of their approach to the front were small, and for 4 – 5 hours of observation they were not captured by the front.

Development of the methods of aerospace remote sensing for the ocean has made it possible to study large-scale ocean phenomena – not only those simulated by introducing artificial tracers but also those emerging from natural contaminants such as turbid waters, accumulations of plankton or floating sea ice. A good indicator of dynamical processes in the ocean could also be the temperature of its waters, which often can be regarded as some passive 'contaminants' being carried around by moving ocean waters. There is by now an immense amount of literature available on the results of such aerospace remote sensing observations that enable the study of the structure of individual currents, turbulent formations, water evaluation zones, frontal zones, etc. An example of recent results of such studies is the Proceedings of the Fifteenth International Liege Colloquium on Ocean Hydrodynamics [141], which is devoted to the remote sensing methods of studying shelf hydrodynamics.

Experiments with dye spots allow us not only to study the transformation of averaged contaminant-distribution pictures in the course of diffusion but also the fluctuation phenomena in the indicator-concentration field. Actually, as was said earlier in this section, each crossing of the contaminant spot by towing fluorimeter give us a record which can be considered the realization of a random function $c(x)$ of one variable. This function, however, is usually highly 'non-stationary' as, on the average, the contaminant concentration in the spot decreases from the center to the periphery. Therefore, it is convenient to subtract from the function $c(x)$ the function $\bar{c}(x)$ averaged in this way or another [e.g. with the help of sliding averaging with the smoothing kernel of the type (2.4)] and obtain the fluctuation of contaminant concentration $c'(x)$. If the fluorimeter measurements were conducted over a long period of time in a fixed point of the spot, than the argument x is naturally substituted for the time t. The function $c'(x)$ [or $c'(t)$] can further be analyzed by the general methods for random functions of one variable. Such an approach makes it possible to study the 'inner structure' of contaminant-concentration fields, allowing a number of problems of practical importance (for example, the problem of amplitudes and duration of 'overshoots' in concentrations that often decisively affect the characteristics of water masses as habitats of living organisms) to be solved.

For the first time, statistical characteristics of dye concentration fields from instantaneous contaminant sources were calculated based on the data obtained in the 1968 diffusion experiments conducted in the Baltic Sea [82]. In the experiment off Rügen Island, 8 kg of dye were used, with the observation of the spot lasting for more than 7 hours. On the termination of the experiment, the horizontal size of the spot exceeded 1 km while its penetration into the depth amounted 10 m. The second spot of dye (10 kg) was released to the nord of Gdansk and watched for 8 hours and 30 minutes. The penetration depth in that case was 7 m with the horizontal size of the spot about 1 km. The discretization of dye-concentration records obtained when crossing the spots was made with the step of 3.75 m. Record smoothing was done by the cosinus-filter with spacial scale of smoothing $L_0 = 2\pi/k_0$ (k_0 is the wave number) equalling 15, 22.5, 45 and 90m. Statistical characteristics of the functions $c'(l)$ (where l is the coordinate along a given direction of the crossing of the spot) were computed using the entire set of data for each spot irrespective of the horizons and the directions of measurement tacks, through which a substantial increase in the volume of the initial information to be used for computation was achieved. The homogeneity of the hydrological conditions in the area of the experiment as well as the results of test calculations, which discovered no dependence of fluctuation characteristics of dye concentration on the direction of the tack or its depth, could serve to justify such a technique.

Using the values for $c'(l)$, the dispersions of fluctuation concentrations were calculated in the first place:

$$D = \overline{c'^2}, \tag{2.11}$$

where the bar means averaging with respect to the entire array of fluctuation data used in the computation.

Since the time of crossing of the contaminant spot (tack) by the ship is small as compared to the characteristic diffusion time (the time for essential changes in spot sizes), a certain moment of time t, equal to the moment when the ship was passing the center of the spot, was attributed to each tack. Dispersion val-

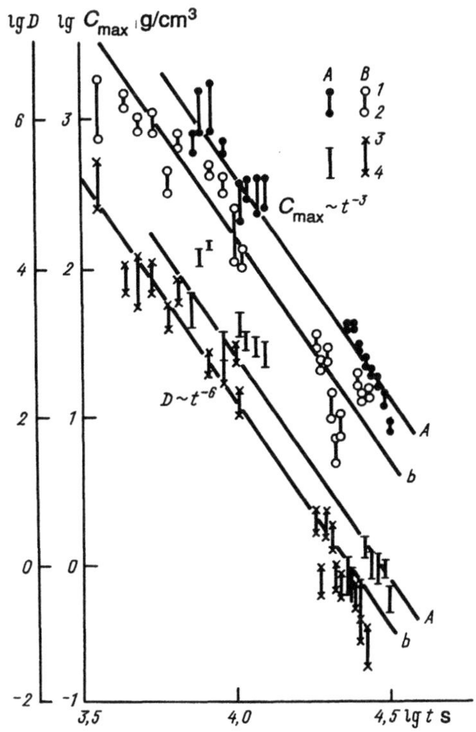

Fig. 2.31. Change in variance D and contaminant concentration C_{max} in the center of spot depending on diffusion time t (after [87]). A – the September 1, 1968, experiment; B – the August 21, 1968, experiment in the Baltic Sea; 1 – absolute concentration maximum; 2 – maximum at L_0 = 45 m; 3 and 4 – variances at L_0 = 45 m and L_0 - 22.5 m, respectively; y-axis: lg D, lg C_{max} g/cm^3; x-axis: lg t s.

ues as functions of the diffusion time (counted from the moment of spot release) for two values of L_0 are plotted in Figure 2.31. The figure also presents the absolute values for the maximum dye concentrations in spots as well as maximum mean concentrations $\bar{c}_{max}$ when smoothing the concentration profiles with $L_0 = 45$ m. As is seen from the figure, the maximum concentration in spots changed proportionally to the minus third power of time, while the dependence of dispersion on t was well described by an expression of t^{-6}. Thus, there was a linear relation observed between D and $\bar{c}^{-2}_{max}$:

$$D = a\bar{c}^{2}_{max}, \tag{2.12}$$

where a is some constant.

The dependence of D on the square of the mean concentration $\bar{c}^2$ for a given cross-section (tack) was also found to be linear, i.e.,

$$D = b\bar{c}^2 \tag{2.13}$$

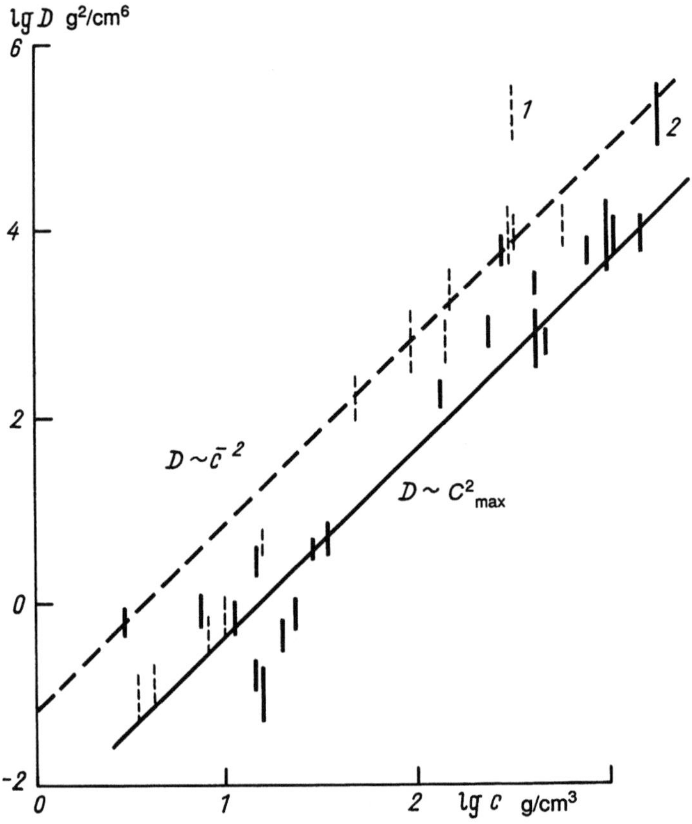

Fig. 2.32. Dependence of variance D on the mean (1) and maximum (2) dye concentrations at $L_0 = 22.5$ m and $L_0 = 45$ m (utmost values of vertical lines), according to the August 21, 1968, experiment in the Baltic Sea (after [87]). y-axis: lg D g^2/cm^6; x-axis: lg c g/cm^3.

The square root of the constant b may be defined as the relative intensity of the concentration fluctuations ($\sqrt{b} = \sqrt{\bar{c}'^2}/\bar{c}$).

Graphs for the dependences (2.12) and (2.13) are given in Figure 2.32 for the dispersion values calculated with the smoothing scales of 22.5 and 45 m. Here, the constants appeared to be the following: $a = 4 \cdot 10^{-3}$, $b = 6.3 \cdot 10^{-2}$. That means that the relative intensity of concentration fluctuations in this case was close to 0.25.

Using the dispersion values determined under different smoothing scales L_0 were are able to get an idea of the relative contribution of fluctuations of a given scale to the total 'energy' of concentration fluctuations, i.e., an idea of the form of the spectral density of fluctuations $E_c(k)$. The computation of values for the function $E_c(k)$ was done according to the formula $E_v(k) = \Delta D/\Delta k$, where ΔD is the dispersion increment when the wave-number of smoothing increase by Δk. Since it is the rate of variation of $E_c(k)$ with the growing k which is the most interesting and not the absolute values of the function $E_c(k)$ (apparently changing from one experiment to another and also in the course of experiment), the values for $E_c(k)$ were normalized by an $E_c(k)$ value at $k = 0.186$ m^{-1} (which corresponds to the mean value for L_0 of 22.5 and 45 m) The thus plotted

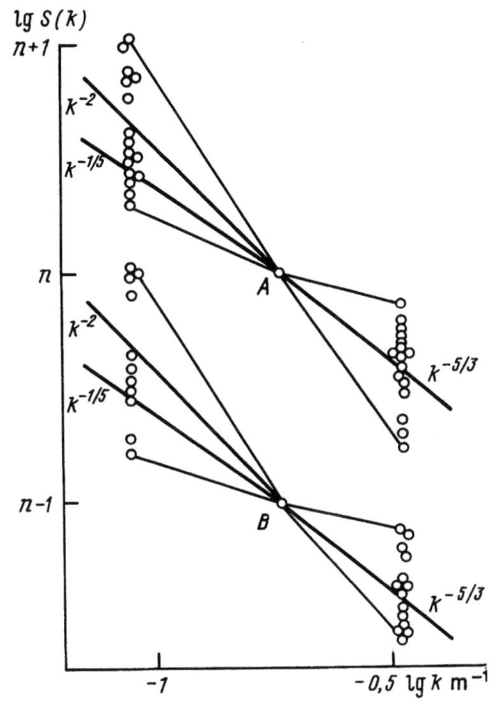

Fig. 2.33. Estimations of spectral densities of contaminant-concentration fluctuations, based on dispersion differences (after [87]). A – the September 1, 1968, experiment; B – the August 21, 1968, experiment in the Baltic Sea. Conditional units are used on y-axis.

graphs presented in Figure 2.33, show the changes in the function $E_c(k)$ as well as the possible values for the exponent n when approximating the function $E_c(k)$ by power expressions. The spread of points in the graph appeared to be rather significant, and the exponent n could be varied within a relatively wide range. For clarity of illustration, the graphs also present the dependences corresponding to the powers –5/3, –7/5 and –2. Mean values for n, calculated with respect to all the data of the two experiments being considered, appeared, respectively, to be the following: $n = 2.10$; 2.12 in the range of k from 0.093 to 0.186 m^{-1}, and $n = 1.5$; 1.30 for $k = 0.186 \ldots 0.334$ m^{-1}.

For measurement tacks longer than 375 m (i.e. with the number of discrete values for $c'(l)$ more than 100), correlation functions of the random functions $c'(l)$ were calculated. The forms of the correlation functions significantly depended, as would be expected, on the scale of smoothing of the original series for $c(l)$. Based on the correlation functions, the functions $E_c(k)$ were determined with the help of the Fourier transform. An example of the thus-calculated function $E_c(k)$ for one of the sections across the spot is given in Figure 2.34. Within the range of k from 0.2 to 0.65 m^{-1}, the function $E_c(k)$ in this example could be approximated by the law of the '–5/3 power'. The decline of $E_c(k)$ in

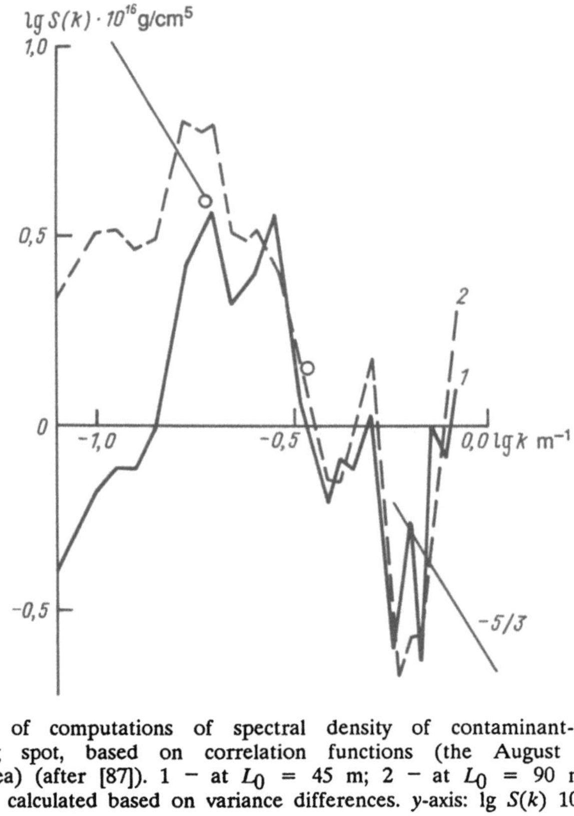

Fig. 2.34. Examples of computations of spectral density of contaminant-concentration fluctuations in diffusing spot, based on correlation functions (the August 21, 1968, experiment in the Black Sea) (after [87]). 1 – at L_0 = 45 m; 2 – at L_0 = 90 m. Light points refer to $E_c(k)$ calculated based on variance differences. y-axis: lg $S(k)$ 10^{16} g/cm^5.

the left-hand side of the graph is a result of the filtration effect, and the increase in $E_c(k)$ in the utmost right-hand part of the graph is caused by the noise effects. If the exponent $-5/3$ for the function $E_c(k)$ can be explained by the presence of a local isotropic range within the above limits for the wave number k, then the inverse square dependence of $E_c(k)$ on k, which in some cases gave better approximation results, would require special consideration. In [87] considering the shear wind friction stress τ (or the so-called friction velocity $u_* = \sqrt{\tau/\rho}$) as a factor determining the diffusion regime in the upper sea was suggested for explaining this dependence. In fact, the energy supply for the surface waters can, generally speaking, occured over the entire range of motion scales and, therefore, the flux of energy ε over the cascade of scales may be, in this case, essentially variable depending on t. The rate of smoothing of the contaminant-concentration fluctuations ε_c must serve as a second parameter determining the form of the function $E_c(k)$. Under such assumptions for $E_c(k)$ we, on dimensional grounds, determine that

$$E_c(k) \sim \varepsilon_c u_*^{-1} k^{-2} \tag{2.14}$$

The dependence (2.14) actually held true in certain ranges of scales. This was proven later by a number of experiments both with dye spots and continuous point sources of contaminants.

Immense calculations of statistical characteristics for contaminant-concentration field from instantaneous point sources were made in the following years using data of diffusing experiments conducted in the Baltic and Black Seas [19, 29, 88]. The series of discrete contaminant concentrations obtained when crossing the spots usually had about 400 terms (with a discretization step 1 – 4 m), which made it possible to calculate spectral densities for the range from several meters to several hundred meters. Examples of graphs for the obtained functions $E_c(k)$ are given in Figure 2.35. The confidence interval for the functions $E_c(k)$ in the calculations was from 0.6 to 2.1 of a calculated value (at 95% confidence probability). The spectral functions has, as a rule, rather 'smooth' forms, with only some peaks in the scales of 10 – 20 m outstanding the confidence interval. Apparently these peaks were caused by surface waves, since one could often notice a decrease in the peaks with the growing depth of submergence of the instrument sensor (e.g. see Figure 2.35b). In some spectral functions, the peaks were observed at larger scales as well (several dozens of meters), which could be accounted for by the Langmuir circulations of waters developing in the upper sea. Such circulations can usually be observed well enough visually by the elongated dye streaks, which are parallel to each other. This allows us to choose tacks for the ship: either parallel or perpendicular to the direction of the circulations. In the case of measurements in tacks parallel to the streaks

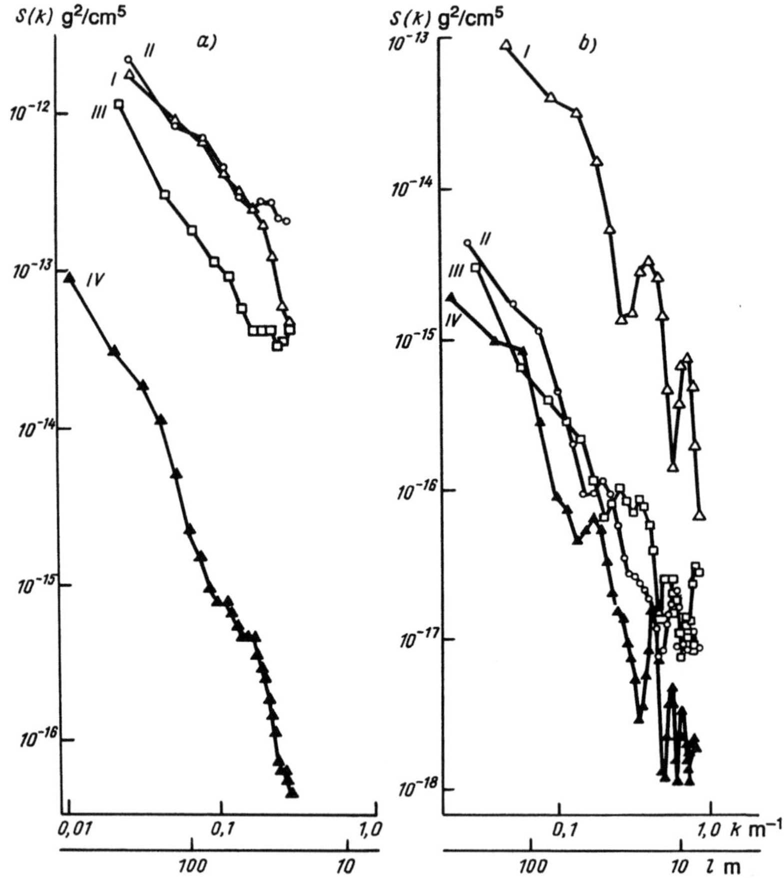

the peaks in spectral functions disappear. An example of such 'peak-free' spectra, well approximated by square dependences, is given in Figure 2.36.

Approximating straight lines were drawn on the logarithmic graphs for all the calculated spectral functions $E_c(k)$. Table II.II presents the dependences of slope tangents of the approximating lines (i.e., of the exponent n when approximating the function $E_c(k)$ by a power expression) on the diffusion time t. Although the approximation of the function $E_c(k)$ throughout the entire range of its definition by a sole line is rather rough it is possible to notice, using the tabulated data, a sharp change in the slope 4–5 hours from the moment of spot release. This phenomenon is apparently related to the fact that since that time new factors such as horizontal current-velocity gradients, in particular, begin to influence the diffusion of the contaminant cloud.

Figure 2.37 presents an example of spectral functions for the fluctuations of dye concentrations in spots with slopes varying significantly with the increasing wave number. These spectral functions are plotted based on the field data

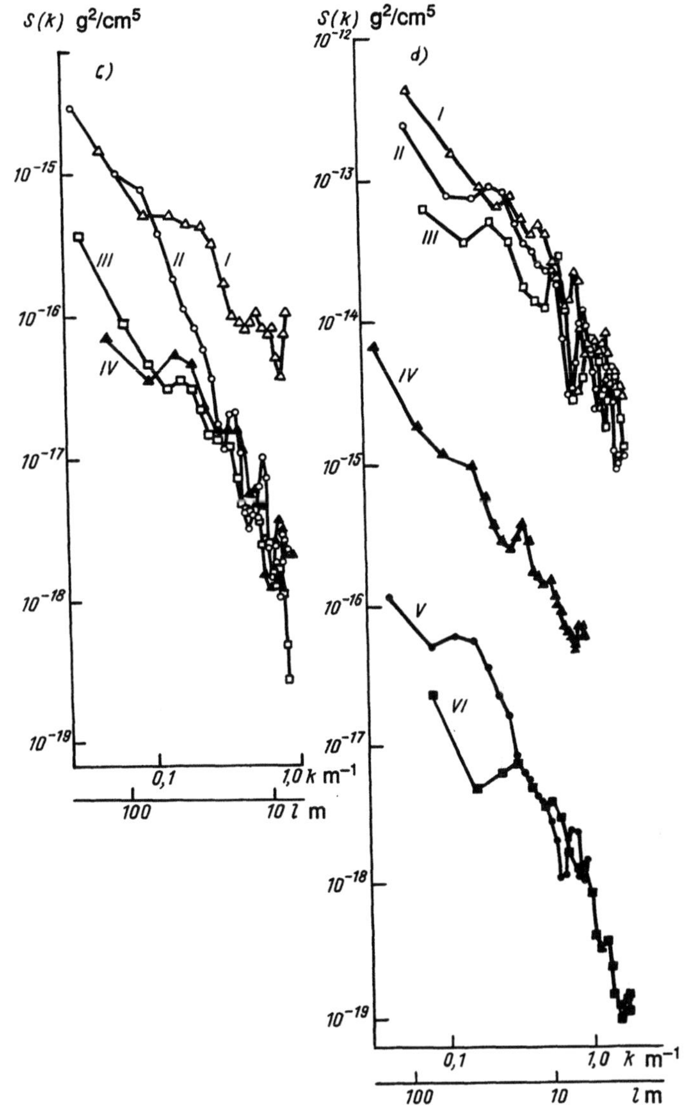

Fig. 2.35. Spectral densities of contaminant-concentration fluctuations, according to the experiments with dye spots in the Black and Baltic seas. 1 – the Black Sea, June 16, 1968; Roman figures denote individual measurements, measurement horizon is 0.5 m; b – the Baltic Sea, August 23, 1968, I, II, IV – measurements at the depth of 1 m, III – measurement at the depth of 4 m; c – the Baltic Sea, August 30, 1968, II, III – measurements at the depth of 7.5 m; d – the Baltic Sea, 1969, I–IV – measurements at the depth of 0.5 m (after [88]). y-axis: $S(k)$ g^2/cm^5.

obtained through measurements in the Black Sea off the Bulgarian shores during the international venture Kamchiya-77. The experiments were carried out in

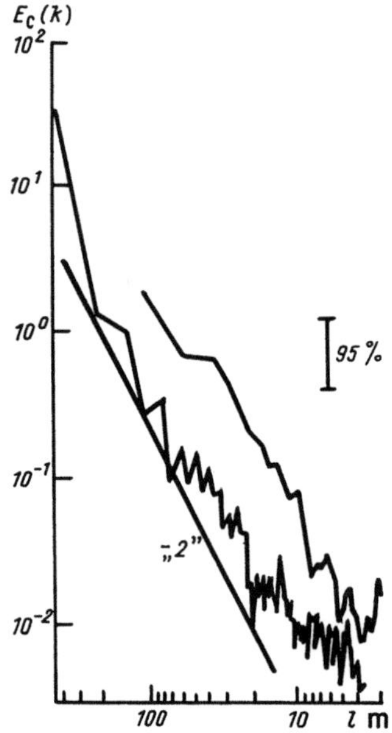

Fig. 2.36. Spectral densities of contaminant-concentration fluctuations, according to the dye-spot experiment in the Baltic Sea in 1972, and their approximation line corresponding to the square law (after Brozin, Kremzer, and Muravyov in [47]).

TABLE II.II

Dependence of the power n from diffusion time t when spectral functions $E_c(k)$ are approximated by power laws

t	n
99	1,4
114	1,4
123	1,4
125	1,4
128	1,4
153	1,4
222	1,4
229	1,4
300	2,2
306	2,2
331	2,2
354	2,2
357	2,2
360	2,2
361	2,5
374	2,2
376	1,7
390	2,5
444	2,2
503	2,2
534	2,2

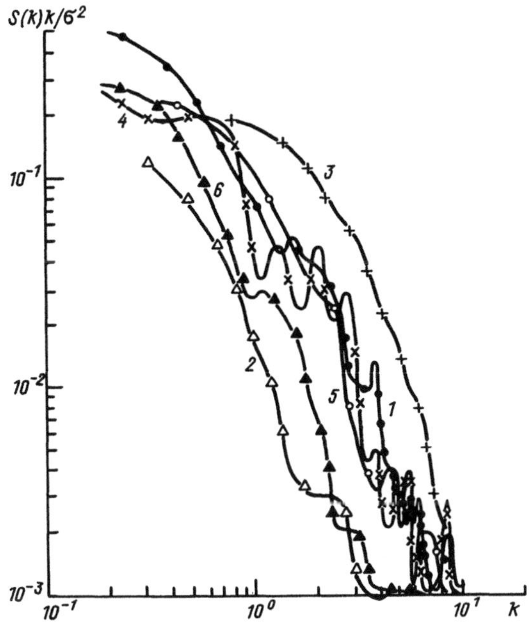

Fig. 2.37. Spectral densities of contaminant-concentration fluctuations, according to the October 7, 1977, experiment off the Bulgarian Black Sea coast (Arabic figures denote individual measurements in spot) (after [29]).

slightly windy weather in a shallow coastal sea area with depths of only 5–7m [29].

The variety of forms of the spectral functions $E_c(k)$ obtained through diffusion experiments with dye spots is naturally explained by the complexity and variability of dynamic processes in the sea that influence the contaminant concentration field. In this regard, when studying the regularities of diffusion in the sea it is necessary to measure not only mean and fluctuation values for contaminant concentrations, but a large complex mean and fluctuation characteristics of natural hydrophysical fields, primarily those of velocity and density (temperature and salinity) of the sea water. It is also necessary to have data available on surface waves and meteorological conditions in the area of the experiment. Kamchiya joint ventures [11, 12], carried out by COMECON scientists, are an example of such complex measurements.

2.3. Diffusion from Continuous Point Sources

As already mentioned in Part 2.1, experiments with stationary point sources of contaminants are very convenient for studying statistical characteristics of the contaminant field; the jet of contaminant from such a source is stationary (certainly, at mean hydrometeorological conditions it is unchanged in the course of an experiment), which makes it possible to obtain, through multiple crossings of the jet by a instrument, vast, statistically homogeneous experimental data suitable or further analysis. Of course, experiments with dye jets can also be used for studying averaged contaminant concentration fields.

Experiments with continuous dye sources began in the Baltic Sea in 1964 [81]. A dye solution with a density identical to that of the sea water flowed out through a hose from a tank on board a small anchored ship onto the surface sea layer. In the first experiments, a 0.05% fluorescein solution was used, and the contaminant-source intensity was equal to 3.3 l/min. This source generated a jet of diffusing indicator which was easily traced up to distances of 200 – 500 m from the source. Measurements of dye concentration distribution in the jet were made with the help of a fluorimeter placed on a second small-tonnage shop that went across the jet different distances from the source.

Altogether, 7 jet crossing at the distances of 25 and 50 m from the indicator release point were made in the 1964 experiments. An example of an individual dye-concentration distribution in the cross-section of the jet (along the axis y) is given in Figure 2.38. The concentration distribution becomes ‘smoother’ after averaging all the curves $c(y)$, obtained by crossing the jet at a fixed distance from the source (along the axis x). These averaged curves are given in Figure 2.39, where the approximating curves of normal distributions are also plotted. Such approximating functions can be obtained by solving the diffusion equation of the form

$$u\partial\bar{c}/\partial x = K_y\partial^2\bar{c}/\partial y^2 + K_z\partial^2\bar{c}/\partial z^2, \tag{2.15}$$

where u is the velocity of constant current along the axis x; K_y and K_z are diffusion coefficients along the horizontal transversal axis y and vertical axis z,

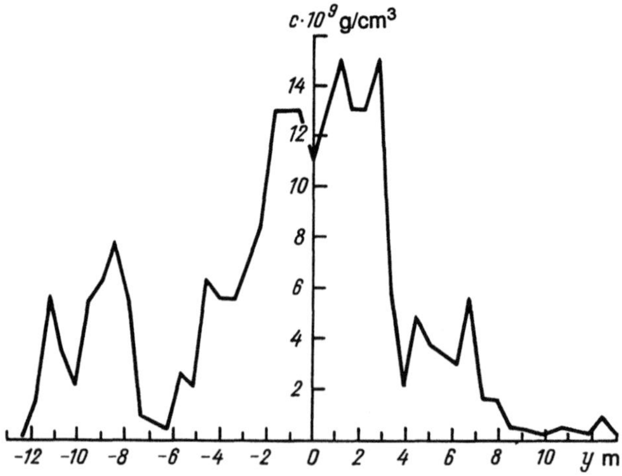

Fig. 2.38. Indicator-concentration distribution in the cross-section of diffusing jet at distance 25 m from the source (the 1964 experiment in the Black Sea). y-axis: $c \cdot 10^9$ g/cm^3.

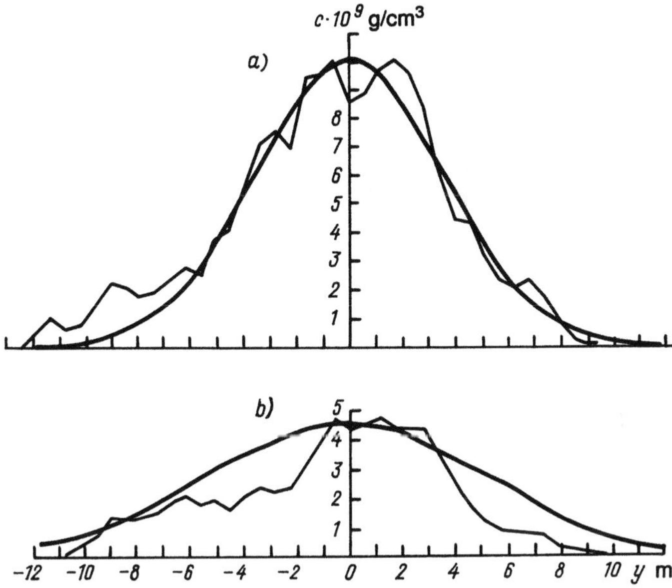

Fig. 2.39. Indicator-concentration distributions, averaged over 7 realizations, in the cross-section of jet (broken lines) and approximating curves. Distances from indicators source are 25 (a) and 50 (b) m (1964 experiments in the Black Sea). y-axis: $c \cdot 10^9$ g/cm^3; x-axis y m.

respectively. The origin of the coordinates was placed on the surface of the sea at the point of indicator release.

The solution of equation (2.15) with the boundary condition of 'reflecting' the contaminant from the plane $z = 0$ and with the condition of boundedness of the function $c(x, y, z)$ at large distances from the source has the form

$$c(x, y, z) = \frac{2q}{4\pi x\sqrt{K_y K_z}} \exp\left(-\frac{uy^2}{4K_y x} - \frac{uz^2}{4K_z x}\right), \tag{2.16}$$

where q is the contaminant source intensity.

The distribution of dye concentration on the sea surface ($z = 0$) has the form

$$c(x, y, 0) = \frac{2q}{4\pi x\sqrt{K_y K_z}} \exp\left(-\frac{uy^2}{4K_y x}\right). \tag{2.17}$$

In the experiment with a dye jet u, q and x were known. From the condition for the best agreement between theoretical curves and experimental data the values for the transport coefficients K_z and K_y can be determined. For example, according to figure 2.39, these appeared to be the following: $K_z = 42$ cm^2/s and $K_y = 7.5 \cdot 10^2$ cm^2/s. Interestingly, the only value obtained for the horizontal-transport coefficient K_y turned out to be very close to that calculated accord-

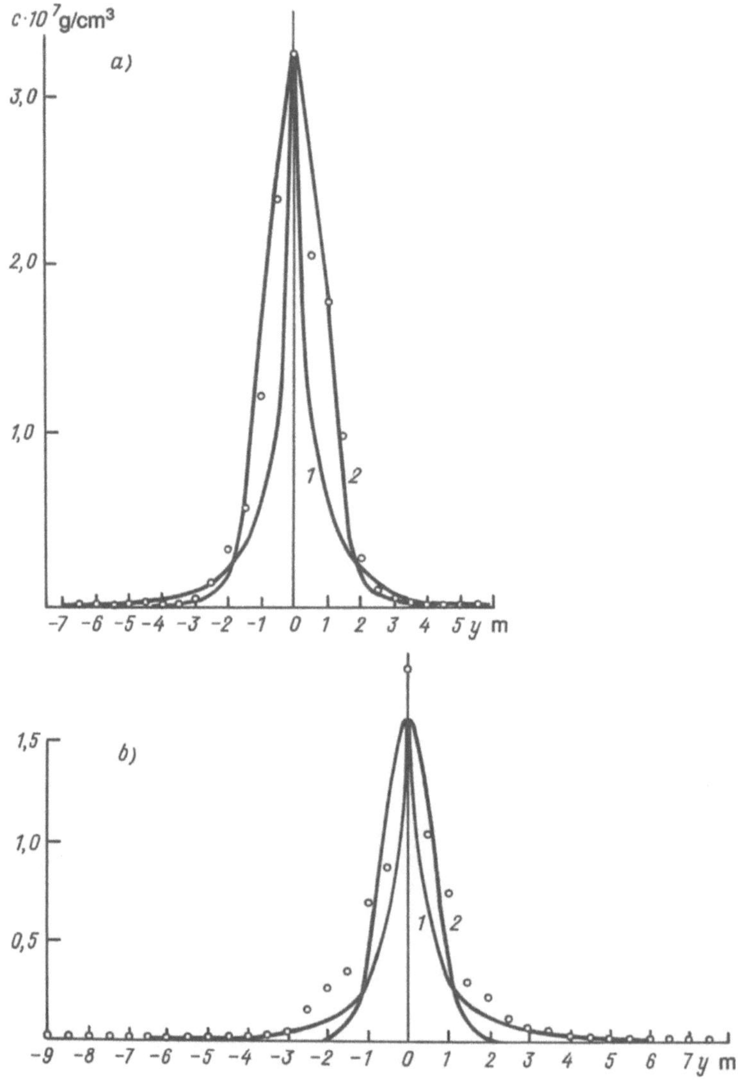

Fig. 2.40. Averaged indicator concentrations in the cross-sections of jets at distances 30 (a), 75 (b), 150 (c) and 250 (d) m (1965 experiments in the Black Sea) and theoretical indicator distribution curves. 1 – according to equation with the "4/3 power law" diffusion coefficient. 2 – according to equation with constant diffusion coefficient. y-axis: $c \cdot 10^7$ g/cm^3; x-axis: y m.

ing to the formula for the '4/3 power law' where the jet width was used as the phenomenon scale.

More representative data on dye-concentration distributions in diffusing jets were obtained in the consequent 1965 experiments in the Black Sea. The intensity

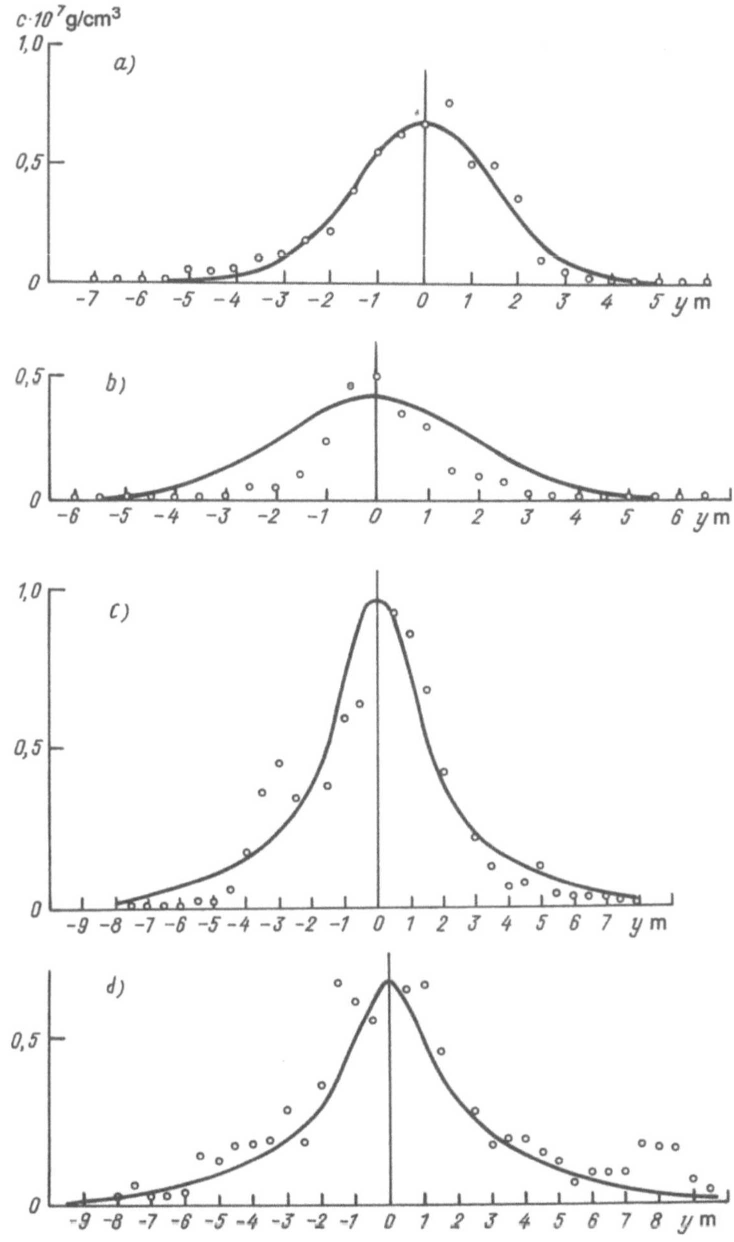

Fig. 2.41. Averaged indicator concentrations in the cross-sections of jets at distances 30 (a), 75 (b), 150 (c) and 250 (d) m (1965 experiments in the Black Sea) and theoretical curves of contaminant distribution, according to equation with constant diffusion coefficient. y-axis: $c \cdot 10^7$ g/cm^3; x-axis: y m.

of the source was enhanced, which made it possible to observe the contaminant jet at distances up to 1500 m from the source. Figure 2.40 gives an example of averaged profiles of dye concentrations in these experiments at the distances of 20 and 50 m from the source, approximated this time by the solutions of diffusion equations not only with constant transport coefficients but with those dependent on the phenomenon scale, too. Averaged dye concentrations at the distances of 30 and 75 m from the source as well as the distances of 150 and 250 m (in the case of greater source intensity) in Figure 2.41 are approximated only by solutions of the diffusion equation with constant coefficients. As is seen from figure 2.40 and 2.41, the approximation of the averaged curves for indicator-concentration distributions in the cross-sector of the jet by Gaussian curves is usually quite satisfactory; however, in some cases noticeable deviations of experimental values from the theoretical curves can be found. For example, the experimental points in the peripheral jet zones in Figure 2.40b deviate distinctly from the normal distribution curve and are better approximated by the solution curve for the diffusion equation with variable coefficient, determined by the '4/3 power law'. Substantial deviations of experimental points from the theoretical curve also found in Figure 2.41b. The greater excess of the experimental values in this case can better be described by the solution of the equation with variable transport coefficients. These facts are more proof of the relation existing between the diffusion characteristics and the scale of the phenomenon, which we discussed earlier in a detailed way.

The 1965 experiments in the Black Sea with dye jets enabled the studying of regularities of falls in dye concentration in the direction of the jet (axis x) within a sufficiently wide range of x. The average values for maximum dye con-

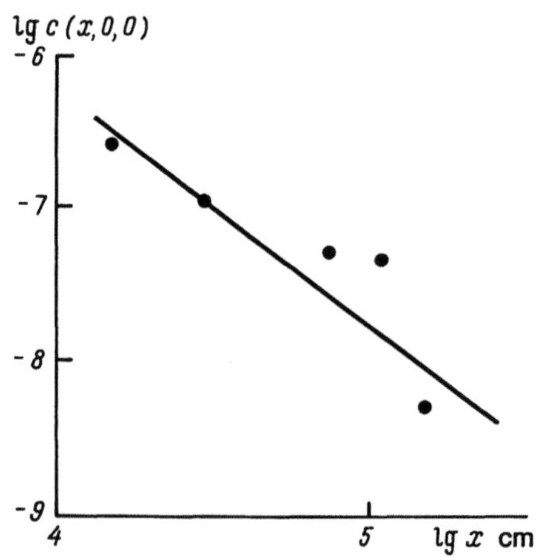

Fig. 2.42. Indicator-concentration variations along jet axis (1965 experiments in the Black Sea) and theoretical dependence derived from the solution of diffusion equation with a "4/3 power law" diffusion coefficient. x-axis: lg x cm.

centrations in the directions of the jet at the distances of 150, 300, 750, 1100 and 1500 m from the source are presented in Figure 2.42. The decrease in the concentration $\bar{c}_{max}(x)$ with the growing x can, with a fair approximation, be considered proportional to x in the power of 2/3, as is seen from the figure. Such a dependence agrees with the solution of the one-dimensional diffusion equation with coefficient proportional of the phenomenon scale (radius vector at the observation point) in the power of 4/3 [81].

Experiments with continuous point sources of dye were started in the Baltic Sea under ventures of COMECON oceanologists in 1969. In that year, dye was released in Gdansk Bay from a anchored ship. Another vessel with fluorimeter made cross-sections through the contaminant jet [115]. The decrease in the indicator concentration along the jet axis in that experiment was found to obey a power dependence but with the exponent $n = 0.82$, which is somewhat higher than 2/3. A close value for n was also found in the Tvären Gulf in the Baltic Sea in an experiment with a radioactive isotope [125]. Such a difference in the value of n in the Black Sea and Baltic Seas experiments may, certainly, be explained by the relatively low accuracy of determining n. However, it was considered quite possible that this difference is also a result of the specific features of each observation area, particularly, the shallowness of Baltic Sea bays. Further experiments with continuous dye sources showed that there really is a dependence of n on the hydrometeorological conditions in the area of observation and on the scale of the phenomenon. For example, in a number of further diffusion experiments the researchers managed to distinguish two phases of the diffusion of a jet with different decrease rates for the maximum concentration along the jet axis, which could be explained by an existing dependence of diffusion characteristics on the phenomenon scale L. Besides, such a dependence on L may even leas to very large variations of n.

In the following years, experiments with dye jets were carried out in the Black and Caspian seas [26, 27, 28] and also in Lake Baikal [99]. Figure 2.43 presents graphs for dependences of maximum dye concentrations on the jet axis, based on the data from several diffusion experiments staged in the Black Sea. Here, the exponents in the approximating dependences varied from 1.1 to 1.6 for distances from the source larger than 100 – 600 m. In a part of the stream closer to the source, the decrease rat for $\bar{c}_{max}(t)$ was much smaller. A point separating the phases of slower and faster decreases in dye concentration in the stream was located, according to a number of experiments, at distances from 50 m to 1 km from the source.

The growth in cross-section sizes of dye jets with the increasing distance from the source can also often be approximated by power dependences. Usually, the concentration-distribution dispersion in the lateral section of the jet $\overline{\delta_y^2}$ is taken for a jet lateral size scale. Evaluation of $\overline{\delta_y^2}$ was achieved with the help of the formula

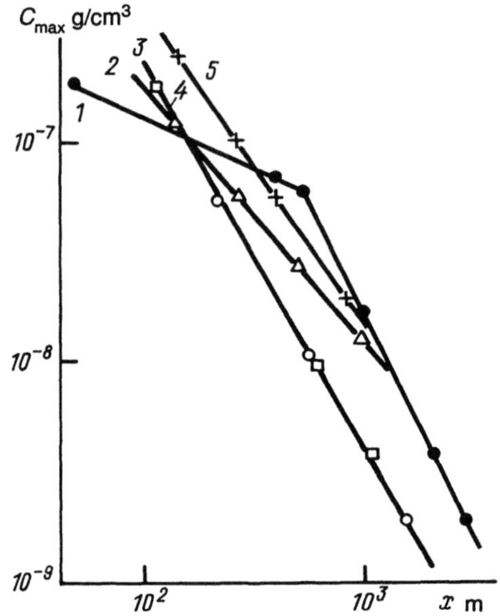

Fig. 2.43. Indicator-concentration variations along jet axis, according to experiments in coastal Black Sea area (after Zats, Nemirovsky and others in [11]). y-axis: C^{m}_{max} g/cm^3; x-axis: x m.

$$\overline{\delta_y^2} = \frac{\int_{-\infty}^{\infty} y^2 c(y)\,dy}{\int_{-\infty}^{\infty} c(y)\,dy}, \tag{2.18}$$

where $c(y)$ is the dye-concentration distribution across the jet at a fixed distance x from the source. The limits of integration in (2.18) are actually determined by jet zones with the minimum $c(y)$ that are registered by instruments. Under discrete sampling of values for the function $c(y)$, integration in (2.18) is substituted by summing with respect to all the points of the cross section.

Examples of the graphs for the dependence of the lateral dispersion $\overline{\delta_y^2}$ on the distance x, based on the Black Sea experiments with dye jets, are given in Figure 2.44. As is seen from the graphs, the lateral sizes of jets grow with the increasing distance from the source and, if approximating the growth by power dependences, the exponents, according to Figure 2.44, will vary, on the average, from 1.1 to 1.5. In some cases (curve I in Figure 2.44) the function $\overline{\delta_y^2}(x)$ is better approximated by two dependences with the exponents n changing at different distances from the source. It is only natural that, it x continues to increase, there must appear a saturation phase when n approaches zero and then becomes negative. Here, the jet width, within the designated dye-concentration isoline, decreases until the jet disappears completely.

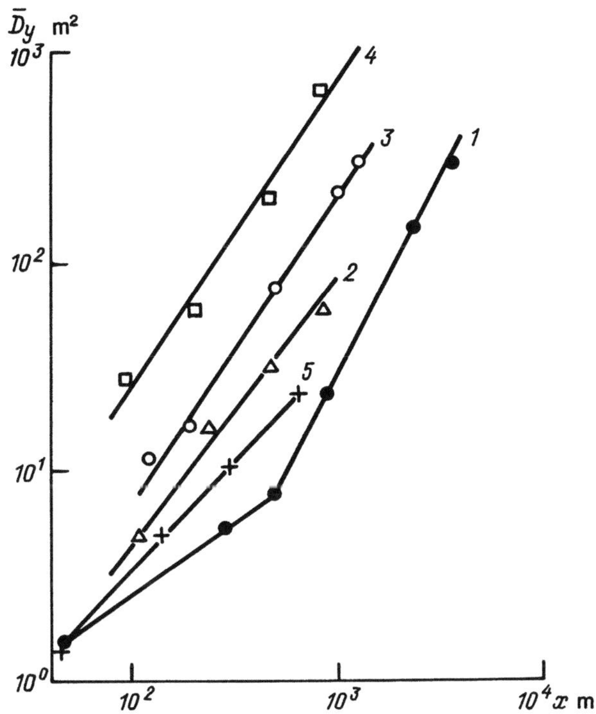

Fig. 2.44. Dependence of lateral variance of indicator jet on distance from source, based on five experiments in the Black Sea (after Zats, Nemirovsky and others in [11]).

An *ad hoc* study on the dependence of the lateral dispersion of contaminant jet on the distance from the source was made by Zhurbas *et al.* [26, 27]. This study provided an analysis of both earlier experimental data and the results of special experiments conducted in the Caspian Sea in 1973 and 1975. The work was done at the Marine Observatory of the Institute of Geography of the Azerbaidzhan Academy of Sciences. The observatory was located 25 km north-east of Artem Island in the basement of a former oil derrick, where the depth of the sea was 40 m. In 1973, the dye jet was being photographed from the height of 30 m with the pictures being further analyzed and processed using the theory of 'visibility through fog' by O. Roberts (see ref.[68, § 24.3]). The dependence of the thus-calculated lateral dispersion of jet $\overline{\delta_y^2}$ (with 95% confidence intervals) on the diffusion time $t = x/u$ (u is the mean current velocity along the axis x) is presented in logarithmic scale in Figure 2.45. The experimental results were analyzed within the framework of the theory of relative diffusion in the field of locally isotropic turbulence. Since the size of the contaminant source l_1 exceeded the Kolmogorov inner scale of turbulence, a square dependence for the

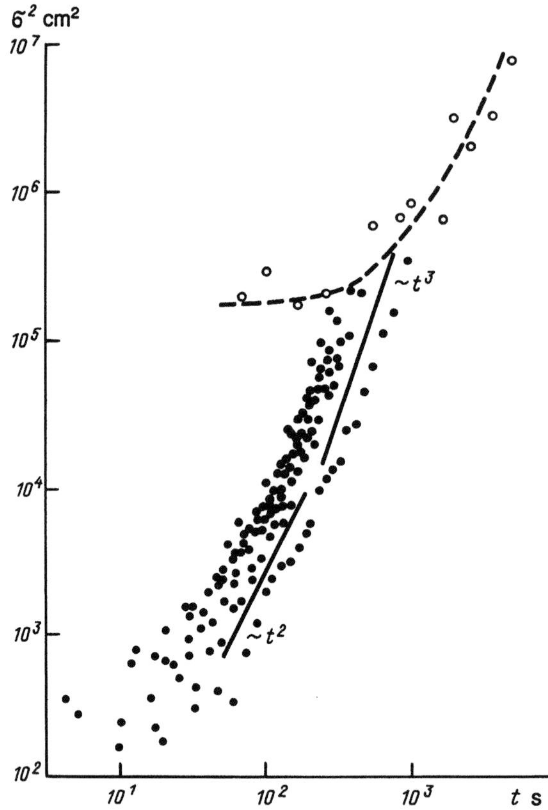

Fig. 2.45. Dependence of lateral variance of indicator jet on diffusion time, according to the 1973 experiment in the Caspian Sea (dots) and the 1965 experiments in the Black Sea (circles) and approximating theoretical dependence (after [27]).
y-axis: σ^2 cm^2; x-axis: t s.

widening of the jet should have been observed during the experiments (see formula (1.162)) when

$$\overline{\delta_y^2}(t) = l_1^2/2 + c_1(\varepsilon l_1)^{2/3} t^2/2, \tag{2.19}$$

and, at diffusion times $t \gg t_*$, where $t_* = [c_1/(2c_2)](l_1^2/\varepsilon)^{1/3}$, l_1 had to drop out of the formula, so for the dispersion of jet we obtain

$$\overline{\delta_y^2}(t) = c_2 \varepsilon t^3, \tag{2.20}$$

where c_1 and c_2 are universal constants close, by the available estimation, to 1.19 and 0.28, respectively [19].

Figure 2.45 shows that the experimental points agree well with the formulas (2.19) – (2.20). Boundaries between the two diffusion regimes at the moment of time t_* move somewhat from one experiment to the other depending on the state of the sea surface (wave heights in the course of the experiment veried from 0.15 to 0.79 m). It is interesting, that in such a case the dispersion of jet for the moment t_*, equalling $\delta_y^2(t_*) = c_1^3 l_1^2/(8c_2^2) \approx 2.7 l_1^2$, practically did not change in the entire series of experiments. Comparing the experimental data with theoretical dependences made it possible to calculate an important characteristic of turbulence – the dissipation rate of turbulent energy ε. It turned out that ε varied, under different surface wave parameters, from $1.71 \cdot 10^{-3}$ to $2.78 \cdot 10^{-2}$ cm^2/s^3. Figure 2.46 presents a graph for the dependence of ε (multiplied by the disturbance period T in the power of 5) on the height of the waves. The approximation of experimental points in the graph by a cubic function agrees with the theoretical dependence yielded in [7].

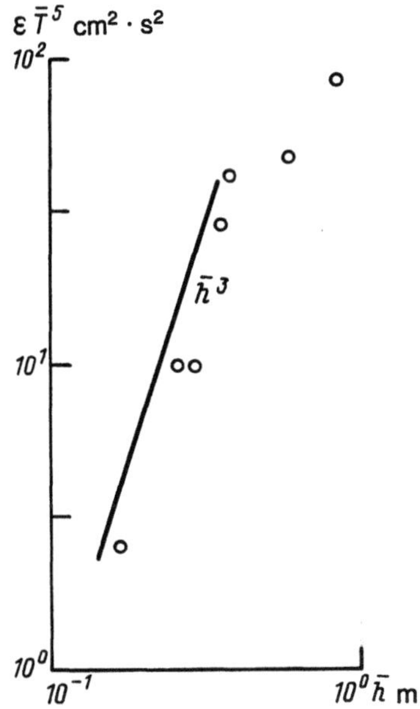

Fig. 2.46. Dependence of dissipation of turbulent energy ε on mean wave height $\bar{h}$ and on mean wave period $\bar{T}$, according to experiments with dye jets in the Caspian Sea in 1973 (after [27]). y-axis: $\varepsilon \bar{T}^5$ cm^2 s^2.

Figure 2.45 also presents points for the dependence $\delta_y^2(t)$ determined based on the 1965 experiments in the Black Sea. These points are approximated by a curve calculated according to the formula (2.19). This approximation yielded the following values for the governing parameters: $\varepsilon = 9 \cdot 10^{-4}$ cm^2/s^3, $l_1 = 6$ m. The obtained large value for the size of the source l_1 is apparently related to the distorting influence of the ship on the initial part of the contaminant jet. The same reason can be used to explain the fact that the experimental values for δ_y^2 in that case did not follow the cubic regime. In the following experiments with jets, dyes were released not in the proximity of the ship's stern but at some distance away from the ship by hose, which resulted in a substantial decrease in the initial 'effective' size of the source l_1.

Based on the 1965 experimental data, the value for ε was estimated at 9×10^{-4} cm^2/s^3. The hydrological conditions in the course of the 1965 experiments were approximately similar to those observed in the Caspian Sea experiments. Different values for ε in those experiments can, therefore, be referred to the larger jet scales in the Black Sea experiments. Thus, there is a possibility to evaluate the additional influx of energy to the ocean turbulence in the intermediate range of motion scales. The mean value for ε is equal to $1.54 \cdot 10^{-2}$

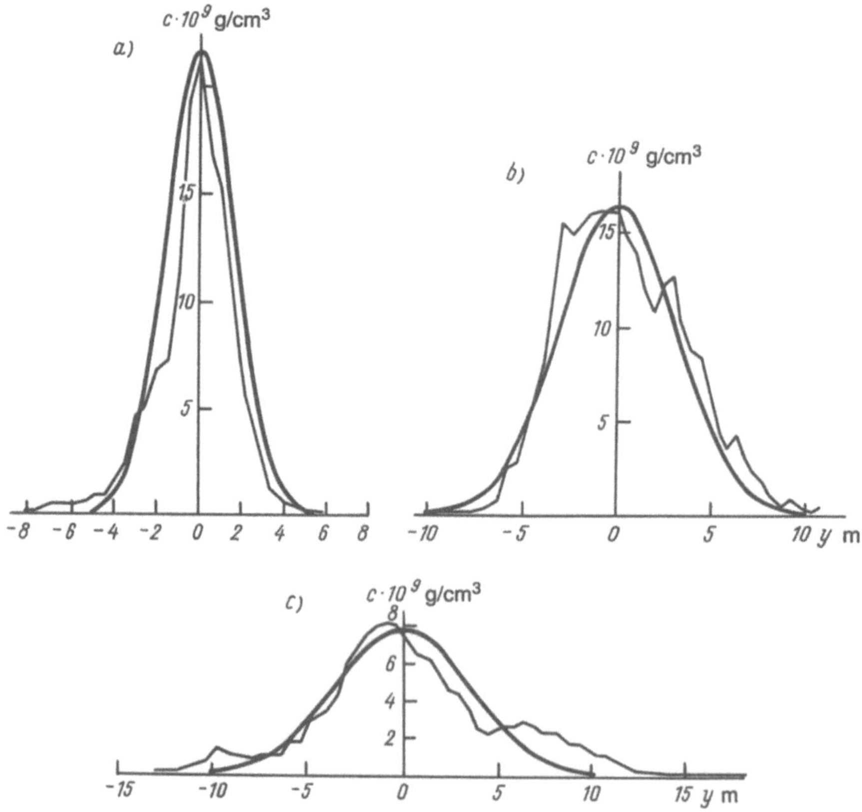

Fig. 2.47. Averaged indicator-concentration profiles in the cross-sections of jet (broken lines) and theoretical curves, based on measurements at the distances of 100 (a), 250 (b) and 300 (c) m from source (1966 experiments in the Black Sea). y-axis: $c \cdot 10^9$ g/cm^3.

cm^2/s^3, based on the Caspian Sea experiments (at wave heights from 0.25 to 0.79 m). Therefore, the energy 'addition' to the sea turbulence for motions with time scales of about 100 s turned out to be equal to $1.45 \cdot 10^{-2}$ cm^2/s^3.

The experiments in the Black Sea with dye streams were resumed in 1966 and 1968 [88]. Figure 2.47 gives averaged dye-concentration profiles at the distances of 100, 250 and 300 m from the source, according to the 1966 experimental data. As is seen from the figure, the profiles are rather well approximated by the Gausian curves. Naturally, the accuracy of such approximation improves with the growing number of 'instantaneous' concentration profiles $c(y)$ used for averaging. In order to study this issue, the value

$$\delta(n) = \frac{1}{c(x,\ 0,\ 0)(N-1)} \sum_{i=1}^{n} (\Delta c_{n,i})^2 \tag{2.21}$$

was calculated based on the 1968 experimental data, where $\Delta c_{n,i}$ is the difference between the theoretical (approximating) and the averaged, over n realizations, concentrations at the i-th point of the cross-section (concentrations were taken at discrete points); N is the number of points in realization in a given cross-section; and $\bar{c}(x,\ 0,\ 0)$ is the mean concentration in the jet axis in a given cross-section (at the distance x from the source). The approximation errors $\delta(n)$, calculated according to the formula (2.21), are given in Figure 2.48. As is seen from the figure, the approximation error quickly decreases with the growing n and becomes small at $n = 8...10$.

The above experiments with dye jets were conducted at very small velocities of jet outflow and with dye-solution densities equal to that of the sea water at

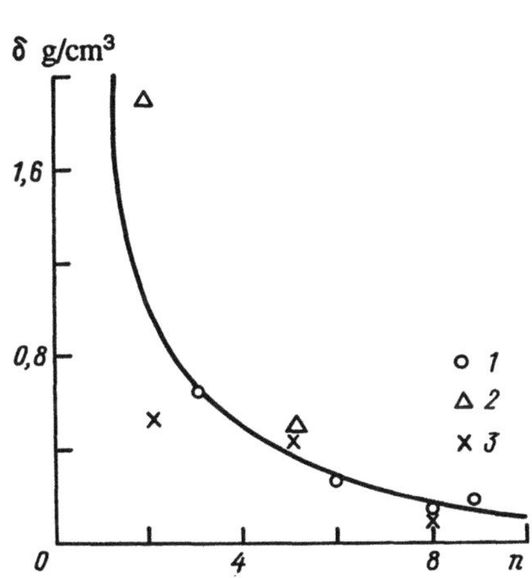

Fig. 2.48. Approximation errors for cross-section distributions of averaged indicator concentrations when approximated by Gaussian curves under different numbers of averaged realizations n (according to the 1968 experiments in the Black Sea). 1, 2, 3 are cross-sections at 100, 200 and 400 m from source, respectively. y-axis: δ g/cm^3; x-axis: x n.

the jet release point. In this case, the main mechanism for the indicator mixing is the natural turbulence of sea waters. The process becomes much more complicated if the contaminant jet has a high initial outflow velocity. Here, the kinetic energy of the jet itself can, certainly, become an important factor determining the initial mixing of the jet with the surrounding fluid. The process becomes even more complicated if the jet density differs from that of the sea water in the jet release region. In that case, the jet will be influenced by the buoyancy forces that cause either deepening or elevation (in the case of deep-sea release) of the jet. Here, additional mixing of the stream with the surrounding water masses will occur. Obviously, studies on contaminant jets with initial non-zero outflow velocities and various densities are of practical importance as these jets simulate various kinds of into-the-sea discharge systems for municipal and industrial sewage.

A larger number of papers have been devoted to theoretical studies of the behavior of contaminant jets in a stable stratified medium (e.g., see [24, 124]. Here, it is often assumed that lateral profiles of the liquid velocity u in the jet and also profiles the differences between the densities of the surrounding waters and the jet are described by the Gaussian curves

$$u(s, r, \varphi) = u(s) \exp(-r^2/b^2); \tag{2.22}$$

$$\frac{\rho_a(s, r, \varphi) - \rho(s, r, \varphi)}{\rho_0} = \frac{\rho_a(s) - \rho(s)}{\rho_0} \exp\left(-\frac{r^2}{\lambda^2 b^2}\right), \tag{2.23}$$

where s is the curvilinear coordinate along the jet axis; r is the radial coordinate; φ is the angle between the vertical plane xz (the origin of the coordinate is at the jet outflow point) and the described direction; $u(s)$ and $\rho(s)$ are the fluid velocity and density at the jet axis, respectively; ρ_0 is the initial density; $b = b(s)$, the characteristic width of the jet; and $\lambda = 1.16$, the empirical constant, taking into account the fact that profiles of density differences in cross-sections of the jet appear somewhat more 'smeared' than velocity profiles.

Usually, theoretical analyses of the problem further assume that the jet always maintains its axis-symmetrical form (i.e., they neglect the effect of the 'flattening' of the jet in a stratified medium) and that the pressure in the medium in the presence of the stream remains close to the hydrostatical one while the difference between the densities of the surrounding medium and jet is small, i.e., $|(\rho_a - \rho)/\rho_0| \leqslant 1$. Usually, the dissipation of energy in the stream and the centripetal forces, aroused due to the curvilinear shape of the jet, are neglected.

If further assuming the constancy of the coefficient of entrainment of the surrounding fluid into the jet we can arrive at the following system of equations describing the variations of flow rates, horizontal and vertical components of the momentum and jet buoyancy flux:

$$\frac{d}{ds}(ub^2) = 2\alpha ub; \quad (2.24)$$

$$\frac{d}{ds}(u^2b^2 \cos\theta) = 0 \quad (2.25)$$

$$\frac{d}{ds}(u^2b^2 \sin\theta) = 2g\lambda^2b^2 \frac{\rho_a - \rho}{\rho_0} \quad (2.26)$$

$$\frac{d}{ds}[ub^2(\rho_a - \rho)] = \frac{1 + \lambda^2}{\lambda^2} b^2u \frac{d\rho_a}{ds}, \quad (2.27)$$

where θ is the angle of inclination of the jet axis to the horizon; α is the entrainment coefficient, close, according to the available data, to 0.057; g is the acceleration due to gravity; and the coordinates s and θ are related to the Cartesian coordinates x and z by the expressions $dx/ds = \cos\theta$, $dz/ds = \sin\theta$. The system (2.24) – (2.27) is solved under the initial conditions

$$u(0) = u_0;\ b(0) = b_0;\ \rho(0) = \rho_1;\ \theta(0) = \theta_0;$$
$$x = z = 0 \quad \text{if} \quad s = 0 \quad (2.28)$$

In the case of constant vertical density gradient $d\rho_a/dz$, it would be appropriate to introduce non-dimensional values of flow rate Q, jet momentum components M_x and M_z and buoyancy flux P as well as non-dimensional coordinates ξ, X and Z. Thes are defined by the formulas

$$Q = \{N^{10}/[F_0^6 \cdot 64\alpha(1 + \lambda^2)]\}^{1/8} ub^2; \quad (2.29)$$

$$M = \{N^2/[(1 + \lambda^2)F_0^2]\}^{1/2} \frac{(ub)^2}{2}; \quad (2.30)$$

$$M_x = M\cos\theta;\ M_y = M\sin\theta; \quad (2.31)$$

$$P = \left(\frac{\lambda^2}{1 + \lambda^2} b^2ug \frac{\rho_a - \rho}{\rho_0}\right)/F_0; \quad (2.32)$$

$$\zeta = [N^6 \cdot 64\alpha^4(1 + \lambda^2)/F_0^2]^{1/8} s; \quad (2.33)$$

$$X = [N^6 \cdot 64\alpha^2(1 + \lambda^2)/F_0^6]^{1/8} x; \quad (2.34)$$

$$Z = [N^6 \cdot 64\alpha^2(1 + \lambda^2)/F_0^6]^{1/8} z, \quad (2.35)$$

where $N^2 = -g/\rho_0\, d\rho_a/dz$ is the square of the Väisälä – Brunt frequency and $F_0 = \lambda^2/(1+\lambda^2)\, b_0^2 u_0 g\, (\rho_0 - \rho_1)/\rho_0$ is a dimensional constant.

When using (2.29) – (2.35), the original system of equations will have the form

$$dQ/d\zeta = M^{1/2} \tag{2.36}$$

$$dM_x/d\zeta = 0; \tag{2.37}$$

$$dM_z/d\zeta = PQ/(2M); \tag{2.38}$$

$$dP/d\zeta = -QM_z/M; \tag{2.39}$$

$$dX/d\zeta = M_x/M; \tag{2.40}$$

$$dZ/d\zeta = M_z/M; \tag{2.41}$$

$$M^2 = M_x^2 + M_z^2, \tag{2.42}$$

with the initial conditions

$$Q(0) = Q_0;\ M(0) = M;\ \theta(0) = \theta_0;\ P(0) = 1,$$
$$X = Z = 0 \quad \text{if} \quad \zeta = 0. \tag{2.43}$$

In the general case, it is difficult to solve the system (2.36) – (2.42) analytically, therefore, numerical methods are often used. These methods, in particular, showed that a jet with $\rho > \rho_a$ released horizontally ($\theta_0 = 0$) begins, after certain elevation, to vertical oscillations with time of variable amplitude. In order to calculate the period of these oscillations it is convenient to introduce [24] the mean velocity across the jet $\bar{u}$, substituting the rectangular velocity distribution in jet for the Gaussian distribution while preserving the previous values of the flow rate and momentum. Here, the velocity $\bar{u}$ turns out to be equal to half the fluid velocity in the jet axis $u(s)$. For the mean time $t = t(s)$, necessary for a jet element to move from the source to a position s, the following relation may, apparently, be true:

$$dt = ds/\bar{u} = 2ds/u. \tag{2.44}$$

If we introduce a non-dimensional time according to the formula

$$T = t\Big/\left[\frac{2\pi}{N}\frac{\sqrt{2}}{(1+\lambda^2)^{1/2}}\right], \tag{2.45}$$

then, (2.24) will have the form

$$\frac{dT}{d\xi} = \frac{1}{2\pi\sqrt{2}}\frac{Q}{M}. \tag{2.46}$$

Further dividing (2.39) by (2.38) we will obtain

$$dP/dM_z = -2M_z/P. \tag{2.47}$$

Integrating (2.47) and using the initial conditions we will arrive at

$$P = (1 - 2M_z^2)^{1/2}. \tag{2.48}$$

Then dividing (2.46) by (2.38) and taking (2.48) into account we will find the relation between T and M_z:

$$\frac{dT}{d(M_z^2)} = \frac{1}{2\pi\sqrt{2}}\frac{1}{[(1 - 2M_z^2)M_z^2]^{1/2}}. \tag{2.49}$$

The solution of equation (2.49) for the case $\theta_0 = 0$ has the form

$$\cos(4\pi T) = 1 - 4M_z^2. \tag{2.50}$$

The jet trajectory, as it follows from (2.50), reaches its extreme heights at $M_z = 0$. Then we obtain from (2.50) the values for T_k at which the above circumstance holds true:

$$T_k = k/2, \; k = 1, 2, 3,\ldots \tag{2.51}$$

At the time, the multiplier $\sqrt{2}/(1 + \lambda^2)^{1/2}$ in the formula (2.45) is close to one, and thus the non-dimensional time T is practically measured in the units of the Väisälä-Brunt period. Thus, fluid elements of the jet in a stable stratified medium undergo vertical oscillations with the Väisälä-Brunt period. Examples of numerical calculations of the trajectories of such oscillating jets are given in Figure 2.49.

As is shown in [24], the system of equations (2.36) – (2.42) can be solved analytically for the case of a quasi-horizontal jet, i.e., when the condition $|M_x/M_z| \gg 1$ for any $\zeta > 0$ holds true. In that case, M_x can be substituted for M and differentiation with respect to ζ is substituted by that with respect to X. Then the first two equations in the system are integrated, and the entire system is reduced to a second-order differential equation relative to M_z:

$$\frac{d^2M_z}{dX^2} - \frac{1}{X + Q_0/M_0^{1/2}}\frac{dM_z}{dx} + \frac{1}{2M_0}(X + Q_0/M_0^{1/2})^2 M_z = 0. \tag{2.52}$$

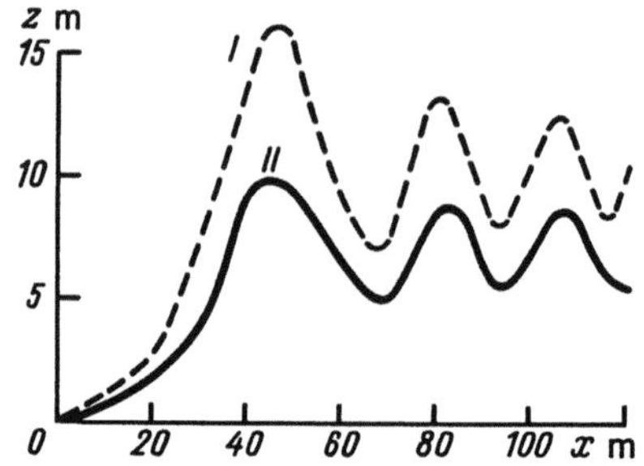

Fig. 2.49. Examples of numerical computations of emerging jet trajectories allowing for (II) and not allowing for (I) the compression effect when jets mix with sea waters (after [24]).

The general solution of equation (2.52) has the form

$$M_z = c_1 \cos\left[\frac{1}{2\sqrt{2M_0}}\left(X + \frac{Q_0}{M_0^{1/2}}\right)^2\right] + $$

$$+ c_2 \sin\left[\frac{1}{2\sqrt{2M_0}}\left(X + \frac{Q_0}{M_0^{1/2}}\right)^2\right], \tag{2.53}$$

where c_1 and c_2 are integration constants.

If further using the initial conditions $M_z(0) = 0$, $\mathrm{d}M_z/\mathrm{d}X|_0 = Q_0/(2M_0)$, then for the vertical-momentum component we obtain

$$M_z = \frac{1}{\sqrt{2}} \sin\left\{\frac{1}{2\sqrt{2M_0}}\left[\left(X + \frac{Q_0}{M_0^{1/2}}\right)^2 - \frac{Q_0}{M_0}\right]\right\}. \tag{2.54}$$

It follows from (2.54) that $|M_z| \leqslant 1/\sqrt{2}$, and thus the condition of jet quasi-horizontally can be written in the form $|M_x| \gg 1$ or in dimensional variables:

$$Nu_0\rho_0/(g|\rho_0 - \rho_1|) \gg 1. \tag{2.55}$$

It follows from (2.55) that a jet can be quasi-horizontal either if it has a large initial outflow velocity or a density only slightly different from that of the surrounding water. A quasi horizontal state will also be observed in cases when the Väisälä – Brunt frequency in the surrounding water is large.

The problem of the influence of the non-linear character of equation of state of sea water on the turbulent jet motions in a basin is considered in [24]. Because of the non-linearity of the equation of state the density of the mixture of two water volumes with different salinities S and temperatures T turns out larger than the weighted mean density of the mixture. For example, if a jet with a temperature much higher than 4°C flows into a fresh water basin with the water temperature lower than 4°C (at which water density is the highest) the jet will float up at the beginning (since the density is less than that of the surrounding water) and then, due to the mixing with the surrounding water, its temperature will approach 4°C, and thus, the water in the jet will become heavier than the surrounding one. In other words, after the jet elevation zone, a submergence zone will be observed which cannot be predicted by using a linear equation of state. The effect of density increase due to mixing can be taken into account by substituting two similar equations for the variation of temperature and salinity fluxes for equation (2.27) in the system (2.24) – (2.27):

$$\frac{\mathrm{d}}{\mathrm{d}s}\,[ub^2(T_a - T)] = \frac{1+\lambda^2}{\lambda^2}\,b^2u\,\frac{\mathrm{d}T_a}{\mathrm{d}s}; \tag{2.56}$$

$$\frac{\mathrm{d}}{\mathrm{d}s}\,[ub^2(S_a - S)] = \frac{1+\lambda^2}{\lambda^2}\,b^2u\,\frac{\mathrm{d}S_a}{\mathrm{d}s}, \tag{2.57}$$

where T_a and S_a are the temperature and salinity of the medium, and T and S are the temperature and salinity at the jet axis.

The thus-modified system must be closed by the equation of state for water:

$$\rho = \rho(T, S). \tag{2.58}$$

In order to estimate the influence of the non-linearity of the equation of state for sea water Zhurbas [24] made a numerical computation, using the above modified version of the system of equations and the original system (2.24)–(2.27), for the following values of the parameters: $\theta_0 = 0$; $u_0b_0^2 = 0.24$ cm^3/s; $u_0^2b_0^2 = 1.25$ m^4/s; $T(0) = 20$°C; $T_a(0) = 5$°C; $\mathrm{d}T_a/\mathrm{d}z = 0.1$°C$/m$ = const; $S = S_a =$ 35%. The calculation results for the two systems turned out to be significantly different (see Figure 2.49). This emphasizes the necessity of taking into account the non-linearity effect of the equation of water stat when describing the spread of different density jets in the ocean. A number of trajectory calculations for contaminant jets were also made for a freshwater basin with typical vertical distributions of water temperatures for different seasons. The contaminant sources were located at the depths of 20 and 40 m, and the parameters of the problem were taken as follows: $\theta_0=0$; $u_0b_0=0.24$ m^3/s; $u_0^2b_0^2=1.25$ m^4/s^2; and $T(0) = 20$°C. The temperature profiles for which the calculations were made are given in Figure 2.50b. As is seen from the figure, different combinations of governing parameters for the problem may lead to the highly varied behavior of the jets. In some cases, they lower, oscillations with the Väisälä – Brunt frequency, while in others they elevate, either with oscillations or without them,

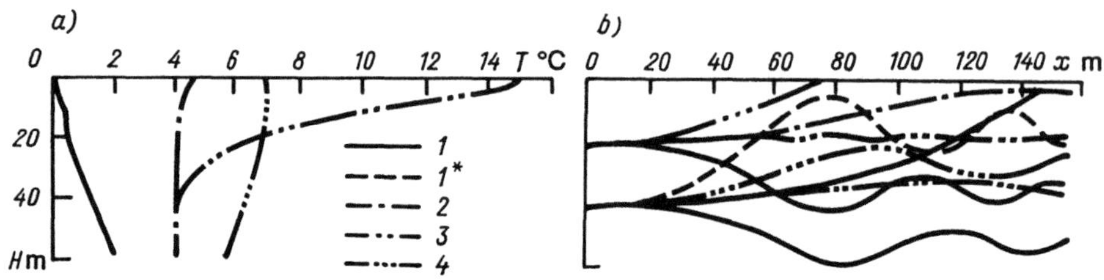

Fig. 2.50. Temperature profiles (a) and computations of jet trajectories (b) in freshwater basin (after [24]). 1 – wintertime allowing for the non-linearity of equation of state for water, 1* – the same without allowing for the non-linearity, 2 – springtime, 3 – summertime, 4 – autumn.

reaching the surface of the water basin or not. It is also seen from the figure that the use of a linear equation of state may result in essential errors when calculating the jet trajectory.

Experiments with submerged dye jets were carried out in Lake Baikal in August of 1975 [99]. The upper 15-m layer of the lake in the area of field works was at that time practically isothermic with T_a = 14...14.5°C. Below 50 m, the water temperature was 4°C. Jets with the dye concentration of 10^{-6} g/ml were released from nozzles located at the depths of 12 and 23 m, at a distance of 140 m from the shore. The currents observed in the area of experiments were in the east-south east direction with the velocities at the depth of 15 m of 3 cm/s at the distance of 0.5 km from the shore and of about 5 – 6 cm/s several kilometers from the shoreline. At larger distances from the shore the current was stronger, and at the depth of 50 m it reached 20 cm/s. At depths exceeding 100 m there was practical no current observed.

Indicator-distribution measurements during the experiments were made at a number of stations with the help of a fluorimeter. The vertical sections of indicator jets, obtained based on the sounding data of 8 stations, is given in Figure 2.51. The jet spread to the east-south east, undergoing vertical oscillations. The indicator from the upper jet came up to the surface at approximately 100 m away from the release point. Here the dye concentration was $10^{-8} – 10^{-9}$ g/ml. At larger distances from the source the jet axis got deepened again, and the dye was not observed at the surface of the lake (within the concentration isoline of 10^{-10} g/ml), but underwent significant vertical oscillations. The spread of the jets was confidently traced up to as far as 1100 m away from the source, and the indicator traces were found at 6.1 km away from the source. Interestingly, both jets did not mix with one other over practically their whole course, which is explained by the large variation of densities (temperatures) of water in the

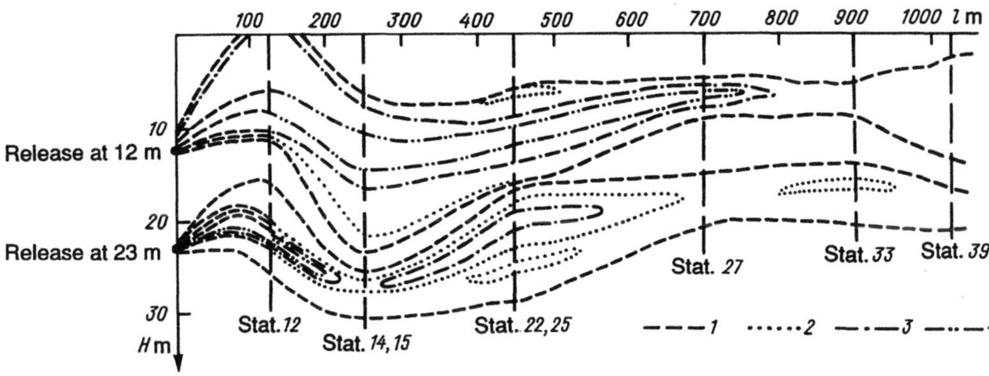

Fig. 2.51. Vertical section of dye-concentration distribution in the experiment with submerged jets in Baikal Lake. Concentration isolines: 1–10^{-10} g/ml; 2–10^{-9} g/ml; 3–10^{-8} g/ml; 4–10^{-7} g/ml.

layer between the sources. This layer of density pump (pinocline) is a kind of screen hampering the vertical transport between the layers of fluid.

In 1975, releases of dye streams in Lake Baikal at depths of 34 and 45 m and at distances of 140 and 240 m from the shore were undertaken. The currents varied substantially in the course of that experiment both in terms of direction and velocity, which resulted in the complicated configuration of dyed water area. In the upper 10-m water layer, the area with noticeable dye concentrations amounted to 110,000 m^2 and dye traces were found at as far as 1000 m from the sources.

In the period from 1972 to 1975, diffusion experiments wiht a radioactive indicator – gold-128 isotope – were carried out in the lake as well [99]. Isotope-concentration measurements in the lake waters were made by a gamma radiometer submerged from a small ship in the summertime and from the ice in winter. Based on the measurement data, the degree of dilution of the isotope cloud in the process of diffusion and the rate of transport of the isotopes by currents was calculated. It was discovered that in low winds and under small current velocities, the isotope spread practically isotropically in the horizontal directions, with the degree of dilution reaching 200–300 at a distance of about 500 m from the source. On releasing the isotope from the same nozzles from which the dye in the 1975 experiment was released, the vertical diffusion was significant – the isotope was spreading inside a relatively thin layer of about 10–15 m thick, with the maximum concentration being registered near the release depth.

Theoretical interpretation of the results of the experiments with dyes and the radioactive isotope in Lake Baikal was made in [99]. Since the contaminant was released from horizontally positioned nozzles with the diameter d = 0.6 m and outflow velocity $u_{outfl.}$ = 2.5 m/s, on leaving the nozzles the jet formed its active zone where $u_{outfl.}$ influenced considerably the behavior of the jet. The Reynolds number for the jet in the proximity of the nozzles was equal to $Re = u_{outfl.}\, d/\nu \approx 1.5 \cdot 10^5$. At the same time, the Richardson number $Ri = (g/\rho)(\partial\rho/\partial z)/(u_{outfl.}/d)^2$ for the starting part of the jet was substantially smaller that the critical Ri_{crit}, if $\partial\rho/\partial z$ is the vertical gradient of water density at the depth of nozzle location. Thus the influence of the density stratification of water at the starting part of jet development can be neglected, and the jet may be assumed axes-symmetrical. In this case, equation (2.24) will hold true for the jet outflow velocity, and for the momentum component along the horizontal axis x, instead of (2.25), we have

$$\frac{\mathrm{d}}{\mathrm{d}x}\int_{\Sigma} u^2(r,\, x)\mathrm{d}\Sigma = \frac{\mathrm{d}}{\mathrm{d}x}\,\frac{\pi u^2 b^2}{2}, \tag{2.59}$$

where Σ is the cross-section area of the nozzle. Evaluating integral in (2.59) we will arrive at

$$u^2 b^2 = (u_0 b_0)^2 + \text{const}, \tag{2.60}$$

where u_0 is the mean velocity of the jet, averaged over the cross-section of the nozzle, equal to $2u_{outfl.}$, and b_0 is the effective initial jet cross-section.

Solving the thus-obtained system for u and b, we obtain

$$u(x) = \frac{u_0}{2\alpha x/b_0 + 1}; \tag{2.61}$$

$$b(x) = b_0(2\alpha x/b_0 + 1). \tag{2.62}$$

At large values of x, these expressions reduce to the simple formulas

$$u(x) \approx u_0 b_0/(2\alpha x); \tag{2.63}$$

$$b(x) \approx 2\alpha x. \tag{2.64}$$

The length of the active zone of the jet x_a can be estimated if we assume that this zone ends up where the characteristic jet velocity $u(\dot{x})$ is comparable with the mean square values of turbulent velocity in the water basin σ_u. In that case for x_a we obtain

$$x_a = u_0 b_0/(2\alpha\sigma_u). \tag{2.65}$$

Assuming, based on the estimations available, that large-scale horizontal variations of horizontal-velocity components in the lake were 4 – 6 cm/s in the period of the experiments, the evaluation for x_a will appear to be 150 – 250 m.

In order to take into account the influence of the density stratification of the lake waters on the jet, let us assume that a stationary jet in such a medium can only exist if there are turbulent eddies with sizes comparable to the thickness of the jet. This is possible in a case in which the local Richardson number for the jet Ri_c is smaller than the critical number Ri_{crit}, i.e., when

$$Ri_c = N^2(\partial u/\partial z)^2 \approx N^2/[u(x)/b(x)]^2 < Ri_{crit}. \tag{2.66}$$

At the starting part of the jet, its velocity $u(x)$ is rather high and the width $b(x)$ is small, therefore, $Ri_c \ll Ri_{crit}$ and the density stratification does not significantly affect the jet. When Ri_c becomes comparable with Ri_{crit} the growth of the vertical sizes of the jet will slow down and, finally, when Ri_c reaches Ri_{crit} this growth will cease completely since the energy of turbulent fluctuations in the jet will already be insufficient to overcome the Archimedean forces. From (2.63) and using (2.66) with the equality sign we will obtain for the maximum width H of the jet

$$H = 2b \approx 2Ri_{crit}^{1/4}(u_0 b_0/N)^{1/2}. \tag{2.67}$$

After reaching the width H, the jet will be widening in the vertical direction much slower, only under the influence of natural vertical turbulence, which is substantially damped in a stably density-stratified water.

Evaluations of H were made for Lake Baikal for typical density gradients in the summer and winter when jet sources were located at depths from 10 to 50 m. The values for H appeared to vary from 9.5 to 24 m, which is in good accordance with the experiment data both for the dyes and the radioactive isotope. In the spring and fall, when the lake water-density gradients become smaller, with even, complete vertical isothermic state of water occurring in certain periods, jet widths may significantly increase. In such situations, the indicator may be observed throughout the entire depth of the lake from surface to bottom in the area of experiment.

During the Lake Baikal experiments, the theoretically predicted floating upward of contaminant jets was observed. The temperature of the indicator solution in jets was about 20 – 22°C, so it greatly exceeded that of the water in the lake. In fall and spring, the upper contaminant boundary was more often located at depths of 5 – 15 m, and the contaminant spread down to the bottom. In summer, the upper boundary was being found at horizons 10 – 20 m, with only a small portion of contaminant reaching the surface by making its way through the sharp thermocline. In the wintertime, under thermal inversion in the lake (the right-under-the-ice water temperature is about 0°C while it grows to 4°C with depth) the interesting phenomenon mentioned above was observed. The warm jet first moved upward, and when its temperature decreased close to 4°C the jet began to descend from the upper layer of the lake as it turned out to be denser than the surrounding water. In this case the jet axis appeared to be lower than the nozzle-location depth, and the upper boundary of the jet was at depths of 20 – 35 m.

TABLE II.III

Lateral dispersions σ_y^2 of indicator jets at various distances from the source x, based on experiments with radioactive indicator in the Lake Baikal (according to [99]).

Winter season		Summer season	
x, m	σ_y^2, m^2	x, m	σ_y^2, m^2
200	$6.8 \cdot 10^3$	200	$2.3 \cdot 10^3$
300	$7.1 \cdot 10^3$	450	$1.5 \cdot 10^4$
400	$1.3 \cdot 10^4$	800	$1.2 \cdot 10^4$
500	$2.6 \cdot 10^4$	1200	$5.2 \cdot 10^4$
650	$3.1 \cdot 10^4$	1700	$6.1 \cdot 10^4$
750	$3.0 \cdot 10^4$		
900	$4.0 \cdot 10^4$		

If the vertical spread of the contaminant jet in the stratified lake was limited by 10 – 20 m, diffusion in the horizontal plane led to much larger sizes of the jet. Table II.III gives data on the lateral dispersion of the jets, based on the experiments with the radioactive indicator in winter and summer in Lake Baikal. As is seen from this table, the horizontal characteristic sizes of the jets reach 200 m at distances from the source of about 1000 m. In other words, at such distances from the release point the vertical and horizontal anisotropies of the jets reached 10 – 20.

Experiments with submerged indicator jets were also carried out in the Black Sea [33]. The experiments were aimed at studying the behavior of contaminant jets created by operational or projected sewers. The submergence of such water sewers is required, first and foremost, in order to ensure better sanitary conditions for the surface and coastal sea waters where most contacts of man with the water medium occur. Besides, a waste-water discharge in the surface sea layer may cause violation of convection mechanisms, and the emerging density stratification of waters in the surface sea layer itself may hamper the normal thermal and gas exchange between the sea and the atmosphere. Certainly, the locating of sewers should be combined with the utmost purification of sewage waters at on-shore facilities and with larger development of waste-free technologies.

Experiments with diffusing dyes near operational sewers in the Black Sea were started in 1969 [31]. An indicator solution was being released at a constant rate into the sewer discharge pipe, with a ship in the release area making fluorimeter measurements. Hydrometeorological and hydrochemical observations were also made in the release area. During 1969 – 1972, seven experiments in the sewage release area near the city of Simeiz were conducted. The release depth was 16 m, with this whole water body being practically density-homogeneous throughout the year. Only during the spring heat wave (May – June) does a short term stratification at depths of 10 – 15 m appear. Without stratification, the sewage waters reached the surface of the sea, and only in the June 26, 1970 experiment (under stratification) did the indicator spread in the bottom layer, not elevating higher than 10 m. During 1971 – 1973, twelve diffusion experiments were conducted near an interim sewer, located at the depth of 37 – 39 m, in the city of Yalta. In the period from October to April, density stratification is not observed in that area. In the summer months, pycnocline is observed that generally deepens to the 20-m horizon or deeper. During storms and surging winds, the pycnocline expands and, in a number of case, the entire coastal water body may become density-homogeneous. Such a variety and changeability of hydrological conditions also resulted in the various forms of sewage-plume behavior. In some cases in which there was a quasi-homogeneous water body, the plume came up to the sea surface while in the case of a prominent pycnocline it usually remained submerged. The height of the plume elevation ΔH_0 also varied from time to time depending on the hydrometeorological conditions. The value ΔH_0 for a contaminant in water with constant vertical-density gradient $d\rho/dz$ can be estimated by the formula (see ref. [31])

$$\Delta H_0^4 = 6340 \frac{Q_0(\sigma_1 - \sigma_a)}{\sqrt{g}\ (d\rho/dz)^{3/2}}, \tag{2.68}$$

where Q_0 is the contaminant flow rate of the source; σ_a and σ_1 are densities *in situ* of sewage and sea waters at the release depth, respectively; and g is the acceleration due to gravity.

For a linear contaminant source, a similar formula can be written in the form (see ref. [31])

$$\Delta H_0^3 = 10.8 \frac{q_0(\sigma_1 - \sigma_a)}{\alpha\sqrt{g}\ (d\rho/dz)^{3/2}}, \tag{2.69}$$

where q_0 is the specific flow rate of contaminant per unit of source length, and α is the parameter of the entrainment of the surrounding fluid into the elevating plume.

The formulas (2.68) – (2.69) are suggested for calculating jet elevation in a stratified sea with no constant currents. If such currents do exist the jet elevation usually decreases since the current accelerates the mixing of the jet with the surrounding fluid, thus, reducing the variance in the jet and sea-water densities. In order to allow for the influence of currents on jet elevations Zats [31] suggested introducing additional empirical multipliers into the formulas (2.68) – (2.69), rewriting the expression for ΔH as

$$\Delta H = K_u \Delta H_0 , \tag{2.70}$$

where the coefficient K_u depends on the current velocity U according to the formulas

$$K_u = 1 - 0.67\, U^{0.6} \tag{2.71}$$

for a case when the variance between jet and sea-water densities is small, and

$$K_u = 1 - 0.93\, U^{0.4} \tag{2.72}$$

for significant differences between σ_a and σ_1.

The results of the Black Sea experiments with submerged dye jets agreed well with the contaminant-jet elevations as calculated by the formulas (2.70) – (2.72). In order to present the outcome of sewage waters up to the sea surface it is very essential, according to these formulas, to enhance the initial dissolving of contaminant jets, which can be done by attaching special diffusers onto the discharge nozzles. It is also worthwhile to spread out the release in order to convert the source from a point one into a linear one or into one distributed over some area.

In the Black Sea experiments with submerged jets Zats [32] happened to observe the interesting phenomenon of the splitting of the elevating contaminant jet into several quasi-horizontal layers stretching along the isopycnical surfaces for large distances from the jet release point. This phenomenon is related to the density 'separation' of the jet substance in a stratified sea. Heavier portions of the contaminant lose their ability to elevate any more on reaching a certain level and begin to spread in the horizontal plane. At the same time, lighter jet fractions still have a velocity directed upward and, therefore, are capable of approaching higher levels where the sea-water density is less.

As was mentioned above, experiments on diffusion with quasi-stationary sources make it possible to gather information suitable for determining statistical characteristics of contaminant concentration fields. Ichiye [128] was, apparently, the first who attempted to determine the spectral structure of the contaminant field in a diffusing jet. Two densitometrical profiles in the lateral sections of the jet, made at the distances of 1900 and 2300 m from the source, were taken from aerial photographs. The profiles were averaged, and fluctuation characteristics were obtained by subtracting the averaged concentration profiles from the 'instantaneous' ones. The spacial spectrum for the distances of 1900 m had a clearly pronounced maximum at the wave number $k = 1.7 \cdot 10^{-3}$ cm^{-1}, and for the second lateral section, the maximum appeared to be far less pronounced and shifted toward lower values of $k(7\cdot10^{-4}$ cm$^{-1})$.

The set of statistical characteristics of contaminant field in the diffusing jet was calculated, based on the data of the experiment with a 'continuous' point source of the indicator in the coastal Baltic Sea area near the settlement of Tsingst [87]. The contaminant source was positioned on a trestle 90 m away from the shoreline and at the distance of 150 m from the source, and at the same distance from the shore a fluorimeter was installed on a special tripod. The registering instruments and electric power supply sources were placed on the shore. The source intensity during the first 3 hours of the experiment was equal to 0.12 g of dye per second, and then it was increased to 0.54 g/s. The velocity of the along-the-shore current in the course of the experiment was 11 cm/s, and the wind velocity approached 6 – 6.5 m/s. The dye-concentration registration lasted continuously for 3.5 hours. The discretization interval when processing the records obtained was 3 seconds. This resulted in 3970 concentration values for most of the experiment in series No. 1, and in 361 terms for the part of the experiment with greater dye concentrations in series No. 2. Calculations of the first four statistical moments of the distributions of concentration fluctuations were made separately for two halves of the first series (series No. 3 and No. 4) as well as for the entire series No. 1 but with the double step of discretion (series No. 5). The thus-obtained estimates of statistical moments are given in Table 2.4. Estimations for $\bar{c}$, dispersion D and also for skewness A turned out, as it follows from the table, very close to each other for series Nos. 3, 4 and 5, which proves the stationary of the main series, No. 1. Somewhat larger variance is seen for the values of kurtosis E. Apparently, longer series are needed for its more accurate estimation.

TABLE II.IV

Estimates of mathematical expectation $\bar{c}$ $(10^{-9}$ g/cm$^3)$, dispersion D $(10^{-18}$ g^2/cm$^6)$, skewness A and kurtosis E as well as confidence intervals j (at the significance level of 0.9) for contaminant concentrations in diffusing jets. Based on the Baltic Sea experimental data – according to [87]

Series Number	$\bar{c}$	$j_{\tilde{c}}$	D	j_d	A	E
2	92.7	1.8	421	52	0.15	−0.17
3	36.9	1.0	695	22	0.61	−0.69
4	34.4	1.0	750	24	0.69	−0.34
5	35.6	1.0	722	23	0.64	−0.51

Using the concentration fluctuations series, probability distribution density functions for c' were plotted (see Figure 2.52). For series No. 2, the empirical distribution is rather well approximated by the normal distribution law, while for series corresponding to lesser source intensity the approximation of contaminant by this law was satisfactory only for concentration fluctuations exceeding the mean value $\bar{c}$. At small values of c' the normal distribution, however, gives underestimated probability values as compared to the experimental ones. The probability distribution density functions for concentration fluctuations, plotted based on the data of a similar experiment with a dye jet in the Black Sea [47], showed satisfactory agreement with the normal distribution law and with the Edgeworth distribution as well.

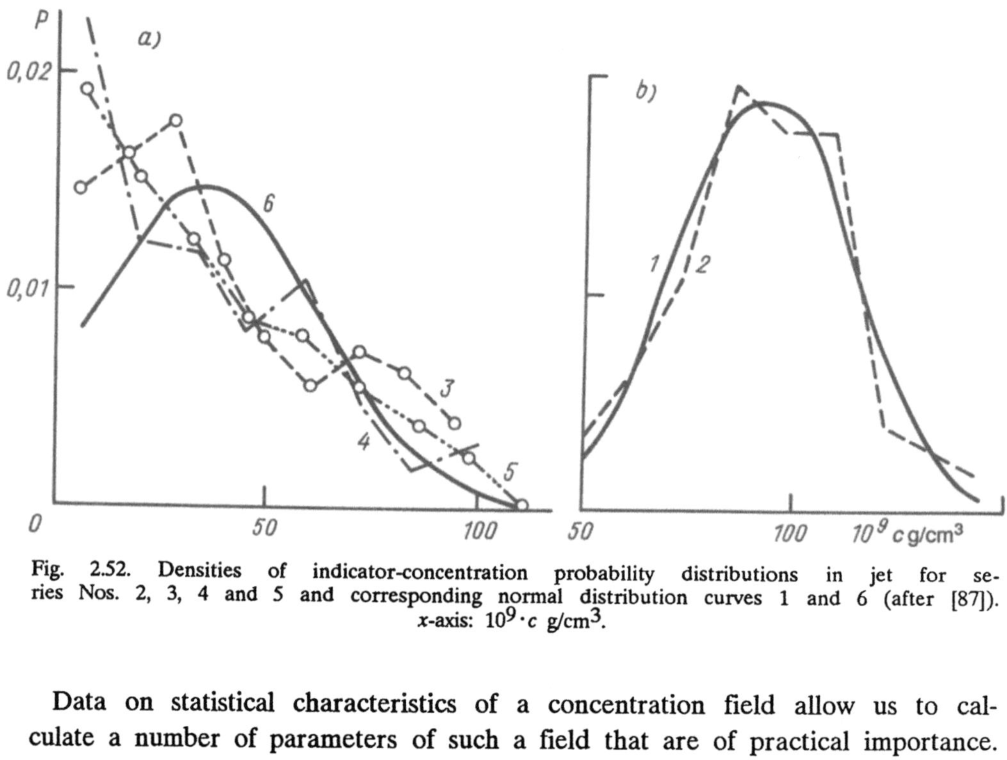

Fig. 2.52. Densities of indicator-concentration probability distributions in jet for series Nos. 2, 3, 4 and 5 and corresponding normal distribution curves 1 and 6 (after [87]). x-axis: $10^9 \cdot c$ g/cm^3.

Data on statistical characteristics of a concentration field allow us to calculate a number of parameters of such a field that are of practical importance. For example, it is often necessary to know how many times n (per unit of time) the contaminant concentration at a point being considered will exceed some value c_0. The function $n(c_0)$ can be determined given the value for contaminant-concentration dispersion and spectral-fluctuation function $c'(t)$. If we assume that the spectral function in the frequency range $\omega_1 < \omega < \omega_2$ is governed by the '-5/3 power law' and dispersion may be determined as the integral of the spectral function from ω_1 to ω_2, then the following expression for $n(c_0)$ can be obtained (see ref. [87]):

$$n(c_0) = \frac{1}{4}\,\pi^{-3/2}\,\frac{\omega_2^{4/3} - \omega_1^{43}}{\omega_1^{-2/3} - \omega_2^{-2/3}}\,\exp\left(-\frac{c_0^2}{2D}\right). \tag{2.73}$$

For the considered series of contaminant concentrations Nos. 2, 3 and 4, ω_1 can be found from the relation $\omega_1 = 2\pi/\tau_{min}$, where $\tau_{min} = 3s$; and $\omega_2 = 2\pi/\tau_{max}$, where τ_{max} shall be considered equal to $l/\bar{u}$. Here l is the phenomenon scale which, in turn, can be estimated by the formula $l = \sqrt{\alpha\varepsilon t^3}$, where α is a universal constant, $t = x/\bar{u}$ is the contaminant 'running' time from the source to a point x. Experimental curves of the function $n(c_0)$ for series Nos. 2, 3, 4 and 5 are presented in Figure 2.53, where the theoretical dependence $n(c_0)$ computed according to the formula (2.73) with the parameters of series No. 3 is also plotted. Figure 2.54 gives the overshoot duration probability distributions $p(\tau)$ above the concentration levels $\bar{c}_1 = 55.4\cdot10^{-9}$ g/cm^3 and $\bar{c}_2 = 42.4\cdot10^{-9}$ g/cm^3 for series No. 5. The theoretical curves in the figure are plotted according to the normal distribution dependence

$$p(\tau) = \frac{\bar{c}R\tau}{4D}\exp\left(-\frac{\bar{c}^2R\tau^2}{8D}\right), \tag{2.74}$$

where

$$R = -\frac{d^2r(\tau)}{d\tau^2} = \frac{1}{2\pi D}\int_0^\infty \omega^2E_c(\omega)\mathrm{d}\omega; \tag{2.75}$$

$r(\tau)$ is the auto-correlational function of contaminant-concentration fluctuations. If $E_c(\omega)$ is governed by the '−5/3 power law' in the frequency range from ω_1 to ω_2 then an expression for R will have the form

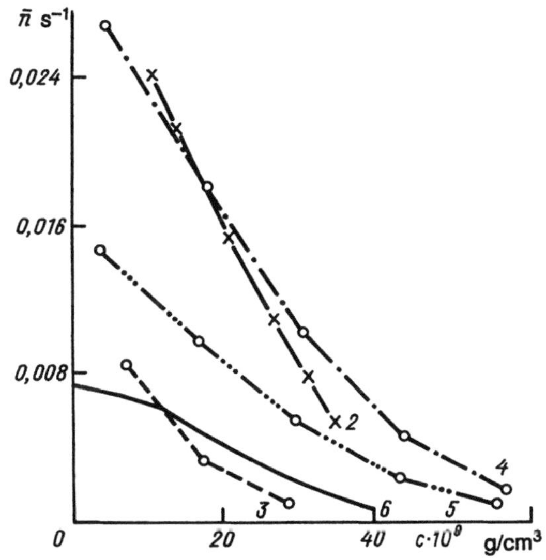

Fig. 2.53. Average number of overshoots vs. indicator-concentration level for observation series Nos. 2, 3, 4 and 5 and theoretical curve 6 as calculated according to formula (2.73) for series No. 3. x-axis: $c\cdot10^9$ g/cm^3; y-axis: $\bar{n}$ s^{-1}

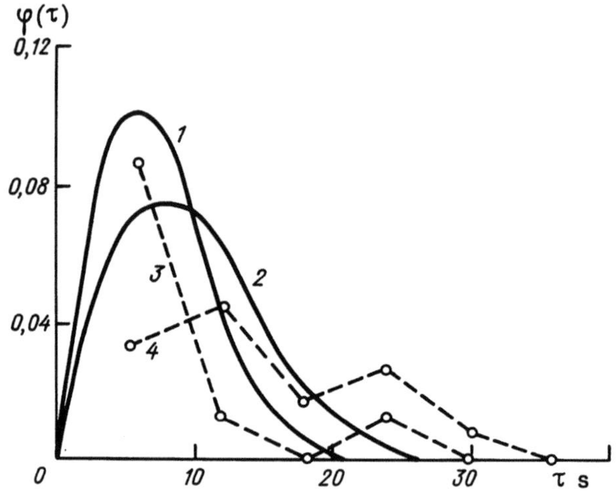

Fig. 2.54. Experimental values of probability distribution density for duration of indicator-concentration overshoots for series No. 5 at two mean concentration levels (curves 3 and 4) and corresponding theoretical curves 1 and 2, plotted according to formula (2.74). x-axis: τ s.

$$R = \frac{3}{8\pi D} B\varepsilon_c \varepsilon^{-1/3} \bar{u}^{-2/3} (\omega_2^{4/3} - \omega_1^{4/3}), \tag{2.76}$$

where B is a universal constant, ε is the turbulent energy-dissipation rate, ε_c is the rate of 'dissipation' of concentration inhomogeneities and $\bar{u}$ is the mean current transport velocity.

Spectral functions of contaminant-concentration fluctuations E_c were plotted for all the series considered. Here, the transition from time arguments to spacial arguments is done according to the hypothesis of 'frozen' turbulence with the use of the known value of the current transport velocity $\bar{u}$. As an example, Figure 2.55 presents spectral densities of concentrations fluctuations, calculated for observation series No. 5 and No. 2. Up to the frequency $f = \omega/(2\pi) =$ 0.03 Hz (which corresponds to the wave number $k \sim 0.02$ cm^{-2}) the experimental points agree well with the '–5/3 power law'. However, for frequencies $f < 0.03$ Hz and up to the maximum value $f = 0.001$ Hz the inclination of spectral function increases to –2. Such a form of spectral function was discovered, as was mentioned in the previous section, also through dye concentration field measurements in diffusing spots and was related to the influence of energy influx to the ocean waters from the wind (formula (2.14)). In experiments with continuous point sources, the form of the spectral function E_c can also be determined by the meandering of the dye jet as a whole under the influence of large-scale turbulent eddies in the velocity field. In that case, a fluorimeter sensor at rest

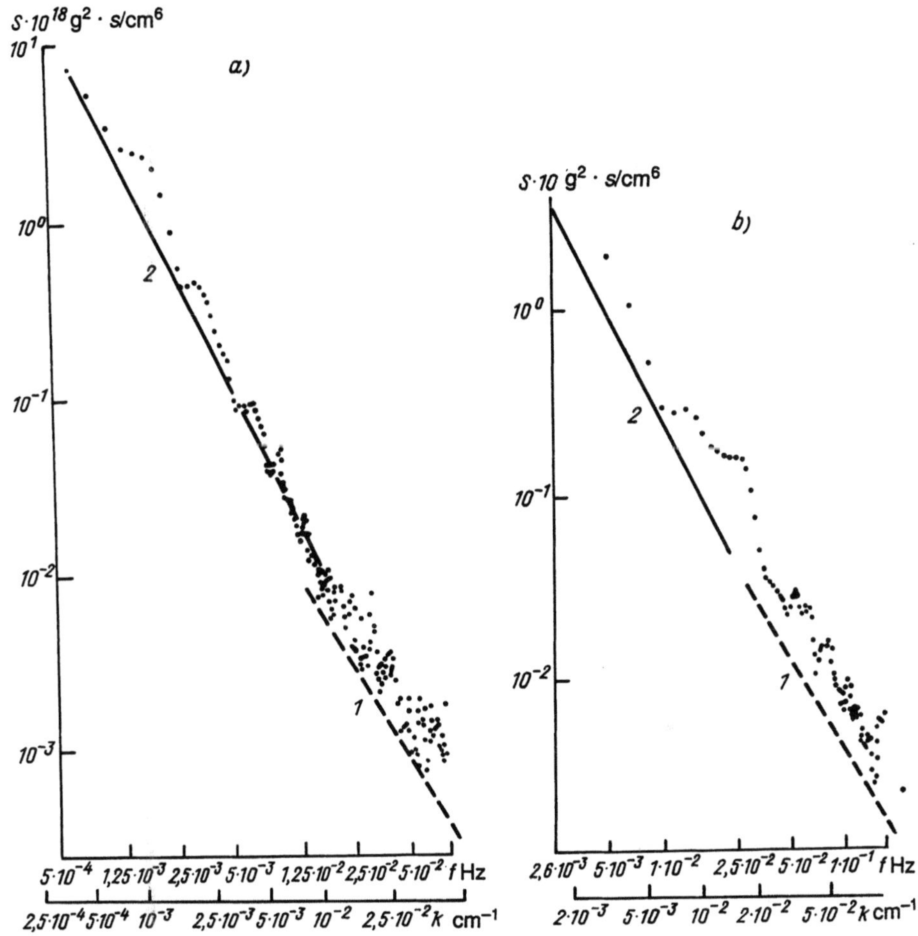

Fig. 2.55. Spectral densities of indicator-concentration fluctuations in jet, calculated for series No. 5 (a) and No. 6 (b). 1 and 2 are dependences with exponents -5/3 and -2, respectively. y-axis: $s \cdot 10^{18}$ $g^2 \cdot s/cm^6$; x-axis: f Hz; x-axis: k cm^{-1}.

will, at different moments of time, enter into the jet zones with a changing mean indicator concentration. If we denote the time spectrum of the jet displacement in a direction perpendicular to that of the averaged flow, by $E_y(\omega)$ the instrument will, under the influence of the jet-axis displacement, register concentration variations whose spectrum $E_c(\omega)$ is related to the mean contaminant concentration in the following way:

$$E_c(\omega) = (\partial \bar{c}/\partial y)^2 E_y(\omega). \tag{2.77}$$

Since the gradient $\partial\bar{c}/\partial y$ is a function of y, the spectrum $E_c(\omega)$ will also be dependent on the amplitude of jet meandering. In order to divide, in a clear-cut manner, the contributions of the jet-axis oscillations inside the jet (under the influence of small-scale turbulence) to $E_c(\omega)$, it is obviously also necessary to register the dye-jet displacement as a whole under the influence of large-scale sea turbulence while simultaneously taking concentration measurements at a fixed point of space.

In the case of multiple crossings of the diffusing indicator jet at a fixed distance from the source x, the problem of excluding the meandering jet influence on the statistical characteristics of the concentration field does not appear. Really, in that case, the lateral coordinate y may be counted off from the maximum point of averaged through smoothing individual realizations of $c(y)$ indicator concentration. Under the thus-chosen system of coordinates, all the consequent profiles $c(y)$ can be brought into coincidence, by points, with the maximum of the concentration and, therefore, make up an ensemble suitable for further statistical averaging. Fluctuation values of the concentration are then obtained by substracting the averaged profile $\bar{c}(y)$ from each 'instantaneous' profile, i.e., $c'(y) = c(y) - \bar{c}(y)$. If the number of crossings undertaken is large enough, the dispersion or mean square deviation σ of concentration fluctuations as a function of the lateral coordinate y can be computed. In order to facilitate further analysis, it is expedient to normalize $\sigma(y)$ by the value of averaged concentration $\bar{c}(y)$, with the non-dimensional distance $\zeta = y/L$, where L is the typical jet width, being taken as the argument for this normalized function. The dependence of $\varphi = \delta(y)/\bar{c}(y)$ on ζ was determined based on the data of the diffusion experiments with dye jets conducted in the Black Sea in 1968. Computations were made for the measurements at three lateral sections of the jet at the distances of 100, 200 and 400 m from the dye source [88]. Similar calculations were made later based on the measurement data for dye jets in the Caspian Sea [26]. Here, the distance between the measurements tacks and the contaminant source reached 2100 m. It turned out that the dependences of φ on the non-dimensional distance from ζ were close to each other for all the tacks, which allowed the presentation of all the data on $\varphi(\zeta)$ as a unique graph (see Figure 2.56). Typical values of the function $\varphi(\zeta)$ in the central part of the jet appeared to be close to $0.3-0.6$, while at the periphery of the jet they increased up to $4-6$. The theoretical curves for the function $\varphi(\zeta)$ presented in the figure were obtained through numerical integration of the second-order differential equation for $\varphi(\zeta)$, derived under the assumption of self-similarity of the solution of the semi-empirical equation for the mean square of contaminant-concentration fluctuations in the jet [121]. The integration was carried out for three values of $\varphi(0)$. The agreement between the experimental curves should be accepted as satisfactory at a first approximation; however, slower growth of

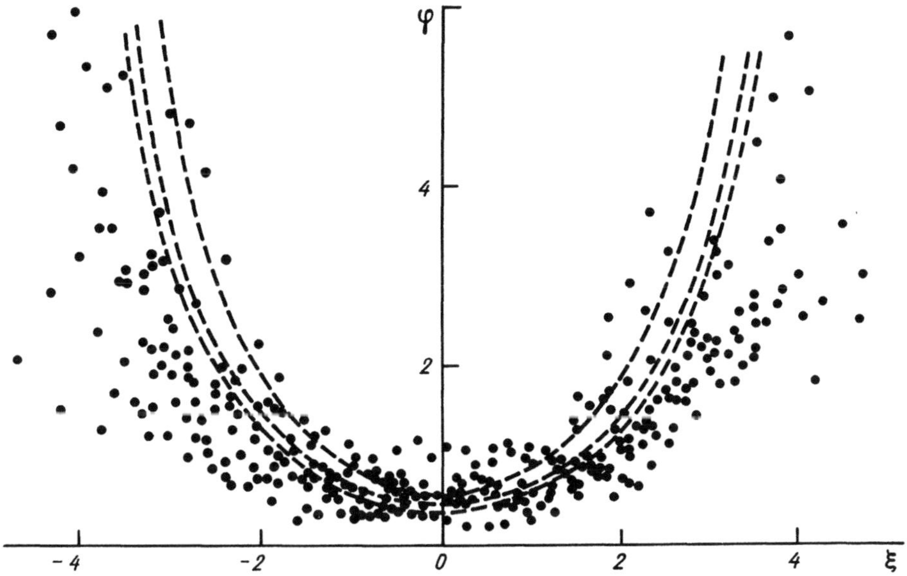

Fig. 2.56. Dependence of relative intensity of indicator-concentration fluctuations on non-dimensional distance to jet axis. Experimental dots and theoretical curves (after [26]).

experimental values for $\varphi(\zeta)$ with the increasing ζ is observed as compared to the dependence predicted by the model.

A more complicated model for the dependence of the relative intensity of indicator-concentration fluctuations on the non-dimensional distance from the source was suggested, as was mentioned in Part 1.8, by Toompuu ([102], also see work by M.S. Nemirovsky in [12]). The value of φ in his model was dependent not only on ζ but on the distance from the indicator source x as well. Figure 2.57 presents the dependences $\varphi(\zeta, x)$ obtained based on the dye-jet measurements taken during the Baltic Sea experiments at the distances from the source of 25, 75, 150, 250 and 350 m [102]. The theorectical curves, plotted with the use of the formula (1.173) under certain assumptions about the mean dye-concentration field, agree rather well with the experimental data displaying a pronounced dependence of φ on the distance x. So it follows that the self-similarity of the solution of an equation for φ does not, generally speaking, hold true as it is violated by the dependence on x. However, Toompuu pointed out, for larger distances from the source this dependence retards and the self-similarity assumptions seems, in that case, more grounded. This fact, obviously, explains why in Csanady's earlier studies [121] as well as in the above experimental results in the Black Sea

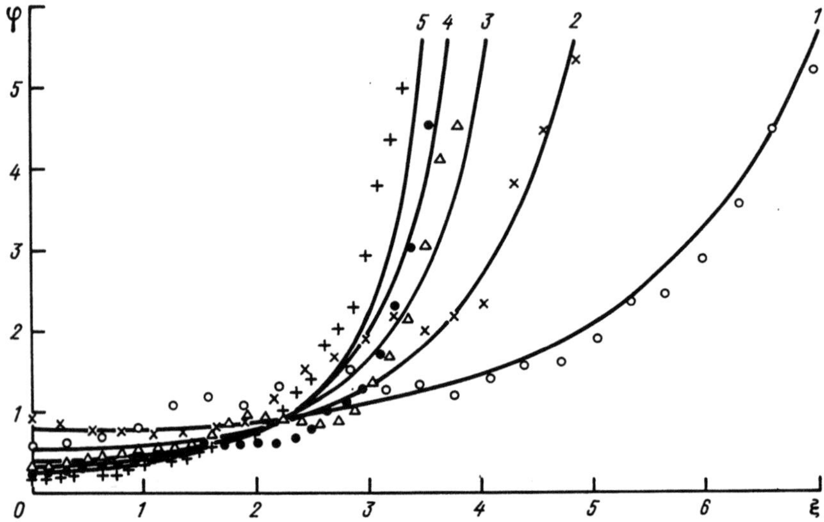

Fig. 2.57. Dependence of relative intensity of indicator-concentration fluctuations on non-dimensional distance to jet axis and on distance to source (1 – 25 m, 2 – 75 m, 3 – 150 m, 4 – 250 m, 5 – 350 m) (after [102]).

and Caspian Sea there were no clear-cut deviations from self-similarity observed.

The influence of the statistical characteristics of contaminant-concentration fields in jets on the distance from the source is also envisaged in the work by Zats, Nemirovsky *et al.* [11]. In that work, the data on diffusion experiments, carried out during the international joint venture Kamchiya-77, in particular the calculations of spectral functions $E_c(\omega)$, were analyzed. The paper presents the values of the functions $E_c(\omega)$ calculated based on the measurements in the dye jet at the distances of 200 and 500 m from the source. It was found that the slope of the spectral function increases with the growing x from –1.5 to –2. In [11] the function $E_c(\omega)$, obtained in another experiment at the distance of 100 m from the indicator source, was also analyzed. Here, the slop varied significantly with the changing argument. Maximum slopes of spectral curves (in the logarithmic coordinates) were observed in the joint venture Kamchiya–77 in stormy weather. Such a variety of spectral forms and their variability are related, naturally, to many a factor affecting the contaminant-concentration field. This was already mentioned when analyzing the spectral functions, plotted based on the experiments with dye spots.

2.4. Diffusion of Discrete Particles in the Ocean

The first experiments on the diffusion of discrete particles in the ocean were started by Richardson and Stommel [153] who, for that purpose, used floats made of Parsnip roots. The experiments were conducted from a pier when the sea depth was only 2 m. The distances between 45 pairs of individual particles were measured at the initial moment of time t_0 and at the moment $t_1 = t_0 + \tau$, where $\tau =$ 30 s. Based on these data, the coefficient of diffusion $F(l)$ in equation (1.164) was calculated by the formula

$$F(l) = \overline{(l_1 - l_0)^2}/(2\tau) = \overline{\Delta l^2}/(2\tau), \tag{2.78}$$

where Δl is the change in the distance between the particles over the time τ, and the bar denotes averaging over the pairs of particles with close initial separations l_0. In the experiment conducted by Richardson and Stommel, averaging was made over two classes of distances between the particles, which resulted in two values for the coefficient $F(l)$, where $l = (l_0 + l_1)/2$. Two values for the coefficient $F(l)$ were also obtained by Stommel [156] from the experiments with discrete particles off the shores of North America and The Bermudas. Sheets of paper were used as indicators in those experiments.

A large serie of experiments on discrete-particles diffusion was conducted by Ozmidov in 1955–1958 [81]. The experiments were carried in artificial ponds, from oil trestles in the Caspian Sea and in the Pacific Ocean. In the first two cases, sheets of paper served as indicators while in the ocean experiments submerged buoys with radar reflectors were chosen. The position of paper sheets in the process of diffusion was registered by photographs and that of the buoys by radar. The distances between the pairs of the diffusing particles varied from dozens of centimeters to several kilometers. When calculating the values of the coefficient $F(l)$, averaging was made over relatively narrow interval of the values l. These computations resulted in discovering a significant increase in $F(l)$ with the growing scale l. In most of the cases, the dependence of F on l was rather well approximated by a power function with the exponent 4/3 (see Figure 2.58). This dependence (the '4/3 power law') can be derived for the diffusion coefficient, as was mentioned in Chapter 1, based on dimensional grounds in the case of locally-isotropic turbulence (formula (1.68)), and empirically it was established by Richardson (formula (1.165)) using atmospheric diffusion data. In the case of diffusion of discrete-particle pairs, the '4/3 power law' must, generally speaking, hold true only within a certain range of the diffusion time τ. Indeed, if we assume that the coefficient $F(l)$ is constant over a small lapse of time τ the solution of equation (1.164) will have the form

$$q(l_1) = \frac{\text{const}}{\sqrt{\tau}} \exp\left[-\frac{(l_1 - l_0)^2}{4\tau F(l_0)}\right], \tag{2.79}$$

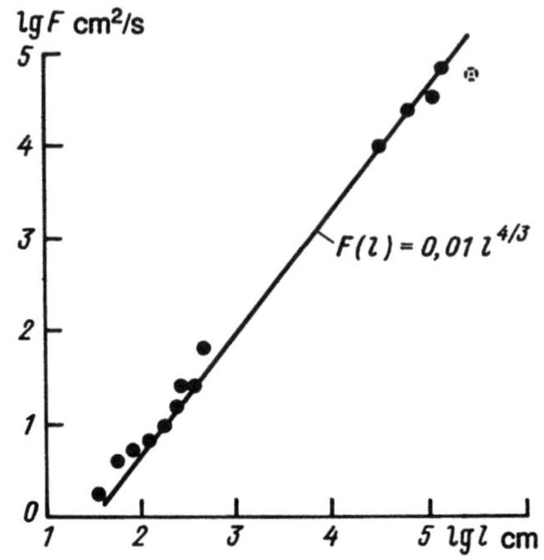

Fig. 2.58. Dependence of horizontal turbulent diffusion coefficient F on scale of phenomenon l, based on experimental data for artificial water basin and for the Pacific Ocean. y-axis: lg F cm^2/s; x-axis: lg l cm.

where l_1 is the distance between the particles at the elapsed diffusion of time τ.

Since the expression (2.79) is the Gaussian probability distribution for l_1, the standard deviation of l_1 from l_0 will be equal to $\sqrt{2\tau F(l_0)}$ from where the formula (2.78) for the coefficient $F(l)$ is derived. However, for comparatively small time-lapses, τ, the relation (1.161), different from the '4/3 power law', should generally speaking hold true for the dispersion of particles. The fulfillment of this law in the ocean and atmospheric diffusion experiments is apparently related to the corresponding choice of τ at different characteristics scales l, as is mentioned in [68].

Experiments with discrete particles in the ocean were carried on in the following years by many researchers (see ref. [68, 81] plus works by Tushinsky in [99], Rozman, Goldberg and Markov in [11], Ozmidov and Zats in [141]). In most of those studies, the dependence of the diffusion coefficient on the phenomenon scale was approximated by a power function of the form $F = kl^n$, with the exponents n sometimes different from 4/3. For example, in the experiments conducted by Ozmidov in 1955, it was notice that n grew, according to some regularity, with the growing thickness of the water layer in the water basins where the experiment was staged. The values of the coefficient of proportionality k also varied. Sometimes, the experimental data were better approximated by power dependences with different values of n and k at different ranges of l. For instance, Zats *et al.* [34] pointed out that in the coastal area of the Black Sea (based on experiments with diffusing floating particles off the cap Ai-Todor) the exponent n did not exceed 0.50 – 0.73, i.e., it was 2 – 2.5 times less than that in the '4/3 power law'. The authors explained this difference by the hydrological situation in the deep-water coastal area of the experiment. complicated sea bed topography and shoreline, the authors believe, impose certain limitations on the horizontal sizes and intensity of turbulent eddies. Relative contributions of horizontal and vertical components of the transport process may also vary. Vertical diffusion, for example, can augment due to surging phenomena, clearly pronounced in area considered, as well as due to convective motions and internal waves.

In the summer of 1974, experiments with discrete particles were conducted by Zhurbas, Mamedov and Tataraev (see ref. [47]) in the Caspian Sea. Small buoys with radar reflectors and underwater parachutes were used as indicators. In each experiment 4 buoys were released. The distances to the buoys and their azimutes were determined with the discreteness of 600 – 900 seconds with the help of a radar system installed in the former oil derrick's basement. Averaging in the formula (2.78) was made when calculating the diffusion coefficient $F(l)$ over groups of values $l = 0-50$; $50-100$; $100-150$; $150-200$; $200-250$; $250-300$ m. The results of the calculations of the function $F(l)$ for all the experimental series are presented in Figure 2.59. As is seen in from the figure, all the experimental points, in that case, will, in the logarithmic scale, fit into a unique straight line, with the slope (i.e., with the exponent n) being equal to 1.14. The coefficient of proportionality turned out to be close to 0.14. The authors account for such a deviation of the experimental results from the '4/3 power law' by the peculiarities of energy supply in the shallow sea areas where the experiment was staged.

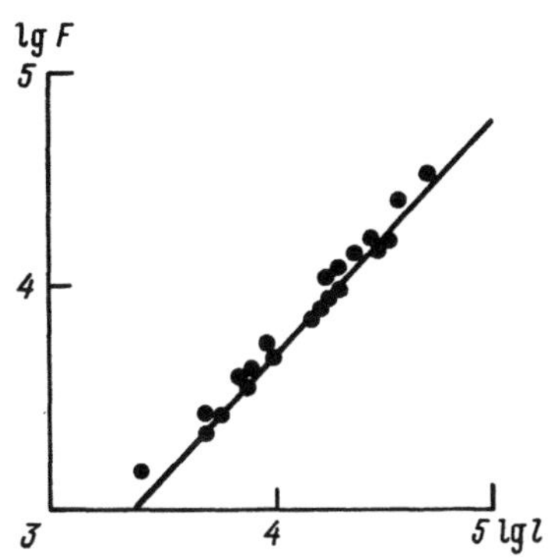

Fig. 2.59. Dependence of horizontal turbulent diffusion coefficient F on scale of phenomenon l, based on the Caspian Sea experiments (after Zhurbas, Mamedov and Tarataev in [47]).

Experiments with discrete indicators diffusing on the water surface were carried out by Tushinsky in Lake Baikal (see ref. [99]). Special floating beacons 2.5 m long with cross-like water sails with cross-sections about 2 m^2 were used as indicators. The position of floats in the course of diffusion was determined by photographing from the ship. Floats were releases near the city of Baikalsk at distances of 0.5 and 12 km from the shore. The phenomenon scale l in the experiments off Baikalsk varied from 6 to 103 m, and the respective values of the diffusion coefficient F from 84 to 5983 cm^2/s. The dependence $F(l)$ obtained through those experiments was approximated by two power expressions, with the exponent $n = 0.98$ for smaller scales and with n close to 4/3 for l exceeding 30 – 60 m. The authors explains this phenomenon by the influence of the pycnocline located during the experiments at the depth of 30 – 60 m on the process of diffusion. The phenomenon scales l in the two first experiments in the Selengin area varied from 14 to 123 m. In that case, $F(l)$ was approximated for all the scales by one power expression with the exponent 1.08. The experiments were conducted in shallow waters and in still weather. In the following experiments, the scale was increased up to 550 m. In these experiments, values of the exponent n for the scales in the range 100 – 500 m were very close to 4/3. These results mainly agree with those obtained by Canadian researchers on diffusion in Lake Huron and Ontario (see ref. [99]) and by Verbolov and Sherstyankin (see ref. [100]) on dye-spot diffusion and current-velocity fluctuations in Lake Baikal. The variability of the parameters n and k depending on the variations of l and current velocities in the lakes was explained by the authors by the influence of the bottom morphology and the coastline of the lakes, the density stratification of waters and their dynamic activity during the observation period.

Similar conclusions were also made by the researchers who studied the diffusion of discrete particles off the Bulgarian coast in 1978 (Rozman, Goldberg, Markov in [12]). According to these experiments, the value of n, for l ranged up to 1.5 km, appeared to be under 0.46. Such a small value of n was explained by the authors by the fact that the observation area was close to the shoreline (not more than 2 km), and the proposed relating n to the non-dimensional ratio l/d, where d is the distance between the observation area and shore. The experimental points for the dependence $n(l/d)$ have a rather significant spread, but nevertheless, it was approximated by a linear dependence in a first approximation. Based on the experiments off the Bulgarian coast as well as on those off the Crimea the anisotropy of the diffusion of discrete particles was also studied [48]. Components of the dispersion of distances between the pairs of particles in two mutually perpendicular directions were calculated separately. The mixed component was also calculated. These made it possible to construct the ellipses of dispersion and to trace their evolution in the course of diffusion. Figure 2.60 presents the calculation results for such ellipses based on the data of the 1980 experiments off the Bulgarian coast. As is seen from the figure, the ellipses were turning during the process of diffusion and increased their anisotrophy α, determined by the formula $\alpha = 1 - \min(\sigma_\tau, \sigma_n)/\max(\sigma_\tau, \sigma_n)$, where σ_τ^2 and σ_n^2 are dispersions along the main axes of the ellipses (index τ referred to the main axis whose direction at the beginning of the experiment was closer to

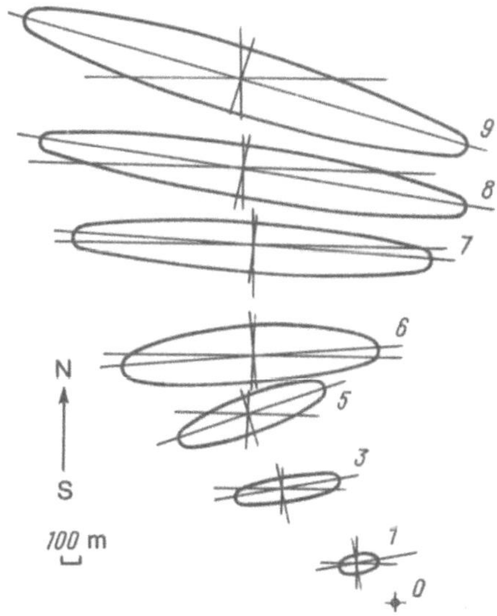

Fig. 2.60. Evolution of diffusing particles variance ellipses with time, based on the Black Sea experiment off the Bulgarian shelf. Figures at ellipses denote hours elapsed since indicator release moments (after Rozman, Goldberg and others in [48]).

that of the mean transport of the indicator 'spot' center of gravity). In most cases, during 4 – 5 hours of diffusion the turn of the axes of the ellipses did not exceed 45°. Sometimes, the turning occured quickly in the initial period of the process and then it slowed down, while in other cases the turning, on the contrary, was going quicker by the end of the observation period. In a number of cases the direction of turning changed in the course of diffusion. The latter regime was observed in the experiments in the proximity of capes where complicated circulation systems of water motions apparently existed. When the indicator transport toward the shore was prevalent, angular acceleration of the dispersion-ellipse axes was usually observed after the indicators reached an area 1 – 2 km from the shore. As a rule, the anisotrophy of ellipses α reached its maximum values 4 – 5 hours after the indicators had been released. Under along-shore indicator transport, maximum values of α were equal to 0.9 while under on-shore transport, α usually was not more than 0.7.

Based on the dispersions σ_τ^2 and σ_n^2, total dispersion $\sigma^2 = \sigma_\tau^2 + \sigma_n^2$ was calculated and, in turn, based on σ^2, the dependence of σ^2 on the diffusion time t was studied. This dependence was sought for in a power form: $\sigma^2(t) \sim t^m$. The experiments were conducted till the diffusion time of 9 hours with varied phenomenon scales from 100 m to 1.5 km. The values of exponent m turned out to be in the range $1 < m < 2$. It was noticed that in the case of along-shore and normal-to-the-shore transport of particles the values of m were significantly distinguished statistically. In the first case, the value m appeared to be close to 2 while in the second one it was within the range $1-1.4$. For small phenomenon scales (from 100 to 400 m), the dependence of m on the depth of observation area was also discovered: for the deep-water shelf area the value of m significantly exceeded those obtained in shallow waters.

The exponent m is in a certain way related to the exponent n in the power dependence of the transport coefficient of diffusion. When $m = 1$ transport coefficient is constant, which corresponds to $n = 0$. For an inertial scale range, the following equalities must hold true: $m = 3$ and $n = 4/3$. According to Goldberg *et al.* (see ref. [48]), the values of m, based on the data available to them for the coastal sea area, mainly lie in the range $1.7-2.2$, and those of n in the range of $1-1.5$. In the case of along-shore diffusion, the values of m closer to 2 more typical, and so are those of n closer to 1. This result can be explained by an analogy between the coastal sea areas and the turbulent boundary layer with a logarithmic profile of longitudinal velocity in that area and a linear profile in the direct proximity of the shoreline. In such a boundary layer, the diffusion coefficient, indeed, may be considered a linear function of the phenomenon scale (distance from the shore). In the coastal sea area, the '4/3 power law' for the diffusion coefficient, apparently, rather seldom holds true; it is more often true for processes with scales of $200-800$ m. An important role in forming the dependence $F(l)$ in the coastal sea area must also be attributed to the effects of vertical shear of the velocity field which, as was shown in Part 1.6, may lead to various forms of dependence of effective horizontal transport coefficients on the depth of the water body, current velocity and other factors.

The measurements described by Ozmidov and Zats (see ref. [141]) may serve as an example of the use of aerial photographic surveys for studying the diffusion of discrete particles at the sea surface. The experiments were carried out close to the Caucasian and Crimean shores in the Black Sea. For better visibility of the diffusing buoys, they were equiped with bags of dye which dissolved in the sea water creating clouds of dyed fluid around the buoys that were clearly visible from the aeroplane. Aerial photographing of floats was done for several hours with intervals of $4-5$ minutes. The distance between the pairs of buoys varied during the experiments from 10 to 474 m, and the values of diffusion coefficients F from 10^2 to 10^5 cm^2/s. The graph of the dependence $F(l)$, based on these experimental data, is shown in Figure 2.61. It follows from the figure that the experiments, held in the deep-water Gaucasian coastal areas, resulted in a dependence very close to the '4/3 power law', while the exponent n for the Crimean coastal sea area was somewhat smaller.

The dependence of the diffusion coefficient F not only on the phenomenon scale but on other factors as well, particularly on the current velocity V, was investigated in the work cited above [34]. Here, one should bear in mind, naturally,

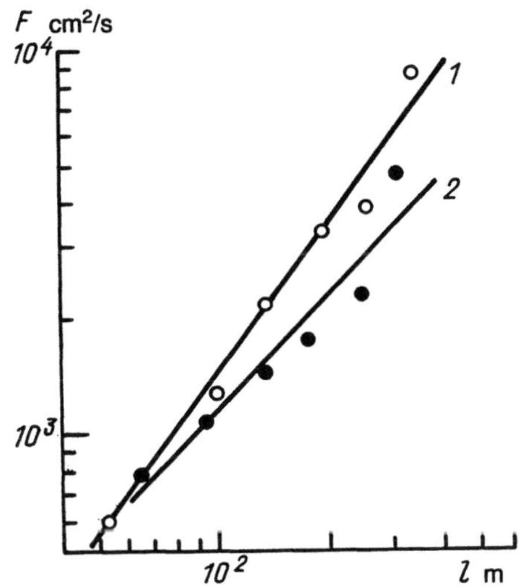

Fig. 2.61. Dependence of horizontal turbulent diffusion coefficient F on scale of phenomenon l, based on serial photographic survey of diffusing discrete indicators in the Black Sea (after [150]). 1 – experiment off the Caucasian shores; 2 – experiment off the Crimean shores; y-axis: F cm^2/s.

that F depends not on the current velocity V itself, but as the authors of that study correctly pointed out, on the velocity gradient V/L, where L is some length scale (e.g. the depth of the sea, thickness of the mixed layer, the drifting current penetration depth, etc.) If the value of L is fixed in the course of an experiment we may envisage the dependence of F or other parameters of diffusion on V itself. Such a dependence was sought for in work [34] in a power form. It was found that when increasing current velocity from 5 – 10 to 50 – 70 cm/s the coefficient of diffusion also increased, approximately, by an order of magnitude. Therefore, based on these data, the dependence of F on V can be, at a first approximation, considered linear. The same conclusion was made by Tushinsky [99] based on the data of experiments in Lake Baikal.

The dependence of the coefficient F not only on the phenomenon scale but on other factors as well is more conveniently described not by a power function with changing exponent n but by introducing some 'correcting' function f, dependent, for example, on the non-dimensional relative depth of the water body l/H, where l is the diffusion scale (the distance between particles), and H is the depth of the water body [81]. In that case, we arrive at the following expression for the diffusion coefficient in the inertial range:

$$F = kl^{4/5}f(l/H), \tag{2.80}$$

where $k = c_1\varepsilon^{1/3}$.

Under large wind velocities (4 – 4 m/s) and current velocities of 8 – 10 cm/s, at which the experiments described in [81] were conducted, k appeared to be close to 0.01 $cm^{2/3}/s$. The value of the function $f(l/H)$ was close to one at large values of the parameter l/H, and increased to several untis at $l/H < 1$. The values of the function $f(L/H)$ obtained from the experiments in water basins and in the coastal area of the Caspian Sea are given in Figure 2.62. One thousand four hundred forty-nine values for the coefficient F, averaged with respect to several ranges of the argument l/H, were used for plotting the graph. For explaining the form of the function $f(l/H)$, a hypothesis on the possibility of changes in the slope of the axes of turbulent eddies with the growing fluid layer depth in the water body and on the influence of this phenomenon on the horizontal diffusion of particles at the surface of the water body was put forward [72]. The hypothesis says that horizontal diffusion is only influenced by the horizontal components of turbulent fluid motions, which should result in a smaller contribution of the turbulent eddy to the diffusion of floating particles when its axis is inclined. Based on simple geometrical plottings, the calculation of such a 'reduction' of the horizontal transport intensity leads to an analytical expression for computing the function $f(l/H)$ [75]. The thus-found graph of the function $f(l/H)$ is given in Figure 2.63. The general form of the theoretical function $f(l/H)$ corresponds to the form of the experimental curve. However, complete quantitative agreement between the two curves is not observed, which can be accounted for by the influence of other factors not allowed for in the theory.

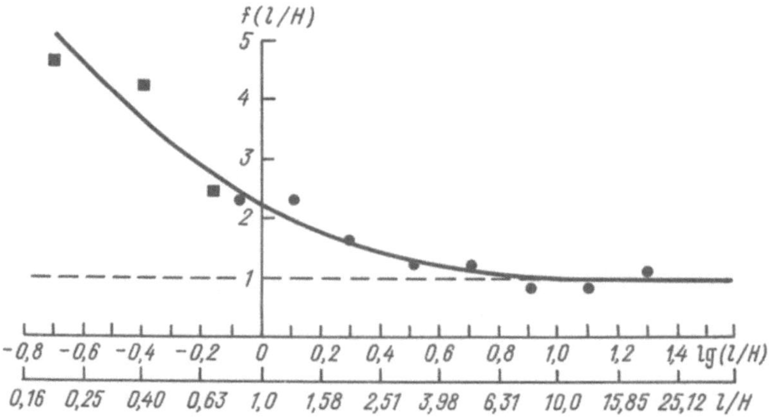

Fig. 2.62. Function $f(l/H)$ 'correcting' the "4/3 power law" at small values of l/h, based on the experiments in basins (dots) and in the Caspian Sea (squares).

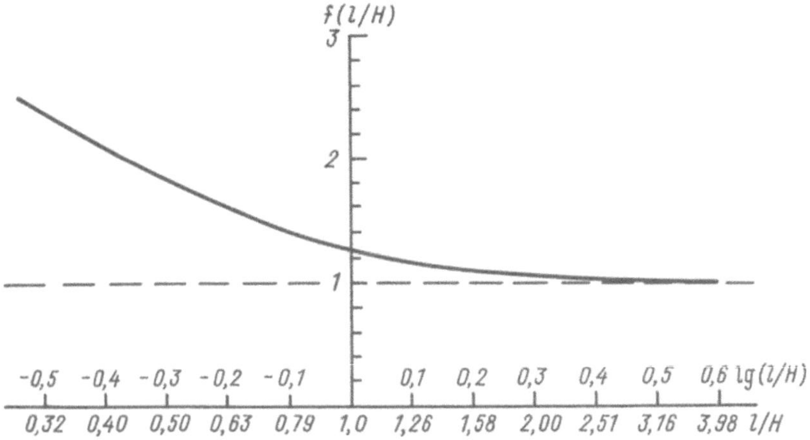

Fig. 2.63. Theoretically obtained function $f(l/H)$ 'correcting' the "4/3 power law".

An interesting question is how the velocity of relative diffusion of discrete particles is influenced by their sizes. It was shown experimentally [73] that small indicators diffuse more quickly compared to larger particles, all other parameters being equal. This can be explained by the fact that in the diffusion of smaller indicators small turbulent eddies were also involved. However, these are completely 'covered' by large indicators. If, for the sake of simplicity, it is assumed that the distribution of energy among eddies in a flow where diffusion indicators are present is described by the regularities of locally isotropic turbulence, then for the ratio of energies of turbulent eddies that are related to the diffusion of a pair of particles with sizes h and of that with point sizes, we will obtain

$$\frac{\overline{u'^2_{h,z}}}{\overline{u'^2_{0,l}}} = \frac{\int_{1/l}^{1/h} c\varepsilon^{1/3}k^{-5/3}dk}{\int_{1/l}^{\infty} c\varepsilon^{1/3}k^{-5/3}dk} = 1 - \left(\frac{h}{l}\right)^{2/3}, \tag{2.81}$$

where l is the distance between the pairs of the particles considered. However, as is shown in Part 1.7 (formula (1.159)), at large diffusion times the relative dispersion of particle pairs become proportional to the mean square of fluctuation velocities $\overline{u'^2}$, therefore (2.81) may be rewritten as

$$\frac{\overline{\Delta l^2_{h,l}}}{\overline{\Delta l^2_{0,l}}} = 1 - \left(\frac{h}{l}\right)^{2/3}. \tag{2.82}$$

The formula (2.82) was compared with the experimental results on horizontal diffusion of indicators of three sizes: 4.6, 8.6 and 17.3 cm [73]. The experiments were conducted in a water basin, with the time interval Δt between subsequent determinations of indicator location equalling 10 seconds. The values of $\Delta l^2_{h,l}$ obtained in the experiment were averaged with respect to five intervals of the ratio h/l: 0.00 – 0.05; 0.05 – 0.10; 0.10 – 0.20 and 0.20 – 0.30. The value $\Delta l^2_{0,l}$ was calculated according to the formula $\overline{\Delta l^2_{0,l}} = 2k\Delta t l^{4/3}$, where k is the coefficient of proportionality in the '4/3 power law'. The five values of the ratio $\overline{\Delta l^2_{h,l}}/\overline{\Delta l^2_{v,l}}$ thus determined are given in Figure 2.64, where the theoretical curve also presented. There is a quite satisfactory coincidence between the theoretical curves and experimental points, which justifies the assumptions made when deriving the formula (2.82).

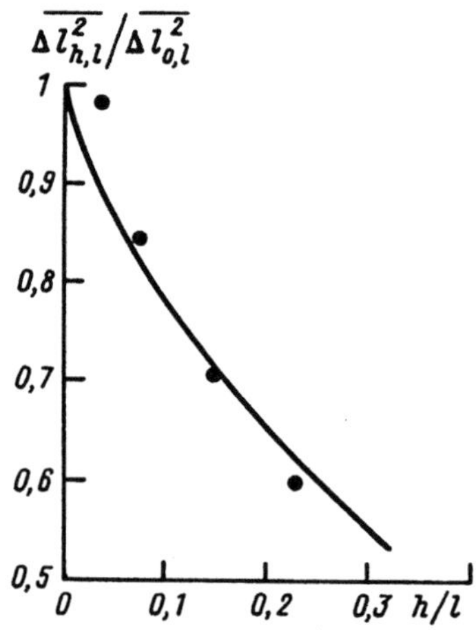

Fig. 2.64. Ratio of variances of indicators of size h to those of point indicators as a function of ratio h/l (l is distance between diffusing indicators). Theoretical curve and experimental dots.

CHAPTER 3

Diffusion of Contaminants from Distributed Sources in the Ocean

3.1. Diffusion of Contaminants Fallen Out of the Atmosphere

The air-sea interaction leads to the exchange of various substances of both natural and antropogenic origin as well as to the exchange of energy, heat and moisture. The exchange of oxygen and carbon dioxide mainly determines the gas composition of the atmosphere while the biogenic elements and various air contaminants are transferred from the atmosphere to the ocean. During the post-war years special attention was paid to the possible contamination of the ocean by radioactive elements that arose after nuclear tests. Naturally, of no less importance is knowing the distribution in the ocean waters of any other substances that can be dangerous either for humans or for ocean inhabitants. Meanwhile an input of such substances and contaminants into the ocean is increasing continually due to the industrial development of the world community.

A large amount of radioactive elements (radionuclides) had been delivered into the atmosphere by nuclear tests conducted intensively from the late 50's to the early 60's. The accumulation of radionuclides in the ocean, soil and biological objects had forced leading nuclear powers to conclude in 1963 a treaty banning atmospheric, space and underwater tests of nuclear weapons. However, the vigorous growth of nuclear energetics in the following years as well as nuclear tests that are still being conducted by China and France do not permit one to relax one's attention to the problems of the contamination of the environment by radionuclides. To resolve those problems the methods of determination of radionuclide concentration in different mediums and objects have been elaborated and the procedures of the data interpretation have been developed, as well as the technique of the prediction of isotope spread and transformation in the atmosphere and in the ocean. Consequently, such new fields of science as nuclear meteorology [98] and nuclear hydrophysics [70] have arisen. Numerous studies of the content of various radioactive isotopes of artificial origin in the ocean have led one to the conclusion that radionuclides may serve as convenient tracers to investigate circulation of the waters and diffusion properties in the ocean. If a long-term atmospheric fallout occurs, a certain equilibrium of contaminant content in the ocean and the atmosphere will be reached, resulting in some 'background' concentration of the contaminant in the ocean. However, the level of this 'background' depends significantly on the properties of a given water body, i.e., its size, dynamic activity, composition of the water, intensity of biological processes, geographical location of the water body, etc.

Global fallout of the nuclear-fission products produced by nuclear explosions appeared to be the main source of the contamination of the open parts of the oceans by the artificial radioactive elements. Of these elements, ^{127}Cs and ^{90}Sr (half-life about 30 years) and the isotopes ^{144}Ce (half-life about one year) are the most dangerous ones for man and for ocean biota. Local increases in the concentration of these elements may be caused either by the discharge of radioactive water waste products of nuclear industry or by the carrying away of fission products into the ocean by rivers flowing close to nuclear plants. The background level of radioactive contamination of the World Ocean open areas was changing in time and had various values for particular areas of the ocean. The Pacific was one of the main nuclear-test sites so the radioactive contamination of its waters was increasing during the late 50's and the early 60's. As has been reported by some researchers, the content of radionuclides in the northern parts of both the Pacific and the Atlantic oceans exceeded significantly the corresponding level of their southern parts. As nuclear-weapon tests were not conducted in the Atlantic and the Indian oceans the global radionuclide atmospheric fallout appeared to be the main source of their background contamination.

As early as the late 1950's the first information on transport of radioactive elements from local fallout caused by American nuclear tests had been obtained in the Pacific by Japanese oceanologists. Radionuclides were proved to be transported up to 2000 km from the explosion site (near Bikini atoll) in no longer time than 4 months and up to 7000 km in 8 months. The distribution in the ocean of radionuclides caused by global fallout from the high layers of the atmosphere rather than by local fallout from nuclear tests became the subject of studies some time later. This was due, first, to the fact that the determination of small concentrations of isotopes in the ocean water turned out to be a rather difficult procedure, and, second, vast measurements were to be carried out to elaborate a 'chart' of the distribution of radionuclides over the large ocean areas.

The largest amount of radioactive products was delivered into the upper atmospheric layers during the powerful thermonuclear explosions. In the stratosphere radionuclides were distributed with height and then they were spread rapidly over nearly all the globe, forming a stratospheric reservoir of radioactive products. Radionuclide fallout from this reservoir onto the Earth's surface has an aerosol form (the 'dry fallout') or occurs simultaneously with atmospheric precipitation (the 'wet fallout'). The intensity of fallout depends not only on the amount of radioactive material brought into the stratosphere but on the peculiarities of the mechanisms accounting for the clearing of the stratosphere as well. The latter, in turn, depend on geographical factors. That is why the amount of the global radioactive fallouts was found to be noticeably dependent on the latitude of the observational site. Temporal variations of fallout intensity, primarily yearly, also occurred (these variations were, naturally, superimposed on the general trend of fallout intensity related with intensification or banning of nuclear tests).

The radionuclides transferred into the ocean from the atmosphere may react intricately with other elements and form chemical compounds that are then transformed by chemical, physical and biological processes. Naturally, the variety of the forms of existence of isotopes in the ocean do make it rather difficult to study their distribution and transformation. Yet isotopes of ^{90}Sr and ^{137}Cs were proved to exist in the ocean water nearly completely dissolved. Consequently, these isotopes turn out to be the most convenient tracers for studying the transport processes in the ocean. These isotopes, on falling out of the atmosphere into the surface ocean layers, 'mark' the waters. The rate of isotope penetration into the deeper layers is related to the intensity and properties of vertical transport in the ocean.

To elaborate a mathematical approach to the diffusion of atmospheric radioactive fallouts, one has, primarily, to analyze observational data on both the intensity of fallouts and the content of isotopes in the upper ocean. As shown by data available till 1965, the averaged yearly fallout of stratospheric ^{90}Sr can be regarded as a constant for large ocean areas. Consequently, the horizontal transport of ^{90}Sr in these ocean areas may be neglected as compared to the vertical one. The Atlantic Ocean, as mentioned above, underwent no influence of the local atmospheric fallouts from nuclear explosions, so this ocean appeared to be the most convenient one in which to study diffusion of the stratospheric ^{90}Sr. Measurements of isotope content in the waters of this ocean were conducted from 1956 (mainly in the surface layers). The results of these measurements for the period from 1960 to 1961 are plotted in Figure 3.1. In this figure all data available are referenced to the same meridional section in order to establish an

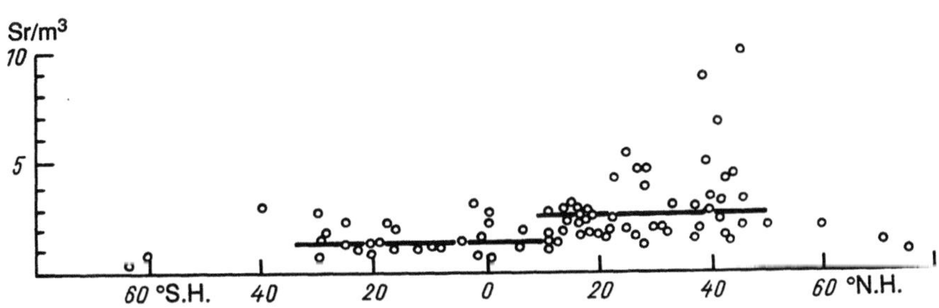

Fig. 3.1. Variations of strontium-90 content with latitude in the surface waters of the Atlantic Ocean. Data obtained during 1961–1962 by different authors.

isotope-concentration dependence on latitude. The experimental data in Figure 3.1 display a rather large scatter that may be accounted for either by inevitable errors of measurements or fluctuations of transport processes and short-period variations of isotope fallout intensity. Nevertheless, the values of concentration averaged over large latitudinal zones can be treated, to a first approximation, as constant ones (horizontal lines in Figure 3.1). The weaker contamination of the southern-hemisphere waters as compared to the northern waters is the only clearly traced feature of these data.

The data on temporal variation of the averaged concentration of ^{90}Sr in the surface waters of the northern Atlantic are summarized in Figure 3.2 for a period from 1954 to 1962. As can be seen from this figure, the isotope concentration in this area varied, on the average, rather slowly and increased by not more than a factor of two from 1955 till 1961.

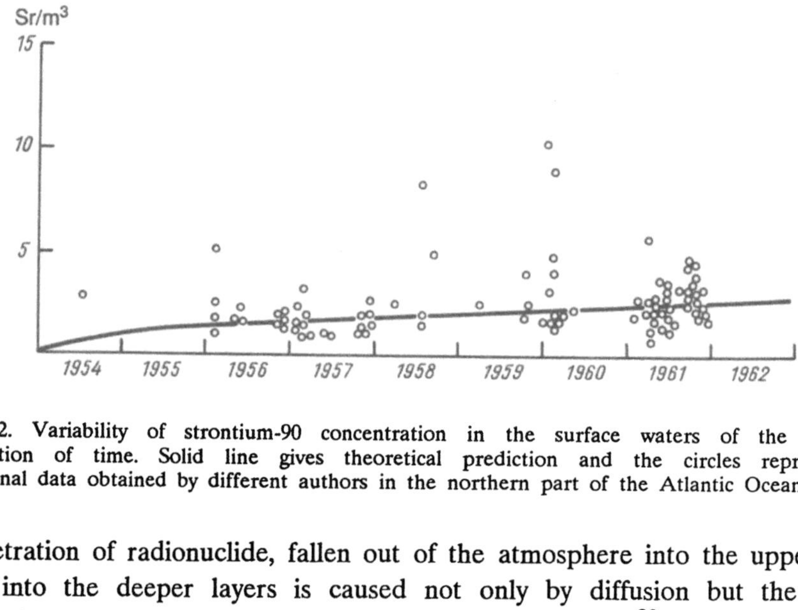

Fig. 3.2. Variability of strontium-90 concentration in the surface waters of the ocean as a function of time. Solid line gives theoretical prediction and the circles represent observational data obtained by different authors in the northern part of the Atlantic Ocean.

Penetration of radionuclide, fallen out of the atmosphere into the upper ocean layer, into the deeper layers is caused not only by diffusion but the vertical averaged motions of the waters as well. Some portion of ^{90}Sr is transferred by hydrobionts assimilating the isotope or by their remains during the process of sedimentation. Yet the latter effects may, apparently, be neglected because a comparatively large amount of a stable isotope of strontium is contained in the ocean waters and is the main source of strontium assimilated by the organisms. As the yearly rate of radioactive decay of ^{90}Sr is as low as about 2.5% of the total content of the isotope, the decay of ^{90}Sr may be neglected as well when considering the diffusion processes with a typical time-scale of about a year (or even of several years). Under these suggestions and neglecting the horizontal effects (if the isotope fallout is homogeneous over the large areas) we arrive at the diffusion equation

$$\frac{\partial c}{\partial t} = \frac{\partial}{\partial t}\left[K_z(z, t)\,\frac{\partial c}{\partial z}\right] - w(z, t)\,\frac{\partial c}{\partial z}, \qquad (3.1)$$

where c is the concentration of the isotope, K_z is the vertical turbulent transport coefficient and w is the vertical component of an averaged velocity of the water motion.

Equation (3.1) was applied for the first time by Beliayev, Kolesnikov and Nelepo in 1964 [3] to estimate the vertical transport of ^{90}Sr in the ocean. In this study the functions $K_z(z, t)$ and $w(z, t)$ were determined by the indirect methods using the data on flow velocity, temperature and salinity obtained at some observational sites in the Atlantic. As stated above, the accuracy of such calculations is usually not high and, moreover, results obtained can be merely applied to a given area and given averaged hydrometeorological conditions only. However, comparison of the predicted diffusion ^{90}Sr fallen out of the stratosphere with field observations could be done only for data averaged over large time intervals and large areas. Consequently, for the problem in question it would be more natural to suppose the diffusion coefficient, K_z, to be a constant. This value of K_z must characterize the intensity of vertical transport in the oceanic layers below the pycnocline because a more intensive mixing occurs above it and, consequently, the isotope fallen out of the atmosphere reaches the pycnocline practically 'instantaneously' (as compared to the typical time-scales of the contaminant diffusion into the deeper layers). Moreover, for large temporal intervals considered in the problem discusses the mean vertical velocity of water motion, w, may be assumed zero over the middle ocean latitudes. Under these suggestions the diffusion equation is reduced to the simple form

$$\partial c/\partial t = K_z\,\partial^2 c/\partial z^2. \qquad (3.2)$$

Equation (3.2) was used to determine the distribution of ^{90}Sr in the ocean by Miyake *et al.* and by Sereda in 1962 (see references in [91]). The problem concerning the isotope diffusion from the point source (explosion) was considered in the first paper, while in the second one the vertical diffusion of atmospheric global fallout of ^{90}Sr in the Pacific and the Atlantic Oceans was studied. Sereda compared the theoretical predictions with the available observational data of vertical distribution of the isotope to estimate the mean values of the vertical diffusion coefficient for the Atlantic (17.2 cm^2/s) and the Pacific (6 cm^2/s) Oceans. Both Sereda and Beliayev *et al.* used isotope concentration at the ocean surface as the boundary condition for solving the diffusion equation. This boundary condition leads, apparently, to agreement between the estimations and experimental data for the upper ocean layer where most of the measurements of isotope concentration were carried out; but, under this boundary condition, it is impossible to predict variations of isotope concentration in the surface waters.

Ozmidov and Popov [90] obtained a solution of equation (3.2) with the most natural boundary condition for the problem in question, imposing contaminant flux through the ocean surface. The boundary conditions were written in the form

$$K_z \frac{\partial c}{\partial z}\bigg|_{z=0} = \begin{cases} 0 & \text{at } t < t_0 = 0; \\ q & \text{at } t \geqslant t_0 = 0; \end{cases} \tag{3.3}$$

$$c(z, t)\big|_{z \to \infty} = 0, \tag{3.4}$$

where t_0 is a time moment of arising of global ^{90}Sr fallout (since 1954); q is the isotope flux through 1 cm^2 area of ocean surface per 1 s. Naturally, the initial condition has the form

$$c(z, 0) = 0. \tag{3.5}$$

The solution of equation (3.2) under conditions (3.3) – (3.5) is

$$c(z, t) = \frac{2q\sqrt{t}}{\sqrt{K_z}} \left\{ \frac{1}{\sqrt{\pi}} \exp(-\zeta^2) - \zeta[1 - \operatorname{erf} \zeta] \right\}, \tag{3.6}$$

where

$$\operatorname{erf} \zeta = \frac{2}{\sqrt{\pi}} \int_0^{\zeta} \exp(-\zeta^2)\, d\zeta, \quad \zeta = \frac{z}{\sqrt{K_z t}}.$$

In accordance with (3.6), the isotope concentration at the ocean surface ($z = 0$) is described by following expression

$$c(0, t) = 2q \sqrt{t}/\sqrt{\pi K_z}, \tag{3.7}$$

i.e., the value of $c(0, t)$ is in direct proportion to the square root of the time elapsed since the start of isotope fallout.

The predicted curve for the concentration of ^{90}Sr in the surface ocean waters is shown in Figure 3.2 by the solid line and indicates a good agreement between theoretical estimation and observational data obtained in the northern part of the Atlantic during 1954 – 1962. The theoretical curves for vertical distribution of the isotope for the beginnings of 1955, 1958, 1961 and 1964, respectively, are plotted in Figure 3.3 together with the horizontally averaged observational data obtained in 1957 – 1958. The best agreement of theoretical curve with experimental values of $c(z, t)$ is achieved for vertical diffusion coefficient

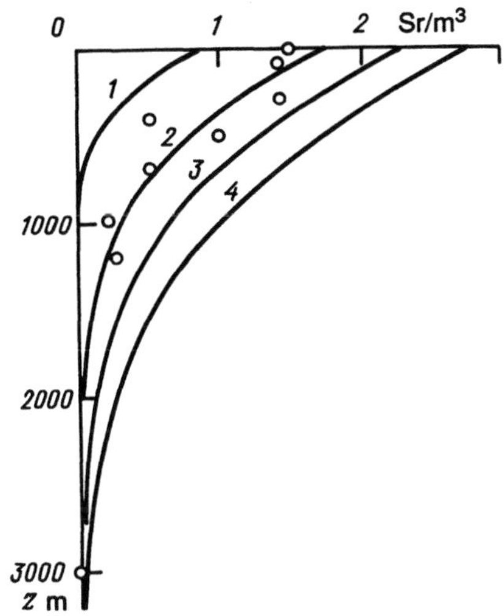

Fig. 3.3. Vertical distribution of strontium-90 concentration in the Atlantic Ocean. Solid curves represent theoretical predictions for the beginning of 1955 (1), 1958 (2), 1961 (3) and 1964 (4). Circles are averaged observational data for 1957–1958.

30 cm^2/s. In accordance with the available data, isotope flux from the atmosphere to the ocean was about $7.4 \times 10^{-10}/(cm^2/s)$ during these years. Figure 3.2 shows that rapid contamination of the surface ocean layers occurred during the first years after the start of isotope input. Afterwards this process was retarded but isotope concentration in the deeper ocean layers increased.

The above-given estimation of the diffusion of contaminant fallen out of the atmosphere is, naturally, a fairly idealized one because the suggestions that flux q and coefficient K_z are constant values and that the horizontal effects may be neglected are both rather severe ones. Therefore a detailed elaboration of estimation procedure has been carried out in the following years. Thus, the flux observed during 1954 – 1970 in different latitudes was taken into account by Nelepo [91] to estimate the theoretical curves for $c(0, t)$ that are shown in Figure 3.4. The maxima and minima of contamination in Figure 3.4 are caused by frequency of explosions and power of individual explosion during the years of nuclear weapon tests. A general decrease in the contamination of surface ocean waters for a period after 1963 – 1964 is due to the nuclear-tests moratorium imposed in 1963.

In 1967 Beliayev *et al.* [4] attempted to take into account a variability of K_z and w with depth. A set of equations describing balance of heat and salt was used to determine the values of K_z and w for different seasons, with pycnocline

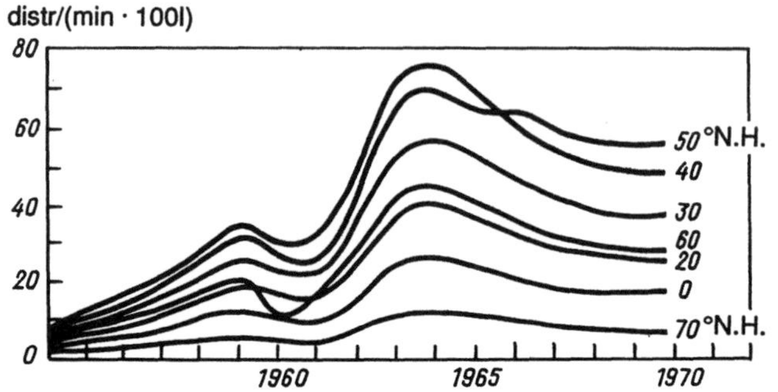

Fig. 3.4. Theoretical concentrations of radioactive isotopes in the surface waters of the ocean. Temporal variability of radionulides' fallout from the atmosphere was taken into account to estimate these curves (after the paper by Nelepo from [91]).

depth H ranging from 60 m (in spring) to 200 m (in winter) prescribed for each of the seasons. The seasonal variations of the pycnocline depth were shown to be responsible for an increase of the rate of isotope penetration in the deeper layers due to the so-called 'lock' effect. Actually, in a thick mixed layer a contaminant is mixed rapidly and reaches depths up to 200 m. When a value of H decreases, radionuclides will diffuse down from a lower part of the former mixed layer, and another 'portion' of isotope will be brought down quickly into these deep layers during the next cycle of pycnocline deepening. This mechanism of contaminant transport can be responsible for the often-observed decrease of isotope concentration in the uppermost layer. The results of computation of the functions $K_z(z, t)$ and $w(z, t)$ are displayed in Figure 3.5, together with the vertical profiles of isotope-concentration distribution determined from diffusion equation with these functions $K_z(z, t)$ and $w(z, t)$. The experimental data on isotope concentration in the Atlantic Ocean that were available at that time are plotted in Figure 3.5c as well.

In diffusion problems, the decay of the long-living isotopes may be neglected in most cases, while the finite decay-time must be taken into account for the short-living radionuclides. Since the mass of an isotope, decaying in unit time, is in a direct proportion to its total amount, the decay of an element can be taken into account in the diffusion equation by adding to its right-hand side a term of the form $-\lambda c$, i.e.,

$$\partial c/\partial t = K_z \, \partial^2 c/\partial z^2 - \lambda c, \tag{3.8}$$

where a constant λ is related to the element half-life, τ, by the relation $\lambda = \ln 2/\tau$.

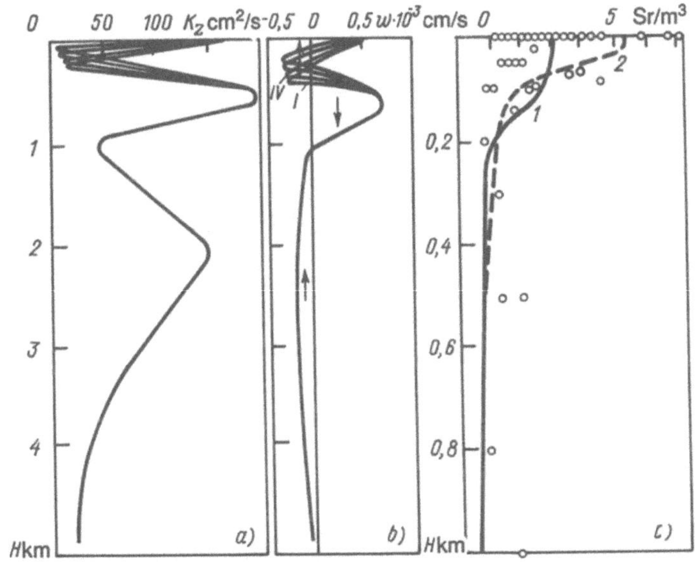

Fig. 3.5. Spacial-temporal variability of vertical turbulent diffusion coefficient, K_z (a), and averaged vertical velocity of water motions, w (b), as estimated by indirect methods in the Atlantic Ocean. Vertical distribution of strontium-90 concentration (c) is determined from diffusion equation in which the dependence of K_z and w on both vertical coordinate and time is taken into account (after [70]). Theoretical curves in Figure 3.5c are determined either taking into account 'lock' effect (curve 1) or neglecting it (curve 2). Circles represent experimental data. Roman numerals denote quarters of a year.

The solution of equation (3.8) under the initial and boundary conditions (3.3) – (3.5) is [13]

$$c(z, t) = \frac{q}{\sqrt{\lambda K_z}} \exp\left(-\sqrt{\frac{\lambda}{K_z}}\, z\right) - \frac{q}{2\sqrt{\lambda K_z}} \times$$
$$\times \left\{\left[\exp\left(-\sqrt{\frac{\lambda}{K_z}}\, z\right)\right]\left[1 - \operatorname{erf}\left(\sqrt{\lambda t} - \frac{z}{2\sqrt{K_z t}}\right)\right]\right\} +$$
$$+ \left[\exp\left(\sqrt{\frac{\lambda}{K_z}}\, z\right)\right]\left[1 - \operatorname{erf}\left(\sqrt{\lambda t} + \frac{z}{2\sqrt{K_z t}}\right)\right]\Bigg\}, \quad (3.9)$$

and the contaminant concentration at the ocean surface is expressed by

$$c(0, t) = \frac{q}{\sqrt{\lambda K_z}} \operatorname{erf} \sqrt{\lambda K_z}\ . \quad (3.10)$$

A decay of radioactive contaminant leads to some interesting phenomenon. Despite the fact that the input of isotope into the ocean from the atmosphere is supposed to be a constant in the problem in question, a steady distribution of isotope concentration may be set. For this steady distribution of radionuclide concentration we arrive at

$$c(z) = \frac{q}{\sqrt{\lambda K_z}} \exp\left(-\sqrt{\frac{\lambda}{K_z}}\, z\right). \tag{3.11}$$

At the ocean surface equation (3.11) reduces to

$$c(0) = q/\sqrt{\lambda K_z} \,. \tag{3.12}$$

Estimations given in [13] show that a steady state (with deviations from it less than 2%) is set after the time $t_c = 4/\lambda$ at the ocean surface, $t_c = 5/\lambda$ at depth $z = \sqrt{K_z/\lambda}$ and $t_c = 9/\lambda$ at depth $z = 6\sqrt{K_z/\lambda}$. Under this steady condition the flux of isotope at the different horizons turns out to be reduced by the factor $\exp(\lambda z/K_z)$ as compared to its value at the ocean surface. This phenomenon can readily be explained. As an isotope is decaying during the process of vertical diffusion, only a certain part of its initial amount is 'brought down' into a layer at horizon z. If so, one has, naturally, to use the value of $h = \sqrt{K_z/\lambda}$ as a depth scale which is characteristic of the vertical distribution of radionuclides in a nearly steady state, i.e., at large diffusion times t. This scale varies significantly for isotopes with different half-lives. Values of h estimated for several isotopes at two values of K_z are given in Table III.I.

TABLE III.I

Characteristic depth scale, $h = \sqrt{K_z/\lambda}$, of some unstable isotopes (after [13])

Isotope	Half-life	h, cm	
		$K_z = 1$ cm²/s	$K_z = 10$ cm²/s
^{7}Bl	53 days	$2.57 \cdot 10^3$	$8.13 \cdot 10^3$
^{22}Na	2.6 year	$1.09 \cdot 10^4$	$3.44 \cdot 10^4$
^{3}H	12.26 year	$2.36 \cdot 10^4$	$7.48 \cdot 10^4$
^{210}Pb	21.4 year	$3.12 \cdot 10^4$	$9.86 \cdot 10^4$
^{137}Cs	29.68 year	$3.68 \cdot 10^4$	$1.16 \cdot 10^5$

As is seen from Table III.I, the ocean (with mean depth 4000 – 5000 m) appears to be 'deep' enough, as compared to depth scale h, for short-living isotopes. If a horizon where the steady distribution of isotope concentration is 100-fold smaller than its value in the surface waters is taken as a depth H of isotope penetration, then from (3.11) we obtain

$$H \simeq 4.6\sqrt{K_z/\lambda} = 4.6\ h. \tag{3.13}$$

The above-given estimations of parameters of diffusion of unstable contaminants and the form of solution of equation (3.9) show that in the present problem it is convenient to use the following non-dimensional variables; non-dimensional concentration c^*, non-dimensional vertical coordinate ζ and non-dimensional time θ, defined by [13] as

$$c^* = c\sqrt{\lambda K_z}/q; \quad \zeta = z\sqrt{\lambda/K_z}; \quad \theta = \lambda t. \tag{3.14}$$

Using these non-dimensional variables, the solution (3.9) can be written in a 'universal' form:

$$c^*(\zeta, \theta) = e^{-\zeta} - \tfrac{1}{2}\left\{\left[1 - \mathrm{erf}\left(\sqrt{\theta} - \frac{\zeta}{2\sqrt{\theta}}\right)\right]e^{-\zeta} + \right.$$
$$\left. + e^{\zeta}\left[1 - \mathrm{erf}\left(\sqrt{\theta} - \frac{\zeta}{2\sqrt{\theta}}\right)\right]\right\}. \tag{3.15}$$

The universal curves calculated accordingly to (3.15) at several values of non-dimensional time θ are given in [13]. In logarithmic scale, steady regime of the process is described by a straight line, while the curves for current distribution of concentration approaches this line asymptotically with increasing diffusion time.

The diffusion equation for unstable contaminant was solved under a boundary condition of zero isotope concentration at $z \to \infty$. In the case of finite ocean depth one has, in a general case, to impose boundary conditions at the ocean floor suggesting either non-penetration of the isotope through the floor or its sorption by a floor substance. Besides that, in some cases one has to take into account the possibility of a chemical reaction of radionuclide with a floor substance as well. But for ocean depths on the order several thousand meters, the ocean as seen from above can be considered, with high accuracy, as infinitely deep even for the long-living isotopes. Consequently, any interaction of isotope with a floor substance can be neglected. Nevertheless, for shallow basins this interaction can play a rather important part in the balance of radionuclides delivered into the water body, and it has to be taken into account in the corresponding diffusion problems.

Three types of isotopes, distinguished from each other by their origin, exist in the ocean. They are, firstly, terrigenous isotopes present in the Earth's substance; secondly, cosmogenic isotopes arising mainly in the Earth's atmosphere under the effect of cosmic rays; and lastly, the isotopes of an artificial origin which have arisen on the Earth due to the human activity. Unlike the artificial isotopes, the cosmogenic ones (^3H, ^7Be, ^{10}Be, ^{14}C, ^{22}N, etc.) are continuously delivered from the atmosphere to the ocean surface so, when considering diffusion of these into the deep ocean layers, an initial condition (3.5) is not satisfied. Besides, pronounced seasonal variations of intensity of fallout are typical for cosmogenic isotopes. As estimated, the intensity of fallouts can be changed by a factor of from 2 to 3 during a year, its maximum being reached in the spring-summer period and minimum in the fall-winter one. This peculiarity of the cosmogenic isotopes' fallout was taken into account by Vinogradov [15] in considering the problem of vertical diffusion of isotopes in the ocean waters. He considered the diffusion equation (3.8) without initial conditions while the boundary condition at the ocean surface was taken in the form

$$K_z \left.\frac{\partial c}{\partial z}\right|_{z=0} = -q + q_1 e^{i\omega t}, \tag{3.16}$$

where q is a 'constant' component of isotope flux, q_1 is an 'amplitude' of its variable part, $\omega = 2\pi/T$ and T is an annual period of flux variability. For the diffusion times at which isotope distribution in the ocean waters caused by constant component of the flux is already steady, the solution of the problem in the non-dimensional variables c^*, ζ and θ takes the form [15]

$$c^*(\zeta, \theta) = e^{-\zeta} - \frac{a}{\sqrt[4]{1+\mu^2}} e^{-r\zeta} \cos(2\pi\theta - s\zeta - \beta), \tag{3.17}$$

where $\mu = \omega/\lambda$ is non-dimensional frequency; $a = q_1/q$ is non-dimensional isotope flux;

$$r = \sqrt{(\sqrt{1+\mu^2}+1)/2}; \; s = \sqrt{(\sqrt{1+\mu^2}-1)/2}; \text{ and } \operatorname{tg}\beta = s/r = \mu/(\sqrt{1+\mu^2}+1).$$

For long-living isotopes $\mu \gg 1$ and the steady part of the solution becomes predominant, while for short-living isotopes amplitude of the non-steady part of the solution approaches $ae^{-\zeta}$ and becomes comparable to the amplitude of the steady component. Maximum values of isotope concentration in the ocean waters are somewhat shifted relative to fallout time moments; the shift is β at the ocean surface and linearly increases with depth z. As in the case of heat-wave propagation, the phase shift relative to the maxima (minima) of concentration at the ocean surface is inversely proportional to the square root of the vertical turbulent transport coefficient. The theoretical curves, calculated using (3.17), for vertical distribution of cosmogenic isotope ^7Be are shown in non-

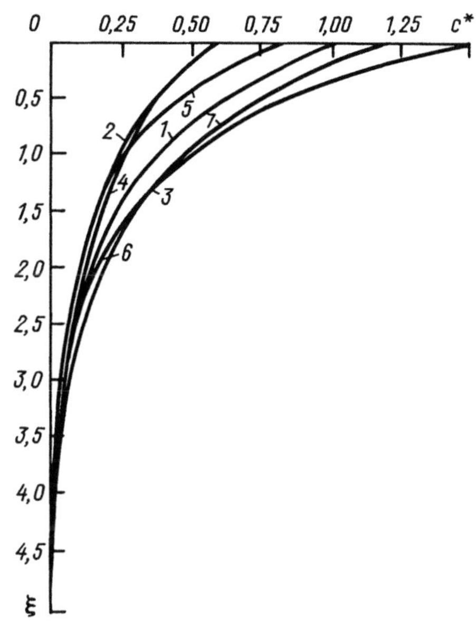

Fig. 3.6. Vertical distribution of non-dimensional concentration of cosmogenic isotope ^{7}Be in the ocean determined taking into account of annual variations of fallout intensity (after [15]). 1 – averaged yearly distribution, 2 – minimum of concentration, 3 – maximum of concentration, 4 – distribution on January 1, 5 – on March 1, 6 – on July 1, 7 – on September 1.

dimensional variables in Figure 3.6, borrowed from [15]. For this radionuclide a value of a was specified as 0.55 and December 1 was taken as the beginning (minimum) of the annual cycle of isotope fallout. The figure shows phase-delay of maximal values of concentration at different depths and decreasing of the contribution of the periodic component of concentration with depth.

A problem of vertical isotope diffusion in a particular case in which the intensity of isotope fallout is exponentially decreasing with time was considered by Vinogradov [16]. This statement of the problem was to simulate a decrease of the global fallout of artificial isotopes after a moratorium on nuclear weapon-tests. The diffusion equation (3.8) was solved with the initial condition at zero and boundary condition (3.4) at infinity, while at the ocean surface the following condition was imposed:

$$K_z \left.\frac{\partial c}{\partial z}\right|_z = -\sum_{i=1}^{n} q_i \exp[-(\lambda + \lambda_c)(t - t_i)], \tag{3.18}$$

where q_i is the initial flux of isotope fallout arisen due to the i-th run of the nuclear tests (or to an individual explosion); t_i is the time of the conducting of this run (or an explosion); λ is the isotope decay constant; and λ_c

is the constant defining clearing of the stratosphere from isotope contamination.

Analysis of the solution of this problem indicates that the isotope concentration in the upper ocean layer increases rapidly at the beginning of the diffusion process, reaching maximal value at time $\theta = \lambda t \simeq 1$. The onset of concentration maximum in the deeper layers is delayed relative to the surface waters' concentration maximum. For $\theta > 10$, an upper layer with a nearly constant concentration of isotope is formed due to the decreasing of fallout intensity with time. The theoretical curves for vertical distribution of cesium-137 concentration in the ocean waters, estimated under boundary condition (3.18) at $i = 1$, are given in Figure 3.7. Taking into account several test runs allows, from consideration of isotope-concentration vertical distributions, the estimating of a particular contribution of any given test run into the forming of the total contamination field. Thus, examining the available observational data on distribution of strontium-90 in the eastern part of the Pacific over the latitudes from 30° to 40° and cesium-137 in the equatorial part of the Indian Ocean, Vinogradov [16] arrived at the conclusion that isotope concentration in the upper ocean layer in 1966 and 1975 was determined mainly by the explosion runs carried out

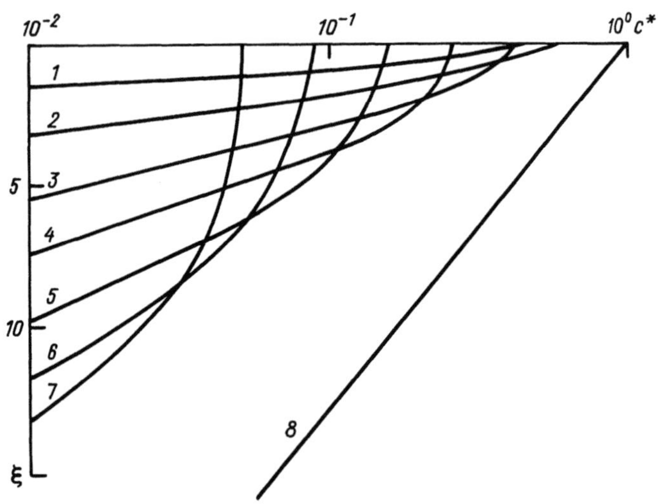

Fig. 3.7. Vertical distribution of non-dimensional cesium-137 concentration, c^*, determined taking into account of decrease of isotope fallout in time (after [16]). 1 – distribution at non-dimensional time $\theta = 0.25$; 2 – $\theta = 1.00$; 3 – $\theta = 2.36$; 4 – $\theta = 4.84$; 5 – $\theta = 9.00$; 6 – $\theta = 16.00$; 7 – $\theta = 25.00$; 8 – steady distribution for constant intensity of isotope fallout.

during 1961 – 1962, while the explosion runs carried out during 1952 – 1958 determined completely the isotope content in the deep ocean layers in 1966 and, to a large extent, in 1975.

Influence of the upper mixed ocean layer on vertical diffusion of contaminant fallen out of the atmosphere was considered by Vinogradov [17] as well. The contribution of turbulent transport in the mixed layer to the vertical-diffusion process was usually neglected because a contaminant was supposed to penetrate through the mixed layer nearly 'instantly' due to a rather intensive turbulent transport within it. Therefore, it was often supposed possible to 'transpose' the boundary condition for contaminant flux from the ocean surface to the lower bound of the mixed layer without essential computational errors, because a thickness of the mixed layer, h, is usually much smaller than the total depth of the ocean. However, as correctly pointed out by Vinogradov, the upper mixed ocean layer can serve as a kind of a reservoir that accumulates and preserves some portion of contaminant fallen out. This effect must correspondingly influence a process of further vertical propagation of a contaminant. If contaminant concentration is supposed as a constant in the upper mixed layer and equal to $c(h_0, t)$, the boundary condition on the lower boundary of mixed layer can be written in the form

$$h \ \partial c(h, t)/\partial t - K_z \ \partial c(h+0, t)/\partial z + \lambda h c(h, t) = q(0, t), \tag{3.19}$$

where the first term describes the variation of total isotope content in the upper layer, the second terms stands for the turbulent flux of a contaminant from the upper layer to the lower one (the differentiating is in a positive z-direction) and the third term describes the decay of a contaminant in the mixed layer. In the right-hand side of the relation (3.19), $q(0, t)$ denotes the specific flux of contaminant from the atmosphere to the ocean surface.

One more non-dimensional variable, $\chi = \sqrt{\lambda/K_z}\, h$, which is the ratio of the depth h to diffusion-decay scale $\sqrt{K_z/\lambda}$, arises in the diffusion problem with boundary condition 3.19. For h = 25 ... 200 m, the value of χ, for instance for isotopes ^{7}Be and ^{3}H, ranges from 0.3 to 24.5 and from 0.03 to 2.6, respectively, while the vertical transport coefficient varies from 0.1 to 10 cm^2/s. Consequently, the depth of the upper mixed layer cannot be considered, generally speaking, as small for these isotopes and it must be taken into account. The existence of the upper mixed layer in the ocean leads, in that case, to some decrease of the surface isotope concentration due to both uniform distribution of radionuclides in the layer and their decay. At the same time the existence of this layer encourages a more rapid transport of contaminants into the deeper ocean layers as well as dampens the annual variations of cosmogenic isotope concentration. One other length scale, $\delta = \sqrt{K_z/\lambda_c}$, may be of significant importance in considering the behavior of radioactive isotopes that arose in the ocean due to nuclear explosions. The value of λ_c, as estimated in [17], can be taken to equal 0.063 month^{-1}, and so the scale δ ranges from 20 to 200 m for K_z = 0.1, 1.0 and 10 cm^2/s. Consequently, the upper mixed layer, which can be considered as a kind of 'depot' for isotopes, can affect significantly the mechanism of propagation of artificial radionuclides into the deep ocean waters. The curves of non-dimensional concentration of the cesium-137 isotope in the surface ocean layer, obtained for different values of non-dimensional parameter δ, are displayed in Figure 3.8 as a function of non-dimensional time θ. The figure clearly shows that if the influence of the mixed layer is taken into account, the resulting values of c^* at large δ can differ significantly from those obtained disregarding this effect.

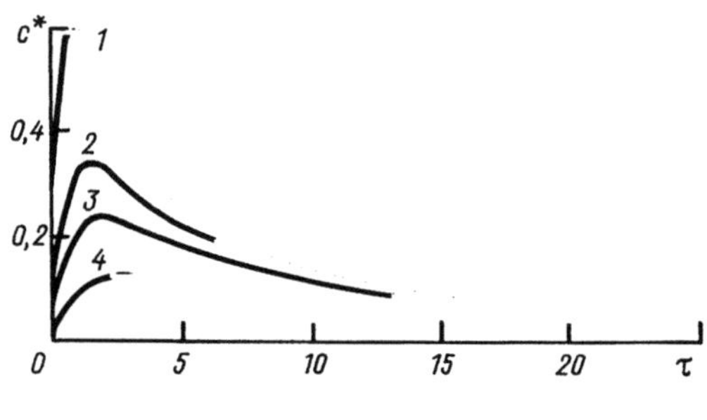

Fig. 3.8. Temporal variation of cesium-137 concentration in the upper ocean layer determined with and without the effect of upper mixed layer. The curves are plotted as a function of non-dimensional time, $\theta = \lambda t$ (where λ is the isotope decay constant), at values of δ = 0 (1), δ = 1 (2), δ = 2 (3), δ = 5 (4) (after [17]).

The integral method of estimation of isotope content in individual parts of the ocean applied by Vinogradov to the upper mixed layer is widely used to study the large-scale features of contaminant transport in the World Ocean as well. The whole water body of the ocean, or of its individual parts, is divided into several 'reservoirs', assuming the distribution of an isotope, or any other contaminant, to be homogeneous over any of them. It is supposed that the interchange of a contaminant takes place through the boundary surfaces separating neighboring reservoirs. Denoting the volume of any of the n reservoirs considered as V_i and the concentration of contaminant in the reservoir as c_i, we can write the equation for balance of decaying contaminant in any of the n reservoirs in the form

$$V_i \frac{dc_i}{dt} = \sum_{j \neq 1} R_{ij}(c_j - c_i) - \lambda c_i V_i, \tag{3.20}$$

where R_{ij} is the water flux from the j-th reservoir to the i-th one, and i, j = 1, 2, ..., n. Besides that, one has to impose additional boundary conditions at the outer surfaces of the surfaces of the reservoirs that are not contiguous with the neighboring reservoirs. Provided any of the reservoirs has common surfaces with the other ones, the number of unknown R_{ij} in equations (3.20) will be equal to the number of combinations of reservoirs taken two at a time, i.e., $n(n-1)/2$. Under typical ocean conditions the number of the unknowns may turn out to be much smaller, exceeding the number of independent equations of contaminant balance (3.20). Therefore, several different substances are usually used in a reservoir method: stable or radioactive isotopes, temperature, salinity, etc. Since for the accepted scheme of defining the reservoirs the values of R_{ij} are the same for any tracer, the number of independent equations required to determine R_{ij} increases. But the fluxes R_{ij} are often represented as a sum of advective and turbulent terms R^a_{ij} and R^T_{ij}. In that case, naturally, the number of unknowns is twice as much and the larger number of independent equations describing contaminant balance is required to determine them. A notion of renovation time needed to renew water content in a given reservoir is introduced in a reservoir method as follows:

$$T_i = \frac{V_i}{\sum\limits_{i \neq j} R_{ij}}, \tag{3.21}$$

where R_{ij} denotes either a total water flux from the i-th reservoir or its advective or turbulent components. Given the values of R^T_{ij}, an effective turbulent diffusivity K_{ij} for two neighboring reservoirs i and j can be defined as well:

$$K_{ij} = R^T_{ij} L_{ij}/(c_j - c_i), \tag{3.22}$$

where L_{ij} is some characteristic length-scale, e.g., thermocline depth, width of a frontal zone, etc.

The reservoir method permitted the estimation of the values of T_i and K_{ij} for the different water bodies and the areas of the ocean. Either stable isotopes or radionuclides of terrigenous, cosmogenic or artificial origin were used as the tracers as well as ocean water temperature, salinity and dissolved oxygen. Especially vast information on the distribution of different elements (tracers) in the ocean has been obtained during the International Program GEOSECS (Geochemical Ocean Section Study).

The data collected during the program made it possible to obtain refined features of the individual reservoirs as well as to improve the very procedure of defining such reservoirs in the ocean. A typical time of water renovation in the reservoirs determined by this method was found to be equal to a hundred or several hundreds of years, the vertical transport coefficients were on the order of several units or even fractions of square centimeter per second, the horizontal turbulent transport coefficients were from 10^6 to 10^8 cm^2/s, horizontal velocities of deep-water currents were units or fractions of centimeters per second, the vertical velocities of water motions were of the order of 10^{-5} cm^2/s. These figures, of course, must be considered as the averaged ones and rough estimates that can vary significantly both in time and space. Some additional information on the reservoir method and the results obtained, as well as extensive references on the subject, can be found in the article by Nelepo in monograph [91] and in Vinogradov's paper [14].

3.2. Diffusion of Biogenic Elements, Oxygen and Carbon Dioxide in the Ocean

Chemical and biological processes essentially influence both the transport and transformation of biogenic elements, oxygen and carbon dioxide in the ocean waters.* Consequently, one has to include the terms accounting for these processes in the mathematical models describing distribution of the above-mentioned substances in the ocean waters, and these terms complicate significantly the estimation procedure. Biogenic elements exist in the ocean waters in either of two forms: dissolved or in the composition of the suspended organic and inorganic matters. The production and destruction of the organic matter determine mainly the variability of the concentration of the biogenic elements. Biogenic elements are transferred into the ocean from the atmosphere due to sedimentation of terrigenous and volcanogenic suspension as well as by river and subsurface runoff.

Biogenic elements in the ocean waters are consumed primarily by phytoplankton, with the intensity of this process depending strongly on the biogen concentration c, the illumination I and the water temperature T. There are a number of empirical and semi-empirical formulae relating a rate of biogen consumption by phytoplankton to c, I and T. The illumination varies with the depth so the dependence of a rate at which the biogenic elements are consumed by phytoplankton on the depth is usually taken as an exponential function with a power $-kz$, where k is the extinction coefficient of the sea water. All the above-mentioned factors are frequently assumed to affect a rate of consumption independently so that their total effect can be represented as a product of the functions describing the individual influence of any of the factors.

The biogenic element that has been consumed by phytoplankton can be released back into the environment in the form of organic or inorganic dissolved compounds. A certain fraction of the biogenic element is returned back into the water medium during the dying of the phytoplankton while the remainder is precipitated in a form of suspension. The rate of the release of biogen by living organisms as well as the rate of these organisms dying off are both usually regarded as being in direct proportion to biomass or to the production of phyto-

* For detailed information on distribution, forms of existence and transformation of chemical elements and compounds in the ocean see [105].

plankton. The biogenic element can be regenerated out of the dead organic matter by bacteria and protozoa. A rate of mineralization of the organic matter caused by these processes can be, apparently, considered to be in direct proportion to the concentration of the organic matter, yet the coefficient of proportionality turns out to be a function of water temperature. The phytoplankton is further consumed by zooplankton and other animals. The food consumed by them is spent on their growth and reproduction and remains partly non-assimilated. Intensity of feeding depends, in its turn, on food concentration, and the rate at which the organisms are dying off and the rate of release of the fraction of biogen dissolved can both be regarded as being in direct proportion to the biomass of the organisms.

To describe the concentration of the biogenic element $c(x, y, z, t)$ in the ocean water we shall use a customary diffusion equation with a term R accounting for the sources (sinks) of element of biochemical origin:

$$\frac{\partial c}{\partial t} + u\frac{\partial c}{\partial x} + v\frac{\partial c}{\partial y} + w\frac{\partial c}{\partial z} = K_L\left(\frac{\partial^2 c}{\partial x^2} + \frac{\partial^2 c}{\partial y^2}\right) + $$
$$+ \frac{\partial}{\partial z}\left(K_z\frac{\partial c}{\partial z}\right) - \frac{\partial}{\partial z}\left(w_0\frac{\partial c}{\partial z}\right) + R, \tag{3.23}$$

where w_0 is a rate of gravitational sedimentation of biogen which is supposed to exist in the water as suspension; all other notations have been introduced previously.

To solve equation (3.23), one obviously has to assign the velocity components, vertical and horizontal turbulent diffusion coefficient and a function R as well as initial and boundary conditions for biogen concentration in a given ocean area. But usually it is impossible to define a unique functional form of R for all forms of biogen existence in the ocean because the biochemical processes are rather complicated. Consequently, block schemes are widely used to estimate any of the individual forms of biogen existence. Thus, for instance, a five-block scheme has been proposed by Yakushev [111] to estimate phosphate distribution in the ocean. In this scheme the first block is used for dissolved inorganic phosphorus, the second block for dissolved organic phosphorus, the third block for phosphorus of phytoplankton, the fourth block for phosphorus of zooplankton and the fifth block for phosphorus of detritus. The phosphorus can pass from the first form of existence to the third one, from the second to the first, from the third to all of the others, from the fourth to the first, the second and the fifth and from the fifth to the first, the second and the fourth. The same diffusion equation (3.23) is used to describe biogen concentration c_i for every block (i.e., for every form of biogen existence), but with a particular form of function R_i and initial and boundary conditions for every block. But as the expressions for R_i are dependent on the values of biogen concentration in other forms of existence (from other blocks), then the individual equations for c_i form a set of i coupled equations. Specific of R_i in this set of equations are determined by the above-mentioned empirical and semi-empirical relations.

As an example, we present the expressions for R_i proposed by Yakushev [111] for the five-block scheme describing the phosphorus balance in the ocean:

$$R_1 = P_3(K_{31} - 1) + K_{41}C_4 + K_{51}C_5 + K_{21}C_2; \tag{3.24}$$

$$R_2 = K_{32}C_3 + P_4(1 - u_4)H_4 + K_{52}C_5 - K_{21}C_2; \tag{3.25}$$

$$R_3 = P_3(1 - K_{31}) = K_{32}C_3 - K_{35}C_3 - K_{34}\frac{c_4}{1+C_4K_3/(c_3-c_3^t)}; \tag{3.26}$$

$$R_4 = P_4u_4 - K_{41}C_4 - K_{45}C_4; \tag{3.27}$$

$$R_5 = K_{35}C_3 + K_{45}C_4 + P_4(1 - u_4)(1 - H_4) - - K_{54}\frac{c_4}{1+C_4K_5/(c_5-c_5^t)} - K_{52}C_5 - K_{51}C_5, \tag{3.28}$$

where $P_3 = K_{13}C_3/[1+(c_3/c_1)K_1]$ is the consumption of inorganic phosphorus by phytoplankton; K_{13} is the maximum rate of the consumption of inorganic phosphorus by phytoplankton; K_1 is a constant determining the consumption of phosphates by phytoplankton; K_{31} is the specific rate of phytoplankton respiration; K_{32} is the specific rate of release of dissolved organic phosphorus by phytoplankton; K_{35} is the specific rate at which phytoplankton is dying off; $P_4 = \{K_{34}/[1+C_4K_3/(c_3-c_3^t)] + K_{54}/[1+C_4K_5/(C_5-C_3^t)]\}c_4$ is the consumption of phytoplankton and detritus by zooplankton; K_{34} is the maximal rate of the consumption of phytoplankton by zooplankton; K_3 is a constant determining the consumption of phytoplankton by zooplankton; c_3^t is a threshold concentration of phytoplankton; K_{54} is the maximal rate of consumption of detritus by zooplankton; K_5 is a constant defining the consumption of detritus by zooplankton; c_5^t is the threshold concentration of detritus; u_4 is the efficiency of the food assimilation by zooplankton; H_4 is the ratio of liquid zooplankton excretations to its total value; K_{41} is the specific rate of zooplankton respiration; K_{45} is the specific rate of zooplankton dying off; K_{52} is the detritus autolysis coefficient; K_{51} is the detritus-phosphatofication coefficient; and K_{12} is the coefficient of phosphatofication of dissolved organic phosphorus.

The available observational data were used by Yakushev to estimate the numerous coefficients in expressions for R_i, and in some cases these coefficients were chosen as achieving the best agreement between the observed distributions of biogen and the predicted ones. Generally speaking, the stability of the set of equations for c_i relative to variations of the coefficients has to be examined before solving the set of equations. This is caused by the comparatively low accuracy of the available estimations of the coefficients and by the difficulties of their field determinations. A study of stability often shows that even small variations of some coefficients can lead to significant changes in the solutions obtained while the solutions can be more stable with respect to variations of other coefficients. Thus, for instance, the set considered by Yakushev has been proved to be rather sensitive to variations of the coefficient K_{13} and the constant K_1 as well as to the coefficients K_{51} and K_{52}, while inaccuracy in other coefficients and constants produces an insignificant influence on the solution.

The sets of equations describing the transport and transformation of biogens in the oceans are, in most cases, too complicated to be solved analytically, so a numerical technique is widely used for such solutions. In some cases, of course, the set of equations can be essentially simplified if some of the phenomena and processes are neglected. Usually, such a simplification is achieved successfully when one is analyzing the transport and transformation of biogen in certain areas or layers of the ocean where some particular processes dominate while others are insignificant. Thus, for instance, Yakushev [112] has considered a model problem of the seasonal change of phosphorus concentration in the photosynthesis layer of the moderate-latitude seas. In this model the removal of phosphorus out of a given zone by the sedimenting suspensions and the compensating input of the dissolved phosphorus into the zone were the only processes taken into account. The intensity of the consumption of the dissolved phosphates by phytoplankton (coefficient K_{13}) was supposed to vary in time with a period of a year. The effects of advection and turbulent transport were not considered in this problem, and the influence of the initial conditions was not accounted for, (i.e., a purely periodic solution was sought). The vertical distribution of phosphorus in the photosynthesis ocean layer has been considered by Yakushev [112] in another model problem, which was supposed to be a steady one but the vertical turbulent transport was taken into account as well as the sedimentation of suspensions containing phosphorus. In this problem the intensity of the consumption of phosphates by phytoplankton was considered as a function of depth depending on the water illumination at the different horizons, and the vertical turbulent diffusion coefficient was approximated by a step function of the vertical coordinate z. The vertical distribution of the concentration of the dissolved phosphorus determined by Yakushev on the basis of this model is displayed in Figure 3.9 along with observational data obtained in the waters of the central Atlantic south of the Azores.

A steady problem on phosphorus distribution in which both the advective and diffusion terms in equation (3.23) were taken into account was studied by

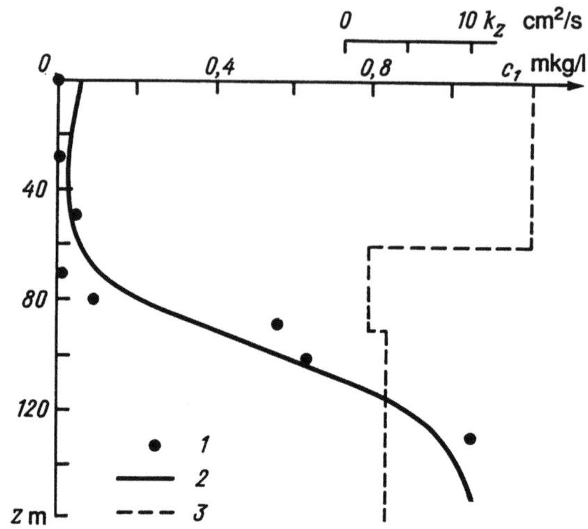

Fig. 3.9. Vertical distribution of dissolved phosphorus in the upper ocean layer (after [112]). 1 – experimental points, 2 – theoretical curve, 3 – coefficients of vertical turbulent transport.

Yakushev [111] as well. The problem was considered for the meridional cross-section of the ocean and, consequently, the terms in the equation with derivatives along the latitude (i.e., along the x-axis) were omitted. The horizontal diffusion coefficient K_L was assumed constant (10^8 cm^2/s), and coefficient K_z was supposed to depend on the Richardson number; its values ranged from 0.3 cm^2/s at 0–5° S to 100–200 cm^2/s at 60–65° S. The values of the flow-velocity components were taken from the study by Zalesny and Ivanov [30] in which the two-dimensional water circulation was predicted for the meridional cross-section of the World Ocean over the latitudes ranging from 70° N to 70° S. The wind shearing stress at the ocean surface and the water temperature of the surface ocean layer were used to estimate the flow-velocity field. The problem was supposed to be symmetrical with respect to the equator, so the meridional flow velocities at the equator were assumed zero as well as the heat and salt fluxes through the equator. The mean ocean depth was assumed equal to 4000 m and the bottom friction was taken into account, while the heat and salt fluxes through the ocean floor were neglected.

The five-block set of the equations describing the transport of the different forms of phosphorus in existence has been considered by Yakushev numerically. The vertical velocities of sedimentation for the third, fourth and fifth blocks were taken as 50, 10 and 5 m/day, respectively, and the functions R_i were used in accordance with equations (3.24) – (3.28). The problem was solved as an initial value one (method of reaching the steady state) provided the initial concentration of biogen in the ocean was uniform for all forms of its existence. A time step of 1 day and spatial steps of 5° along the meridian and of 200 m along the depth were taken to carry out numerical computations. Photosynthesis was supposed to occur at the upper points of the considered grid domain only. A steady state was reached after approximately 3500 days elapsed since a time moment that was taken as the initial one. The steady vertical distributions of the concentration of the dissolved inorganic phosphorus c_1 determined at the different latitudes for the three oceans are plotted in Figure 3.10 along with the observational curves of the distribution of biogen. The predicted curves describe satisfactorily the latitudinal changes of some typical features of the vertical distribution of phosphates. Thus, for instance, the concentration of c_1 in the surface waters decrease from 1.7 – 1.8 micrograms of atomic phosphorus per liter of the water at the latitudes 60 – 70° to 0.010 g/l in the tropics. A layer with the maximum content of phosphates in the tropical and equatorial waters, with the characteristic ascent of this layer near the equator, is observed as well. The determined meridional distributions of concentrations c_3, c_4 and c_5 make it

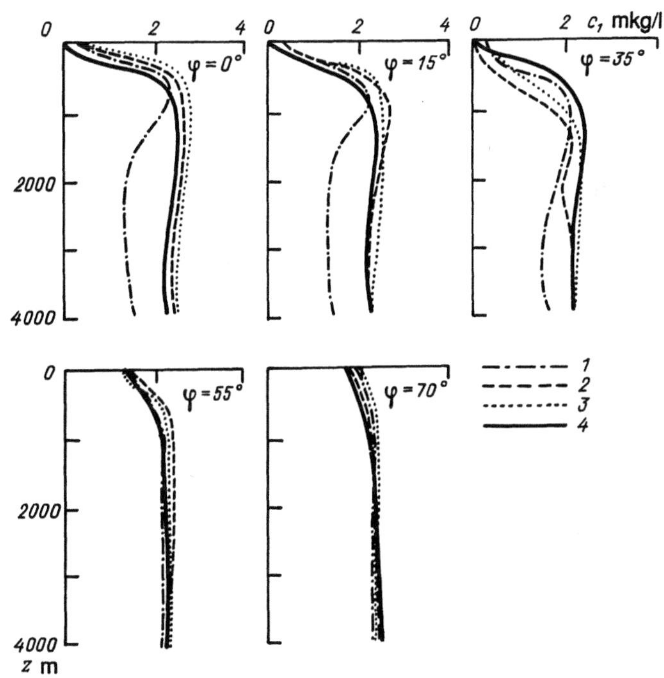

Fig. 3.10. Vertical distributions of dissolved phosphorus at different latitudes of the southern hemisphere. The curves are for observations in the Atlantic (1), the Indian (2) and the Pacific (3) oceans and for theoretical prediction (4) (after [111]).

possible also to estimate the distribution of phytoplankton, zooplankton and detritus in the meridional cross-section of the oceans. The predicted maximum of phytoplankton concentration occurs at the latitude of about 55°, near the equator its concentration is twice as small, and at latitudes 10 – 25° the concentration is twice as small again. The analogous decrease of zooplankton concentration is even more sharp – by four- and eightfold, correspondingly. The agreement between the predicted and observed distributions of concentration of phosphorus is better for the Pacific and the Indian oceans than for the Atlantic. This phenomenon can be explained by the neglection of the transport of the water masses through the equator in the model considered. This transport is especially prominent in the Atlantic where, in the Greenland area, deep waters poor in phosphates are formed and afterwards propagated far southwards, where they give rise to the water tongue of low biogen content. Except for this phenomenon, all the other main mechanisms determining the formation, transformation and transport of phosphorus in the ocean waters are taken into account well enough in the model in question. These mechanisms may be characterized briefly as follows. The concentration of phosphates increases in the surface waters over the high latitudes of the southern hemisphere due to the intensive vertical transport of waters as well as to the reduced average yearly production of phytoplankton. The waters with the increased content of phosphates sink in the area over 52.5 – 57.5° under influence of the vertical transport, and afterwards the water layer rich in biogen arises due to horizontal diffusion. The content of biogens in this layer increases even more in the tropical and equatorial areas where oxidation of the organic matter occurs and gives rise to the additional amount of phosphates. Naturally, seasonal changes of all the forms of phosphorus existing in the ocean waters are superimposed on these processes. This seasonal variability of the biogen-concentration fields can be taken into account by imposing the dependence of some model coefficients on the water temperature. Thus, for instance, Yakushev [112], suggesting that the coefficient K_{13} is a function of latitude and temperature, has obtained seasonal variations of the phytoplankton and zooplankton concentrations in the ocean that were close to the observed ones.

We have considered above some examples of estimation of the phosphorus concentration in the ocean waters. The transformation and transport in the ocean of any other biogenic element, primarily nitrogen and silicon, must be estimated in a similar way. In order to develop the models for transport of these biogens one naturally has to take into account the specific features of their forms of existence in the sea water, as well as the peculiarities of the interaction of a given biogen with other chemical compounds and its assimilability and excretion by the living organisms. The constants and coefficients in the expressions for functions R_i may, obviously, have specific values for any biogenic elements as well.

Besides the problems concerning the distribution of biogenic elements either in the World Ocean or in its individual large parts, the more local problems of diffusion type concerning the concentration of the different forms of the biogen in existence may be of some interest. Thus, for instance, patches of phytoplankton are frequently observed in the ocean. One can assume that the size of these patches as well as the concentration of phytoplankton organisms therein (i.e., the concentration of biogen) are related in some way or another to the hydrophysical processes in a given water body. A simple problem on the possible forms of such a relation had been considered by Ozmidov [81] as early as 1968. Phytoplankton has a rather low ability to move actively in the horizontal direction and, consequently, it can be considered as a passive tracer to a fairly accurate approximation. During the spring sea 'bloom' phytoplankton can reproduce itself rather intensively so that daily increase of its biomass may be as high as 50%. Yet, during the autumn reproduction the rate of the biomass increasing is, as usual, at a significantly lower level. The rate of variation of the phytoplankton quantity (or the concentration) c_3 during the spring 'bloom' can be described by the equation

$$\partial c_3/\partial t = (k_1 - k_2 - k_3)c_3, \tag{3.29}$$

where k_1 is the specific rate of photosynthesis, k_2 is the rate of respiration of the organisms, and k_3 is the rate of their consumption by animals.

The value of $(k_1 - k_2 - k_3) \simeq \gamma$ can be regarded as a constant when considering the transformation of the phytoplankton accumulation within small time intervals (of the order of several days) during the spring 'bloom', when many of the factors restricting the reproduction of phytoplankton are of no importance. If so, the diffusion equation for the phytoplankton accumulation will differ from the equation for a contaminant only by an additional term of the form γc_3. Since phytoplankton is developed only in the upper illuminated layer of the ocean and its vertical diffusion is rather weak, one can neglect all the vertical processes as well as suppose the diffusion process to be horizontally isotropic when studying the size of 'blooming' phytoplankton patches. Introducing a system of coordinates moving with mean flow we arrive at the following diffusion equation for a phytoplankton patch:

$$\frac{\partial c_3}{\partial t} = \frac{1}{r^2}\left[\frac{\partial}{\partial r}\left(K_L(r)\, r^2\, \frac{\partial c_3}{\partial r}\right)\right] + \gamma c_3, \tag{3.30}$$

where r is the radius vector of an observational point, and $K_L(r)$ is the horizontal diffusion coefficient which, in a general case, is a function of r.

In order to take into account the dependence of the horizontal turbulent diffusion coefficient on the phenomenon scale (i.e., on the size of phytoplankton patch), $K_L(r)$ can be presented in the form of a power function. Ozmidov [81] has proposed using in the present problem an expression for $K_L(r)$ following from the "4/3 power law", i.e., letting $K_L(r) = kr^{4/3}$, where k is a coefficient of proportionality. Provided an instantaneous point source of phytoplankton is located at the origin of coordinates at time $t = 0$, we arrive at the solution of equation (3.30) of the form

$$c_3(r, t) = \frac{Q}{6\pi(4k/9)^3 t^3} \exp\left(-\frac{r^{2/3}}{4kt/9} + \gamma t\right), \tag{3.31}$$

where Q is the amount of diffusing phytoplankton.

It follows from the equation (3.31) that the phytoplankton concentration c_3 will be increasing within a circle of radius $r < (4k\gamma/9)^{3/2}/t^3$ and decreasing out of this circle. The radius of such a 'blooming' patch of phytoplankton will increase with time, reaching 20 or 30 km in 10 days at coefficients γ close to 0.4 or 0.55, respectively. Such sizes of phytoplankton patches are in good agreement with observations as regards the order of magnitude. The coefficient γ must be decreased after a sufficiently long diffusion time because the increasing of the phytoplankton concentration leads, apparently, to a worsening of the conditions necessary for the existence of phytoplankton and, consequently, reduces a value of γ which, in turn, will retard or even cease further extending of 'bloom' patch size.

Fennel and Lass [122], analyzing the conditions necessary for the development of phytoplankton patches in the ocean, proposed the following criterion. If the initial size of a plankton patch, L, is larger than the dispersion of a patch, σ, caused by the horizontal diffusion, a given patch will exist and develop. But if $L < \sigma$ a patch will disappear rather quickly due to the diffusion. Provided the horizontal diffusion of a patch is described by an equation with a constant horizontal transport coefficient K_L, the dispersion of a patch is $2\sqrt{K_L/\gamma}$. If the patches rich in biogen are initially scattered over a large ocean area and there are patches sizes that are greater and smaller than σ, the patches of 'blooming' sea will be developed and extended within the former, while the latter will disappear rapidly due to the diffusion process and give no rise to the bursts of the phytoplankton concentration. Thus there can exist a kind of 'natural selection' of the zones of either increased or decreased productivity of phytoplankton and, consequently, selection of the zones of accumulation of the higher trophic-level organisms. As for an initial 'patchness' of the field of the biogen concentration, it either can be caused by the intermittency of ocean turbulence (see Section 3.4) or, in Fennel and Lass's opinion [122], can arise due to the modulation of the velocity field in the ocean by Rossby waves.

The estimation of the diffusion of oxygen and carbon dioxide in the ocean involves the imposing of boundary conditions for concentration of gases at the surface of the ocean, which turns out to be the most intricate part of the problem. The exchange of these gases between the ocean and the atmosphere is a fairly active process, so one naturally cannot neglect their fluxes through the interface surface. Provided concentration of the dissolved gases in the surface ocean layer is imposed as a boundary condition, the distribution of gases in the deeper ocean layer turns out to be strongly predetermined. The problem concerning the distribution of oxygen in the ocean waters under such boundary conditions was studied by a number of authors (see references in [96]). A more interesting diffusion problem was considered by Ryabchenko [96], who imposed oxygen flux at the ocean surface as a boundary condition. He integrated the oxygen-diffusion equation with respect to the vertical coordinate from the ocean surface $z = \zeta$ to the ocean floor $z = H$ and wrote the result in the form

$$\frac{\partial \hat{c}}{\partial t} + \operatorname{div}(\mathbf{v}\hat{c}) = K_L \Delta \hat{c} - K_z \frac{\partial \hat{c}}{\partial z}\bigg|_{z=\zeta} + \int_\zeta^H R_1 dz + \int_\zeta^H R_2 dz, \tag{3.32}$$

where $\hat{c} = \int_\zeta^H c dz$; c is the concentration of oxygen dissolved in the ocean waters; $\mathbf{v}$ is the horizontal component of the flow-velocity vector averaged over the ocean depth; K_L and K_z are the horizontal and vertical turbulent diffusion coefficients, respectively; z is the vertical coordinate measured downward from the undisturbed ocean surface; Δ and div are the Laplace and divergence operators in the spherical system of coordinates; ζ is the ocean-surface level; $H = H(\lambda, \theta)$ is the ocean depth; λ is the longitude; θ is the latitude component; and t is time. Function R_1 determines a rate of oxygen production during the photosynthesis process and function R_2 determines a rate of biochemical oxygen consumption in the ocean waters.

The boundary condition at the ocean-atmosphere interface is given by

$$-K_z \frac{\partial \hat{c}}{\partial z}\bigg|_{z=\zeta} = \alpha[c_p - c(z)|_{z=\zeta}], \tag{3.33}$$

where c_p is oxygen solubility (equilibrium concentration) in the surface waters; α is the exchange coefficient. A value of α is a function of a number of factors but, primarily, it depends on the wind velocity above the surface as well as on the temperature and salinity of the water.

To convert from the values of concentration of dissolved oxygen averaged over the depth H to the values of $c(z)_{z=\zeta}$, Ryabchenko supposed the value of $c(z)$ to be presented in the form of product

$$c(z) = c(z)|_{z=\zeta}\, f_i(z), \tag{3.34}$$

where $f_i(z)$ are non-dimensional universal functions, different for each of the three oceans ($i = 1$ for the Pacific, $i = 2$ for the Indian Ocean, $i = 3$ for the

Atlantic). Under this suggestion we arrive at the equation for the averaged concentration,

$$\hat{c} = c(z)\big|_{z=\zeta} \int_{\zeta}^{H} f_i(z)\, dz. \tag{3.35}$$

One can suggest, to a first approximation, the absence of dissolved oxygen flux at the ocean floor, i.e., suppose

$$\partial \hat{c}/\partial n\big|_{\Gamma} = 0, \tag{3.36}$$

where n is the normal to the rigid boundary surface Γ of the ocean. The normal component of flow velocity becomes zero at this surface as well.

For numerical solution of the problem it is convenient to suppose no dissolved oxygen to be presented in the ocean waters in a time moment taken as an initial one, i.e.,

$$\hat{c}(\lambda, \theta, t=0) = 0. \tag{3.37}$$

Besides that, for some simplification of the problem, the integrals R_1 and R_2 in the right-hand side of equation (3.32) were supposed to be equal in magnitude, which means the equality of production and consumption of oxygen at any vertical in the ocean.

The diffusion problem (3.32) – (3.37) was solved by Ryabchenko numerically. The flow-velocity field was taken from papers of other authors who used the averaged yearly atmospheric pressure field at sea-surface level and the averaged yearly zonal distributions of radiation balance as well as the temperature of the subsurface water layer to determine the stream function of the ocean waters. The available data on atmospheric pressure also presented the possibility of estimating wind velocity over the ocean, which is necessary to determine the values of α in a boundary condition (3.33). The dependence of α on wind velocity as well as on the temperature and salinity of the subsurface waters was determined from the data available from the literature (see references in [96]). Numerical calculations were performed with a time step of five days and a spatial step of five degrees along both latitudinal and longitudinal directions. The calculations were carried out until a steady state was established (to an accuracy of 1%) 43 years after some initial moment. The horizontal turbulent transport coefficient was taken to be 10^8 cm^2/s (coefficient K_z is not required for the present problem).

The chart of the distribution of the vertically averaged concentration of oxygen in the World Ocean waters obtained by Ryabchenko was, on the whole, in good agreement with observational data. In particular, the maximum of O_2 was found in the cold waters of the North Atlantic and in the high-latitude part of the north-western part of the Pacific. It should be noted that this is a satisfactory agreement of estimated and observed values of dissolved oxygen concentration averaged over a given ocean. The maximum oxygen content was observed in the Atlantic and the minimum one in the Pacific. Yet the values of $\hat{c}$ were overestimated for the ocean zones with high depth gradients, which can be caused by the too-large spacial step of the computational grid and indicates the limitations of a given study.

The diffusion problem considered above is, of course, rather simplified for the estimation of dissolved oxygen concentration in the ocean waters. For a more realistic approach one has to consider a three-dimensional problem and take into account both sources and sinks of oxygen in the bulk of the ocean. These functions, in their turn, can be determined by resolving proper diffusion problems on the distribution of phytoplankton, zooplankton, biogenic elements, etc., in the ocean. Finally, a system of blocks describing the distribution and transformation of any organic or inorganic matters that account for the generation and consumption of oxygen has to be involved in a complete problem to describe the distribution of O_2 in the ocean waters. Naturally, the creation of such an adequate model and its solution are left to the future, but progress is being made practically yearly.

Similar problems and difficulties arise in the calculation of carbon dioxide (carbon) content in the ocean waters. The exchange of carbon dioxide between the ocean and the atmosphere is a fairly active process whose intensity, like the exchange of oxygen, depends significantly on the difference between gas solubility and its concentration in the sea water. The solubility, in its turn, depends on CO_2 concentration in the atmosphere as well as on the temperature and salinity of the water. In the upper ocean CO_2 is consumed by phytoplankton in its vital activity, which leads to the increasing of the biomass of photosynthesizing organisms. Some amount of the carbon that is assimilated by phytoplankton is then partly transferred to zooplankton and other organisms and partly, together with dead organisms, sedimented onto the ocean floor or transferred into the water solution again. Carbon dioxide is released into the sea water due to the respiration of hydrobionts as well.

The first attempts to determine the content and transformation of carbon in the ocean waters were based, naturally, on rather simplified model descriptions of a phenomenon. In particular, the fluxes of carbon were taken as constant at the surface of the ocean and at the lower boundary of the photosynthesizing zone, a thickness of this zone and water temperature were considered as constant, and so on. A more realistic model of the carbon cycle was presented by Kagan and Ryabchenko [41]. According to their model two regions are distinguished in the ocean, the first one accounting for the formation of cold deep waters and the second one for the rise of the waters (upwelling) to compensate the delivering of the former. It is supposed, naturally, that the velocity of the water's rise in the upwelling zones, w, must be of the order of $w = W/S$, where W is an intensity of source of cold deep waters and S is an area of upwelling. The two layers are defined within the latter region, which are the upper mixed layer (UML) and the deep one. These layers as taken together with the region of the formation of deep waters form a system of three interconnected reservoirs, or boxes. For these boxes one can write down two similar systems of three equations describing the evolution of water temperature, T, and carbon concentration, c, as follows:

$$\frac{\partial}{\partial t} X_0 H = q_{s0}^x + \frac{S_1}{S_0} w(X_1 - X_0) + (Q_1^x + Q_2^x) + B_0; \tag{3.38}$$

$$\frac{\partial}{\partial t} X_1 h - X_1\left(\frac{dh}{dt} + w\right) = q_{s1}^x - q_{h-0}^x - wX_1 - Q_1^x + B_1; \tag{3.39}$$

$$\frac{\partial}{\partial t} X_2(H-h) + X_2\left(\frac{dh}{dt} + w\right) = q_{h+0}^x + wX_0 - Q_2^x + B_2, \tag{3.40}$$

where X stands for either T or c; q_s^x is the turbulent flux of substance through the ocean surface; q_{h-0}^x is a flux at the lower boundary of UML; q_{h+0}^x is a flux at the upper boundary of the deep layer; Q^x is the turbulent transport between the region of formation of the cold deep waters and the region of upwelling (which will be neglected hereafter); B are sources and sinks of inorganic carbon; H is the ocean depth; h is the UML thickness, t is time; and subscripts 0, 1 and 2 refer to the region of the formation of the cold deep waters, UML and deep layer of the upwelling zone, respectively.

Turbulent fluxes of heat at the air-water interface are functions of time (within a yearly cycle) and are supposed to be given in the problem in question. The carbon dioxide fluxes at the ocean surface can be presented, similarly to the equation for the oxygen flux (3.33), in the following form:

$$q_s^c = K[c_p - c(z)\,|_{z=\zeta}], \tag{3.41}$$

where c_p is gas solubility (equilibrium concentration) in the subsurface water layer, which depends on CO_2 partial pressure in the atmosphere as well as on water temperature T and salinity S; $c(z)\,|_{z=\zeta}$ is CO_2 concentration in this layer, which depends on T and S as well as on the total content of inorganic carbon in the water; and K is the gas exchange coefficient, which depends primarily on wind velocity over the surface.

When dissolving in water carbon dioxide undergoes a chemical reaction with the water to form carbonic acid, which afterwards dissociates. As a result the total concentration of inorganic carbon in the sea water is represented as a sum of the concentrations of dissolved carbon dioxide and of bicarbonate and carbonate ions.

In the present model the heat and carbon transport through the lower UML boundary is supposed to be cause only by the UML deepening. If the lower UML boundary rises there occurs the transfer of heat and carbon contained previously in the lower part of the UML into the deep layer. Under these assumptions the expressions for fluxes q^x_{h-0} and q^x_{h+0} take the form

$$q_{h-0} = \begin{cases} (X_1 - X_2)\left(\frac{dh}{dt} + w\right) & \text{if } \left(\frac{dh}{dt} + w\right) > 0; \quad (3.42) \\ 0 & \text{if } \left(\frac{dh}{dt} + w\right) \leqslant 0; \quad (3.43) \end{cases}$$

$$q^x_{h+0} = \begin{cases} 0 & \text{if } \left(\frac{dh}{dt} + w\right) > 0; \quad (3.44) \\ -(X_1 - X_2)\left(\frac{dh}{dt} + w\right) & \text{if } \left(\frac{dh}{dt} + w\right) \leqslant 0; \quad (3.45) \end{cases}$$

For determining UML thickness Kagan and Ryabchenko suggested supplementing the relations (3.42) – (3.45) by an equation of turbulent energy balance in a layer integrated over the UML. Due to this technique a value of $(dh/dt + w)$ is represented as a function of flux $q^T_{s_1}$, the Richardson number and the difference in the temperature of the water above and below the UML boundary.

An unstable fraction of organic matter as well as dissolving humus are the main sources of inorganic carbon in the upwelling zone. From these sources the amount of carbon that is used by phytoplankton to increase its biomass has to be subtracted. The values of B_i may be taken, in accordance with the available estimates of these components of carbon balance, as follows: $B_1 = -34$ g/(m^2/year), $B_2 = 28$ g (m^2/year) and $B_0 = 0$. Zero value of B_0 means that the organics produced in the cold deep waters is decomposed completely in the same waters.

For solution of this problem of the carbon transport and transformation one has to include, as input information, data on heat flux at the ocean surface, on CO_2 partial pressure in the atmosphere, on wind velocity, on power of the source of cold deep waters and on the ratio of the area of upwelling region to the area where the deep waters are forming. The temperature and carbon concentration in any of the three reservoirs under consideration, the thickness of the upper layer of the upwelling region, the concentration of dissolved CO_2 and the concentration of bicarbonate and carbonate hydrogen ions, the intensity of exchange of a gas with the atmosphere as well as the intensity of exchange of heat and carbon between UML and the deep ocean layer are all the values to be determined in solving the described system. Provided any of the input parameters (for example, heat flux through the ocean surface) is given as a function of time with a seasonal variation, the values to be determined appear, naturally, to be time-varying functions with a seasonal period of variability.

The stated problem was considered by Kagan and Ryabchenko numerically. The set of equations reached periodic solutions (the coincidence of the values of water temperatures for two sequential yearly cycles was within the accuracy of 0.01%) after 560 years from an initial moment.

The results simulated the main features of the thermal state and carbon cycle in the ocean. Thus, some effects which have been known well enough from field observations were obtained. They are, for instance, the increase of gas exchange between the ocean and the atmosphere during the autumn-winter period, the change of the direction of the gas exchange during a year-long period in moderate-latitude and low-latitude regions (in the upwelling region), the constancy of CO_2 uptake in the region where the deep waters are forming, weak if any seasonal oscillations of temperature and the concentration of carbon in the deep layer. The thermal features of the upper ocean layer and the seasonal variation of UML thickness were in good agreement with observational data as well. Besides that, the characteristics of the carbon cycle in the ocean were proved to be influenced mainly by the variations of both phytoplankton production and a rate of the sedimentation of organic matter.

Naturally, no detailed knowledge on either the vertical or horizontal structure of the fields corresponding to the different forms of carbon existing in the ocean could be obtained with the help of a reservoir model that was explored by Kagan and Ryabchenko to study the transport and transformation of CO_2 in the ocean. Yet the three-dimensional differential model of carbon diffusion in the ocean is currently believed to be even more tedious to resolve due, for instance, to some difficulties of imposing the spatial distribution of sources and sinks of carbon in the ocean. To achieve the most drastic approach to the problem in question, one has, naturally, to consider a coupled set of equations describing the transport and transformation of biogenic elements, oxygen and CO_2 accounting for all the forms of existence of these substances and their transformation caused by chemical and biological processes. There is no question that this very approach will be developed as a main one during the further studies in this field of oceanology.

A study by Makkaveev and Yakushev [60] gives an example of the use of a turbulent diffusion equation rather than reservoir models to describe the transport of CO_2 in the ocean. They considered a steady problem in which both horizontal diffusion and advection were neglected. Under these assumptions the diffusion equation for the concentration of carbon dioxide, c, was reduced to the simplest form,

$$K_z \partial^2 c / \partial z^2 = R, \tag{3.46}$$

where the source (sink) function for carbon dioxide, R, accounts for the biochemical processes of synthesis and decomposition of organic matter.

The consumption of carbon by phytoplankton is determined by its mass and the intensity of photosynthesis P, the latter, in turn, depending on the illumination of water by sunlight I. A value of P can be related to I by the expression

$$P(z) = k_1 I(z) \exp[1 - I(z)/k_2], \tag{3.47}$$

where $I(z)$ is illumination at horizon z; and k_1 and k_2 are some coefficients. A value of $I(z)$, in turn, is exponentially related to depth:

$$I(z) = I_0 \exp(-k_3 z), \tag{3.48}$$

where I_0 is illumination at the ocean surface and k_3 is the extinction coefficient.

In this way we arrive at a relation for the consumption of CO_2 caused by photosynthesis,

$$R_1 = k_1 I_0 \exp\left[1 - \frac{I_0}{k_2} \exp(-k_3 z) - k_3 z\right] B(z), \tag{3.49}$$

where $B(z)$ is a distribution of phytoplankton biomass with depth.

Inorganic carbon in the bulk of the water arises mainly during the oxidation of dead organic matter. This process also can be described by an exponential function of depth:

$$R_2 = -k_4 \exp(-k_5 z), \tag{3.50}$$

where k_4 and k_5 are coefficients independent of z.

Thus, the total function R in equation (3.46) takes the form

$$R = k_1 I_0 \exp\left[1 - \frac{I_0}{k_2} \exp(-k_3 z) - k_3 z\right] B(z) - k_4 \exp(-k_5 z). \tag{3.51}$$

Coefficients $k_1 - k_5$ in equation (3.51) are determined by a number of physical and biological phenomena, in particular, by the assimilation ability of organisms and the content of photosynthesizing chlorophyll in them, by illumination conditions optimal for photosynthesis, by some features of biological consumption of oxygen in the bulk of the water, etc. As far as the theoretical description of these phenomena appears to be rather difficult, all the coefficients mentioned above are supposed to be empirical constants whose values range within certain limits.

Makkaveyev and Yakushev used a finite-difference technique to estimate the distribution of total concentration of CO_2 (total content of inorganic carbon) from equation (3.46) using the equation (3.51) for R. The value of concentration of CO_2 measured at horizons 0 and 100 m in summer 1980 in the Atlantic, south of the Azores, were imposed as the boundary conditions. The observations of the vertical distribution of elements of carbonate system and phytoplankton biomass in the ocean waters were carried out in the same area as well. The comparison of the observational data with estimated curves of the vertical distribution of total CO_2 is shown in Figure 3.11. As one can see from this figure, the coincidence of theoretical and experimental data is rather good (one, certainly, has to keep in mind that the coincidence of the boundary observational points with the theoretical curves has been predetermined by imposed boundary conditions). The shape of the obtained vertical distributions of total CO_2 is in good agreement with the observed distributions of phytoplankton biomass. Either position on magnitude of maximum content of total CO_2 in the water was found to vary depending on the depth and magnitude of the maximum concentration of phytoplankton. Variations of the other parameters of the model permitted the examination of their influence on the content of CO_2 and the shape of its vertical distribution in the ocean areas, for which the model equations (3.46) – (3.50) are supposed to describe the processes occurring therein realistic enough.

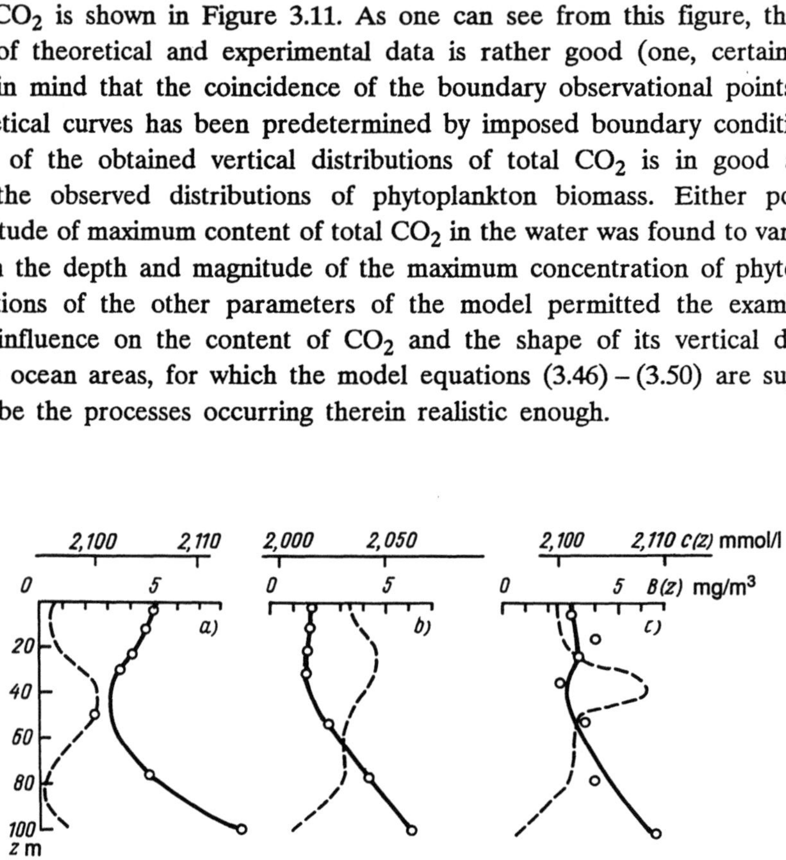

Fig. 3.11. Vertical distribution of concentration of total carbon dioxide $c(z)$ observed at two stations (a, b) south of the Azores (circles) and estimated from diffusion model (solid lines). Dotted lines are for observed distribution of phytoplankton biomass $B(z)$. Data in fig. 3.11c are averaged for several stations (after [60]).

3.3. Some Aspects of Oil-Spot Diffusion

Growing oil pollution in the World Ocean has caused an increased interest not only in techniques of preventing this pollution but also in methods of predicting oil spread on the ocean surface. Sources of oil pollution in the oceans can be of different origin. Thus, the highest oil pollution is observed near the coasts of countries with developed industry, along the busy ocean waterways and in the areas of off-shore oil-production operations. Discharges of oil and its products caused by oil-tankers accidents can lead to pollution in large ocean areas as well. Some information on oil input into the World Ocean may be found in UNESCO publications (see, for instance, [136]), and data on the amount of oil input into the ocean are presented in the survey [25] and the article [97].

Oil input into the ocean spreads on it according to complicated laws whose details are still unknown. In a simplified form oil spread can be outlined as follows. When a significant amount of oil is input onto the ocean surface (for instance, due to oil-tanker accidents), oil spreads first under the action of gravity. As soon as the oil layer becomes thin enough, the forces of viscosity and surface tension will be important in balance of the forces determining oil spread. The forces of viscosity and surface tension depend on oil properties which can vary noticeably during a spreading process because both density and viscosity of oil are usually increasing due to its evaporation and dissolving in the water. As a result, in some moment of surface tension can reverse its sign and further spreading of an oil spot stops under action of this force. Afterwards, not unlike a spot of ordinary passive contaminant, an oil spot can increase its size under the action of the mechanical forces of moving water mass. Expanding of a spot is stopped as soon as the thickness of the oil firm approaches that of monomolecular film. While the processes of oil spread are proceeding, the physical and chemical transformation of oil pollutant are taking place as well, and water-oil emulsions or conglomerates are resulting, their density being higher or lower than that of the sea waters. Consequently, the former are sinking while the latter, in contrast, are floating on the ocean surface for a long time.*

One usually distinguishes three regimes in a process of oil spread on the ocean surface, namely [25]: inertial, gravity-viscous and surface-tension regimes. In the first regime, for a round oil spot of radius R and thickness h a pressure gradient is balanced by inertia, i.e.,

$$\frac{1}{\rho_0}\frac{\partial p}{\partial r} = u\,\frac{\partial u}{\partial r}, \tag{3.52}$$

where ρ_0 is oil density, r is the radius vector with the origin in a center of a spot, u is the radial component of velocity and p is pressure.

The right-hand side of equation (3.52) can be written approximately in the form

* Some more details on the transformation of oil and other pollutants in the ocean waters as well as on methods of their control can be found in the monograph [105] (see the chapter "Chemical Pollution in the Ocean and Its Control").

$$u \frac{\partial u}{\partial r} \approx \frac{R}{t} \frac{R}{Rt} = \frac{R}{t^2}, \tag{3.53}$$

and the left-hand side, correspondingly, in the form

$$\frac{1}{\rho_0} \frac{\partial p}{\partial r} = g \frac{\rho_w - \rho_0}{\rho_w} \frac{\partial h}{\partial r} \approx g \frac{\rho_w - \rho_0}{\rho_w} \frac{V}{R^3}, \tag{3.54}$$

where ρ_w is water density.

Equating (3.53) to (3.54) we now arrive at the following expression for R:

$$R = K_{2u} \left(\frac{\rho_w - \rho_0}{\rho_w} gV \right)^{1/4} t^{1/2}, \tag{3.55}$$

where $V = hR^2$ is a characteristic volume of oil spot, and K_{2u} is a constant of order of unity.

If oil pollution has the form of a stretched strip, rather than that of a round disk, similar considerations give a relation describing the increase of the pollutant-strip width, l, during the inertial regime of its evolution:

$$l = K_{1u} \left(\frac{\rho_w - \rho_0}{\rho_w} gA \right)^{1/3} t^{2/3}, \tag{3.56}$$

where A is a volume of oil contained in a strip of unit length and K_{1u} is a constant of order of unity.

The dependence of the oil-strip width on time, prescribed by equation (3.56), was verified in laboratory experiments. The non-dimensional width, $l/A^{1/2}$, is plotted, together with corresponding experimental points, against the non-dimensional time, $\{[(\rho_w - \rho_0)/\rho_w]gA^{-1/2}\}t^{1/2}$, in Figure 3.12.

In the gravity-viscous stage of the spread of an oil-pollutant spot one has to take into account viscous friction forces in a film. Friction force F per unit film mass can be written in the form

$$F_1 = \rho_w \nu \frac{\partial u}{\partial z} \Big/ (\rho_0 h), \tag{3.57}$$

where z is the vertical coordinate, and ν is the kinematic molecular viscosity of water.

Then, replacing derivative $\partial u/\partial z$ in equation (3.57) by the approximate expression $R/(t\delta)$ (where δ is a thickness of viscous boundary layer in the water which, as is known, is equal to $(\nu t)^{1/2}$) and using the approximation $\rho_w \simeq \rho_0$ we arrive at

$$F_1 \simeq \nu \frac{R}{t\delta} \frac{1}{h} = \nu^{1/2} R^3 V^{-1} t^{-3/2}. \tag{3.58}$$

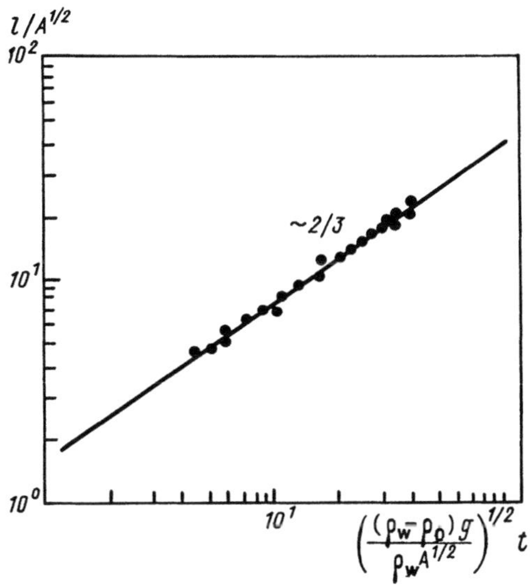

Fig. 3.12. Dependence of non-dimensional width of oil strip on non-dimensional time for inertial stage of spread (after [25]).

Equating pressure gradient to friction force, we can obtain a relation of R with t for the gravity-viscous stage of spot spread in the form

$$R = k_{2w}\left[\left(\frac{\rho_w - \rho_0}{\rho_w}\right) gV\nu^{-1/2}\right]^{1/6} t^{1/4}, \tag{3.59}$$

while for the one-dimensional case, correspondingly, we arrive at

$$l = k_{1w}\left[\left(\frac{\rho_w - \rho_0}{\rho_w}\right) gA^2\nu^{-1/2}\right]^{1/4} t^{3/8}, \tag{3.60}$$

where k_{2w} and k_{1w} are constants of order of unity.

The theoretical dependence of the non-dimensional width of an oil-pollutant strip is plotted in Figure 3.13 against non-dimensional time for the gravitational stage of spread along with experimental points confirming this dependence. A transition from the inertial stage of oil-pollutant spread to the gravity-viscous one takes place when the increasing thickness of the viscous boundary layer becomes equal, by an order of magnitude, to a thickness of oil layer, decreasing during the spreading process, i.e, when $\delta \sim h$.

The surface-tension force becomes a predominant one during the third stage of oil-pollutant spread. Surface-tension force F_2 acting on unit oil mass can be written as follows:

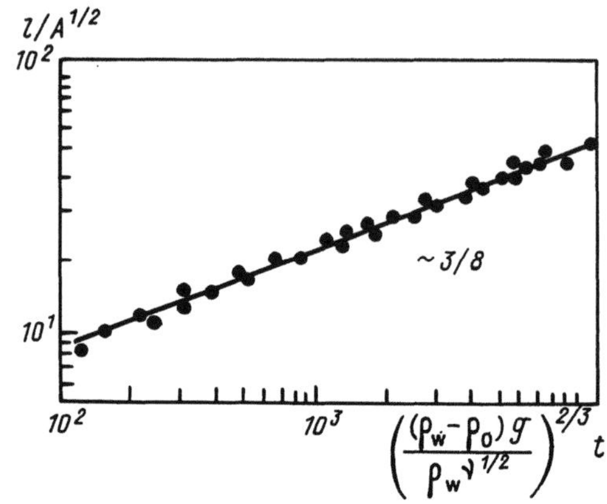

Fig. 3.13. Dependence of non-dimensional width of oil strip on non-dimensional time for the gravity-viscous stage of spread (after [25]).

$$F_2 = \sigma R/(\rho V), \tag{3.61}$$

where $\sigma = \sigma_1 - \sigma_2 - \sigma_3$ is the total surface tension; σ_1, σ_2 and σ_3 are the surface-tension coefficients at the water-air, water-oil and air-oil interfaces, respectively, and ρ is a typical value of density (usually, the difference between ρ_w and ρ_0 is within 10%).

A surface-tension force becomes significant when it is equal, by an order of magnitude, to a pressure gradient. Equating (3.61) to a right-hand side of (3.54) we find that this condition is satisfied when oil thickness h becomes equal to

$$h_k = \left[\sigma\left(\frac{\rho_w - \rho_0}{\rho_w}\right)^{-1} g^{-1}\rho^{-1}\right]^{1/2} \tag{3.62}$$

in the oil-spreading process. A typical value of σ is, in accordance to available data, about 0.025 N/m, and $(\rho_w - \rho_0)/\rho_0$, as was mentioned above, has a characteristic value of 0.1. For these parameters, a typical oil-film thickness h_k, at which the third stage of spread begins, is close to 0.5 cm.

Then, equating the right-hand sides of (3.58) and (3.61), we obtain the relation for oil-spot spread in the surface tension stage:

$$R = k_{2t}\left(\frac{\sigma^2 t^3}{\rho^2 \nu}\right)^{1/4}, \tag{3.63}$$

where K_{2t} is a coefficient of order of unity.

For the one-dimensional case of oil-strip spread in the surface-tension stage one can derive a formula for $l(t)$ that is similar to (3.63) but has another proportion coefficient. The laboratory experiments on oil-film spread during the surface-tension stage are compared to the theoretical relation in Figure 3.14,

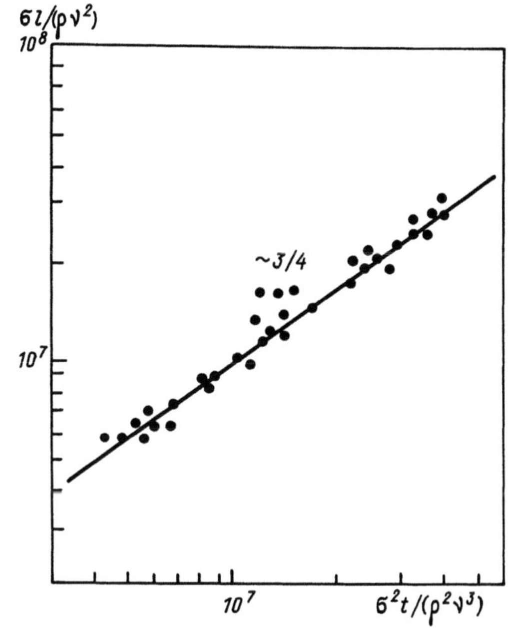

Fig. 3.14. Dependence of non-dimensional width of oil strip on non-dimensional time for regime of spread caused by the surface-tension forces (after [25]).

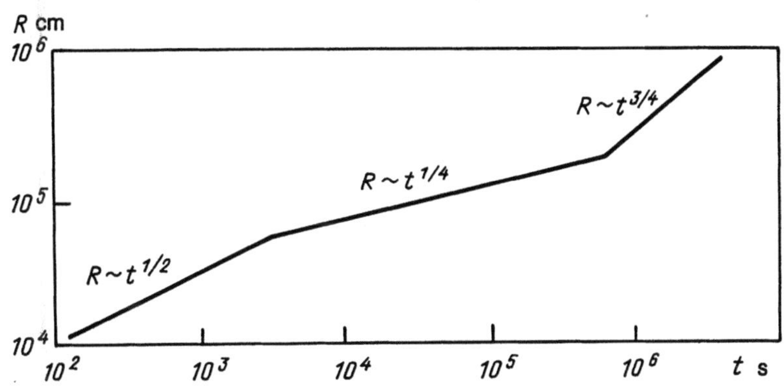

Fig. 3.15. Generalized dependence of radius of a spreading oil spot on time. A spot arose due to a powerful (10^4 tons) instant point source of oil (after [25]).

while a generalized plot for all stages of oil-spot spread from a point source of power 10^4 tons (which simulated the oil spill after the accident of the oil-tanker Torry Canyon off the coast of Great Britain) is shown in Figure 3.15.

The expansion of an oil-pollutant spot in a surface-tension stage is retarded in some time moment. Observations show that this happens when oil-film thickness is diminishing to $(2-3)\times 10^{-3}$ cm. This phenomenon is seemingly due to evaporation of the most volatile oil fractions. As far as evaporation of the volatile fraction occurs from the surface of oil film, a typical time of the process must be the same as a typical time of molecular diffusion of oil fractions across the film. And the latter time, as it follows from the solution of the one-dimensional molecular diffusion equation, is equal to $t = h^2/D$ where D is the molecular diffusion coefficient. Substituting this expression for t into (3.63) we obtain the following expression for the maximum radius of spot:

$$R_{\max} \sim \left(\frac{\sigma^3 h^6}{D^3 \rho^2 \nu}\right)^{1/4}, \tag{3.64}$$

and, correspondingly, for the maximum area of spot:

$$S_{\max} \sim \left(\frac{\sigma^2 V^6}{\rho^2 \nu D^3}\right)^{1/8}. \tag{3.65}$$

A general diagram of the different stages of spreading of an oil spot of volume V is shown in Figure 3.16. The boundaries between the stages are drawn in accordance with the above-given estimates of forces acting on an oil spot.

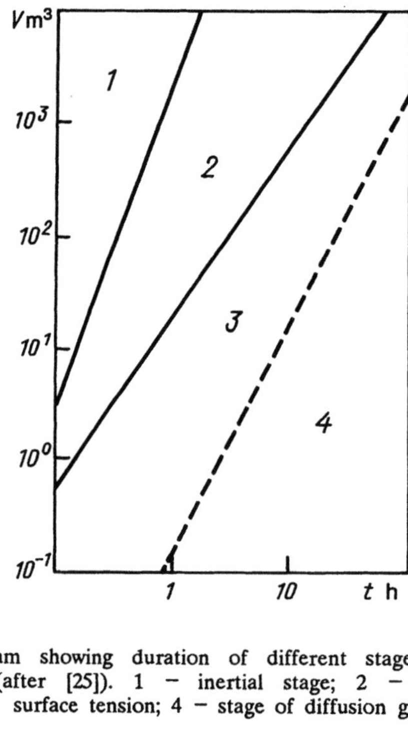

Fig. 3.16. Diagram showing duration of different stages of oil-spots spread as a function of its volume (after [25]). 1 – inertial stage; 2 – gravity-viscous stage; 3 – stage of surface tension; 4 – stage of diffusion growth of a spot size.

As was mentioned above, an oil film that has stopped its expanding caused by the surface-tension forces can further increase it area due to currents and tangential wind stress. Drifting under the action of wind and currents the oil spot usually deviates from the wind direction; a value of deviation reaches up to 20° clockwise in the northern hemisphere and counterclockwise in the southern one. The velocity of wind drift of an oil spot is on an average from 2 to 4% of the wind velocity. If the wind is changing rapidly both in time and space, an estimation of wind drift of a spot becomes difficult, and various statistical approaches can be used for solving the problem (see a review [25]).

When an oil film drifts under action of wind it extends along the drift direction or at some angle with respect to it. As shown in [38], the intensity of extending is directly proportional to both wind stress and film thickness and inversely proportional to the molecular oil-viscosity coefficient ν. The rate of this extending can reach several kilometers per day for oil-film thicknesses ranging from 1 mm to 1 cm and winds from 5 to 10 m/s [38].

Wind waves can affect oil-spot drift as well. As is known, the waves of finite amplitude can cause transport of either fluid particles or passive tracers in the direction of wave propagation. This transport, known as Stokes' drift, is usually small enough to be neglected in most of the sea diffusion problems. Nevertheless, as shown by laboratory experiments (see [25]), drift of the oil films caused by the surface waves can exceed the value of transport predicted by Stokes' theory. This excess is, seemingly, proportional to the product of wave steepness and wave-phase velocity with the coefficient of proportionality close to 0.445. It is interesting to note that if a spot is affected jointly by wind and waves, their individual effects are not merely superimposed but rather intricate interactions take place. Thus, laboratory experiments showed (see [25]) that for gentle winds (up to 80 cm/s) growth of wave steepness leads to increase of the spot drift due to growth of wave transport. But for stronger winds the sea waves can cause decrease of the spot drift; for small steepness the drift velocity decreases with the growth of steepness but for larger steepnesses the effect can become the opposite. The dependence of oil-spot drift velocity on wave steepness can be explained either by the increase of the rough sea surface covered by oil or by the more intricate effects of the interaction between wind and sea surface behind the steep crests of the waves that are blown along by the wind.

It should be noted, however, that the contribution of sea waves into total spot-drift velocity is, as usual, below 20% of wind drift and, consequently, his effect can be neglected in many cases.

During the final stages of oil-spot spread, especially after termination of the surface-tension regime, the turbulent-diffusion effects become significant in increasing the area of oil pollution. The parameters of this process, naturally, depend on the form of existence of the oil spot. Oil film can exist in the ocean in the form of a single spot, be broken into some individual and smaller slicks or present itself as a congestion of suspended particles of emulsion, the solid conglomerates, etc. A rather sharp boundary of pollution area is an important special feature of oil pollution in the ocean. Consequently, an approximation of the distribution of oil-pollutant concentration in the marginal zones of the spots by solutions of the parabolic diffusion equations can give rise to errors. At present all these questions are quite insufficiently studied, so the characteristic properties of oil-pollutant behavior is usually accounted for in the first stages of oil spread only, while the diffusion stage is considered within the scope of the routine diffusion models.

Experiments on oil-film diffusion were carried out in the Caspian Sea by Mamedov and Shirinov (see reference in [25]). In these experiments, some amount of oil (from 12 to 15 l) was spilled on the sea surface and the growth of polluted area S with time t was studied. The dependence $S(t)$ obtained in these experiments for the final (diffusion) stage of spot spread is shown in Figure 3.17. As one can see from the figure, for times exceeding approximately 10^3 s after the oil spill, the dependence $S(t)$ fits satisfactorily the cubic law which, as mentioned above, is the characteristic one for the inertial range of turbulence as well. It is interesting to note that the value of $t_0 = 10^3$ s is in rather good agreement with the time of termination of the surface-tension regime that can be estimated from Figure 3.16.

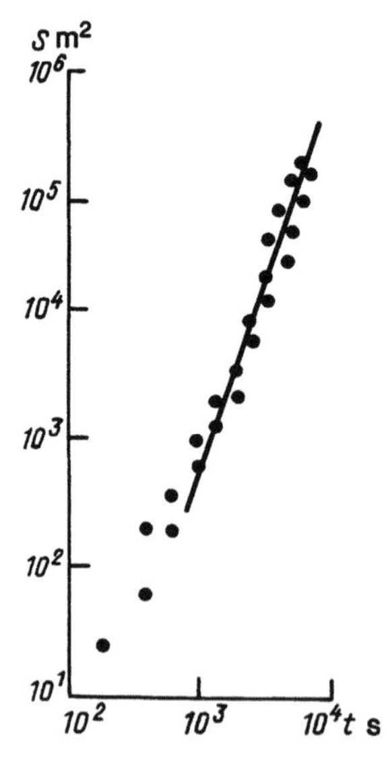

Fig. 3.17. Time variation of the oil-film area under action of turbulent diffusion according to the experiments by Mamedov and Shirinov in the Caspian Sea (after [25]).

If the oil-pollutant spot consists, mainly, of suspended oil particles, then vertical processes can play a significant part in the diffusion of oil as well. Intensity and the other features of these processes are dependent, naturally, on particle sizes or, in general case, on an oil-particle size distribution law. Under the suggestion that the downward turbulent flux of the particles is compensated by their floating up caused by a difference in density between oil particles and the water, we have

$$K_z \frac{\partial c_0(l, z)}{\partial z} = -w(l)\, c_0(l, z), \tag{3.66}$$

where $c_0(l, z)$ is the volume concentration of the oil particles of size l at depth z; and $w(l)$ is the velocity of floating up of the particles that is dependent on l.

It is natural to describe the velocity $w(l)$ by Stokes' law. In accordance with the latter, a spherical particle of size l is floating up in a more heavy medium at a velocity $w(l) = 0.086 \times [(\rho_w - \rho_0)/\rho_w](g/\nu)l^2$ cm/s. It can be suggested as well that the function $c_0(l, z)$ is represented approximately through a total volume concentration of the oil inclusions

$$c_c(z) = \int_{l_{min}}^{l_{max}} c_0(l, z)\mathrm{d}l,$$

maximum sizes, l_{max}, of the particles entering into the bulk of the water and size l_{min} of the minimal oil inclusions existing in the sea water (the particles of even smaller size rapidly dissolving in the water). Under these suggestions we can write $c_0(l, z) = c_c(z)/(l_{max} - l_{min})$. In this case equation (3.66) is reduced to the form

$$K_z\, \partial c_c(z)/\partial z = Ac_c(z), \tag{3.67}$$

where A is a constant with the dimension of velocity.

A solution of equation (3.67) has the form

$$c_c(z) = c_c(0) \exp(-z/z_0), \tag{3.68}$$

where $c_c(0)$ is the value of concentration $c_c(z)$ at the ocean surface, and $z_0 = K_z/A$ is the typical scale of variation with depth of oil-inclusion concentration. In accordance with available estimates (see [25]), this scale has a value from 15 to 20 m and, consequently, the suspended oil inclusions can penetrate down into the ocean to those depths. As for the oil fractions dissolved in the water, they probably can diffuse down into the bulk of the ocean waters to the much deeper layers.

3.4. Diffusion of Contaminant from Distributed Sources in the Ocean with Vertical Fine-Structure

Some interesting effects related to the vertical fine-structure of density and velocity fields in the ocean can be observed during vertical diffusion of contaminants fallen out on the ocean surface. The fine-structure was discovered not so long ago and, according to observations, it is widely distributed in the oceans, significantly affecting many of the hydrophysical, chemical and biological oceanic processes. Provided vertical probing is carried out with a low-inertia meter in the fine-layered ocean, some 'steps' are observed in the records, their vertical sizes ranging from centimeters to dozens of meters. Vast experimental information has been collected until now on the main parameters of these step-structures [67], and the main mechanisms of their generation have been studied as well. The most important of the latter are the following: kinematic effects of internal waves, local effects of turbulent mixing, effects of molecular salt diffusion and molecular heat conduction ('double-diffusion' effects). The forms of the step-structures are related to the above-mentioned mechanisms in some way. In particular, if local events of turbulence generation occur in the ocean body, the 'steps' arise due to turbulent mixing, and their sizes are related to the amount of turbulent energy released during the event of turbulence generation [85].

As, in most cases, the stable density stratification of the waters exists in the ocean body, some amount of energy has to be expended on the mixing of the water masses to give rise to a 'step'. As far as the thickness of a step is comparatively small, an initial gradient of water density $\gamma = \partial\rho/\partial z$ can be considered as a constant within a step. The energy expended on mixing is, evidently, equal to the change of potential energy needed for the mixing of a water column of unit cross-section and height h:

$$E = g\gamma h^3/12, \tag{3.69}$$

while the expenditure of energy needed for the mixing of a unit volume of fluid with a mass of 1 g is

$$E_0 = g\gamma h^2/(12\ \rho) = h^2N^2/12, \tag{3.70}$$

where $N = (g\gamma/\rho)^{1/2}$ is the Väisälä – Brunt frequency, ρ is the mean density of fluid in a mixed volume, and g is the acceleration due to gravity.

Thus, the turbulence-generation mechanism has to release at least $g\gamma h^2/(12\rho)$ energy units per one gram of fluid to give rise to a mixed layer of thickness h in a stratified medium. Generally speaking, some larger amount of energy is released in reality, which is partly expended overcoming the buoyancy forces and is partly converted into kinetic turbulent energy. It is natural to suppose that the energy is distributed in equal shares, on the average, between these two processes during generation events. If so, specific turbulent kinetic energy in the just-arisen turbulent layer can be determined from equation (3.70). The turbulent energy dissipation rate in that layer, ε, can be determined from the following considerations. As shown in [81], the maximal vertical size of eddy in a turbulent layer of stratified fluid can be determined from the expression

$$h = c^{3/2} Ri_{cr}^{3/4} \varepsilon^{1/2} N^{-3/2}, \tag{3.71}$$

where Ri_{cr} is the critical Richardson number at which turbulent motions are capable of overcoming the buoyancy forces in a layer of thickness h; and c is a universal constant in the "–2/3 power law" for structural functions of a locally isotropic turbulence.

From equation (3.71) we obtain

$$\varepsilon = c^{-3} Ri_{cr}^{-3/2} h^2 N^3. \tag{3.72}$$

At the same time, the following relation holds true by definition: $\varepsilon = \overline{a\nu(\partial u/\partial z)^2}$, where ν is the kinematic molecular viscosity of water, a is some numerical coefficient, and the overbar denotes averaging. If the derivative $\partial u/\partial z$ in the latter expression is replaced by the derivative of velocity averaged over the scale h, $\partial u/\partial z = \Gamma$, then for lower estimate of ε we obtain

$$\varepsilon = a\nu\Gamma^2. \tag{3.73}$$

If Γ^2 in equation (3.73) is then replaced by $Ri_{cr}^{-1} N$, we arrive at

$$\varepsilon = a\nu N^2 Ri_{cr}^{-1}. \tag{3.74}$$

Having divided the expression (3.70) for E_0 by the turbulent energy dissipation rate, ε, determined by the formulas (3.72) – (3.74), we obtain the three expressions for the typical time of existence of turbulence in a layer:

$$\tau = c^3 Ri_{cr}^{3/2}/(12\ N); \tag{3.75}$$

$$\tau = h^2 N^2/(12\ \varepsilon) = h^2 N^2/(12\ a\nu\Gamma); \tag{3.76}$$

$$\tau = Ri_{cr} h^2/(12\ a\nu). \tag{3.77}$$

Employing the formulas (3.73) and (3.74) for ε, two more expressions for the typical vertical scale of a turbulent layer can be derived from (3.71):

$$h = \sqrt{a}\ c^{3/2}\ Ri_{cr}^{3/4}\ \nu^{1/2}\ \Gamma\ N^{-3/2}; \tag{3.78}$$

$$h = \sqrt{a}\ c^{3/2}\ Ri_{cr}^{1/4}\ \nu^{1/2}\ N^{-1/2}. \tag{3.79}$$

To determine typical values of the vertical turbulent transport coefficient, K_z, in the turbulent layers, we employ the "4/3 power law" of locally isotropic turbulence:

$$K_z = c_1\ \varepsilon^{1/3}\ h^{4/3}, \tag{3.80}$$

where c_1 is a universal constant.

Using the expression (3.71) for h, we reduce (3.80) to the form

$$K_z = c^2\ c_1\ Ri_{cr}\ \varepsilon\ N^{-2}. \tag{3.81}$$

If the value of ε in the formula (3.80) is substituted by its expressions from (3.74) and (3.72), we obtain two more equations for K_z:

$$K_z = \sqrt[3]{a}\ c_1\ \nu^{1/3}\ Ri_{cr}^{-1/3}\ h^{4/3}\ N^{2/3}; \tag{3.82}$$

$$K_z = c^{-1}\ c_1\ Ri_{cr}^{-1/2}\ h^2\ N. \tag{3.83}$$

So the values of ρ and γ (or N) are, as usual, known from observations in the ocean, while the fine-structure velocity gradient, Γ, is not so frequently measured; it is more convenient to employ the formulas (3.72) and (3.74) to estimate the value of ε, the formulas (3.75) and (3.77) to estimate τ and the formulas (3.83) and (3.82) to estimate K_z. The first of the mentioned formulas involve no molecular parameters while the second ones do involve these parameters. Consequently, the first formulas describe the initial stage of development of the turbulent layers when the buoyancy force is the main one in the balance of forces. The second formulas correspond to the viscous stage of development of turbulent formations in the ocean, when the molecular forces are of significant importance.

The values of ε, τ and K_z estimated at several values of N and h are given in Table III.II, borrowed from [85]. Estimations were carried out either employing formulas with no molecular coefficients (the first figures in each of the lines) or taking into account viscous forces (the second figures). The following physical parameters and universal constants were used in the calculations $\rho = 1$ g/s^3, $g = 980$ cm^2/s, $\nu = 1.5 \cdot 10^{-2}$ cm^2/s, $Ri_{cr} = 0.25$, $a = 15$, $c = 1$ and $c_1 = 0.1$. The values of turbulence parameters given in a table are in agreement with available experimental data [67]. As one would expect for the viscous stage of

TABLE III.II

The parameters of turbulence in the turbulent layers of different thickness, h, in stratified fluids with different Väisälä–Brunt frequency, N

Parameter	N, cm^{-1}	h, cm: 10	10^2	10^3
ε, cm^2/s^3	10^{-4}	$8 \cdot 10^{-10}$	$8 \cdot 10^{-8}$	$8 \cdot 10^{-6}$
		$9.0 \cdot 10^{-9}$	$9.0 \cdot 10^{-9}$	$9.0 \cdot 10^{-9}$
	10^{-3}	$8 \cdot 10^{-7}$	$8 \cdot 10^{-8}$	$8 \cdot 10^{-3}$
		$9 \cdot 10^{-7}$	$9 \cdot 10^{-7}$	$9 \cdot 10^{-7}$
	10^{-2}	$8 \cdot 10^{-4}$	$8 \cdot 10^{-2}$	8
		$9 \cdot 10^{-5}$	$9 \cdot 10^{-5}$	$9 \cdot 10^{-5}$
τ, s	10^{-4}	10^2	10^2	10^2
		9.3	$9.3 \cdot 10^2$	$9.3 \cdot 10^4$
	10^{-3}	10	10	10
		9.3	$9.3 \cdot 10^2$	$9.3 \cdot 10^4$
	10^{-2}	1	1	1
		9.3	$9.3 \cdot 10^2$	$9.3 \cdot 10^4$
K_z, cm^2/s	10^{-4}	$2.0 \cdot 10^{-3}$	$2.0 \cdot 10^{-1}$	$2.0 \cdot 10$
		$4.5 \cdot 10^{-3}$	$9.6 \cdot 10^{-2}$	2.1
	10^{-3}	$2.0 \cdot 10^{-2}$	2.0	$2.0 \cdot 10^2$
		$2.1 \cdot 10^{-2}$	$4.5 \cdot 10^{-1}$	9.6
	10^{-2}	$2.0 \cdot 10^{-1}$	$2.0 \cdot 10$	$2.0 \cdot 10^3$
		$9.6 \cdot 10^{-2}$	2.1	$4.5 \cdot 10$

the 'life' of the turbulent layers, the values of ε and K_z appear as a rule to be smaller, and of parameter τ larger, than those for the initial stage of the layers' existence. That rule is violated only for τ and K_z at a rather small value of the Väisälä – Brunt frequency (10^{-4} s^{-1}), in the thin turbulent layers (h = 10 cm). For $N = 10^{-2}$ s^{-1} and h = 10 cm, the values of K_z become practically equal to each other for both stages of the turbulent layers. For $N = 10^{-4}$ s^{-1} and h = 10 cm, a value of K_z turns out to be smaller than molecular viscosity which, to all appearances, gives some evidence of the inapplicability of the derived expressions under those extreme conditions.

Korotenko [49] has used analytical methods to study the features of contaminant diffusion in the vertically layered ocean. He supposed that the ocean body can be presented as consisting of n quasi-homogeneous layers of thickness h_J, in each of which the turbulence is different and is characterized by the coefficients K_z^J and K_x^J. Each of the layers is of infinite extent in the x-direction (see Figure 3.18) and has its own value of horizontal current velocity, u_J. It is supposed that the velocity is discontinuous at the boundaries of the layers while at the interface separating two neighboring layers the value of velocity is equal to a half-sum of its values in adjacent layers. The contaminant diffusion equation for each of the layers (provided the effects of contaminant transport in the other horizontal y-direction are neglected) has the form

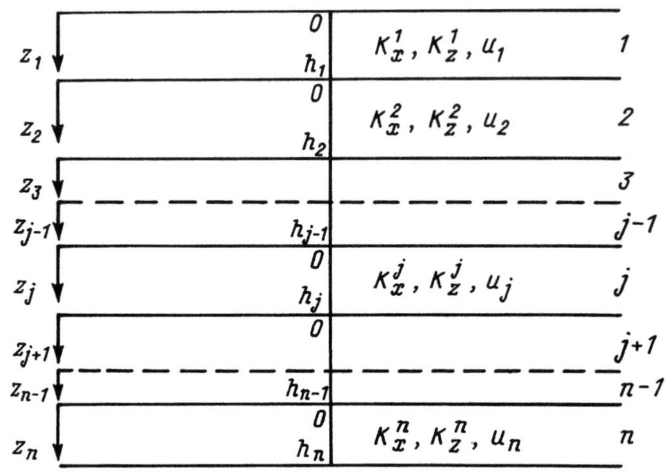

Fig. 3.18. Schematic picture of the layered structure of the ocean waters. Each of the layers has its own current velocity, u_J, and diffusion coefficients K_x^J and K_z^J (after [49]). J is the layer number, h_J is the thickness of the J-th layer, z_J is the vertical coordinate in the J-th layer.

$$\partial c_j/\partial t + u_j \partial c_j/\partial x = K_x \partial^2 c_j/\partial x^2 + K_z^j \, \partial^2 c_j/\partial z_j^2, \tag{3.84}$$

where $j = 1, 2, ...; n$ is the number of a layer; and all other notations are specified above.

Then, the initial distribution of contaminant concentration in the layers, $c_j(x, z_j, 0)$, is supposed to be presented in the form

$$c_j(x, z_j, 0) = \Phi(x, z_j), \tag{3.85}$$

where $\Phi(x, z_j)$ is a known function.

The boundary conditions for each of the layers are written as follows:

$$K_z^j \left. \frac{\partial c_j}{\partial z_j} \right|_{z_j=0} = -\mu_j(x, 0, t); \tag{3.86}$$

$$K_z^j \left. \frac{\partial c_j}{\partial z_j} \right|_{z_j=h_j} = -\mu_j(x, h_j, t); \tag{3.87}$$

where μ_j are contaminant fluxes through the boundaries of the layers. Contaminant fluxes through the outer boundaries of both the first and the last of the layers are supposed to be absent, i.e., $\mu_1(x, 0, t) = \mu_n(x, 0, t) = 0$.

To resolve the problem (3.84) – (3.87), Korotenko employs the integral Fourier transform technique which, under certain conditions imposed on the kernel of the transformation, makes it possible to eliminate the derivatives along the vertical coordinate in the equation. Finally, the general solution describing the concentration of substance in the j-th layer was obtained in the following form:

$$c_j(x, z_j, t) = \frac{2}{h_j} \sum_{l=0}^{\infty} \exp\left[-K_z^j\left(\frac{\pi l}{h_j}\right)^2 t \cos\frac{\pi l}{h_j} z_j \times\right. \tag{3.88}$$

$$\times \left\{\int_{-\infty}^{\infty}\int_0^{h_j} \frac{\exp[-(x-u_j t-\xi)^2/(4K_x^j\, t)]}{(4\pi K_x^j t)^{1/2}} \cos\frac{\pi l}{h_j}\, \xi_j\, \Phi(\xi, \xi_j)\, d\xi d\xi_j + \right.$$

$$+ \int_{-\infty}^{\infty}\int_0^{t} [(-1)^l \mu_j(x, h_j, \tau) - \mu_j(x, 0, \tau)] \exp\left[K_z^j \left(\frac{\pi l}{h_j}\right)^2 \tau\right] \times$$

$$\times \left.\frac{\exp\{-(x-x'-u_j(t-\tau))^2/[4K_z^j(t-\tau)]\}}{[4\pi K_x^j(t-\tau)]^{1/2}}\, dx'd\tau\right\},$$

where ξ, ξ_j, x' and τ are integration variables.

The first double integral in the expression (3.88) for the concentration of substance in the j-th layer describes the effect of the initial conditions on the forming of concentration field, while the second one describes the influence of interchange of substance between the layers. The shape of the vertical profile of contaminant concentration over all the layers depends on the individual contribution of each of these terms to the total balance of substance in a given layer. At certain values of the parameters involved in the solution (3.88), profiles of the 'inversion' type may arise, in which the concentration along the z-direction has maxima and minima. To demonstrate the influence of individual factors on the formation of the vertical profiles of contaminant concentration, Korotenko [49] has considered a problem with an instantaneous linear (in the x-direction) source of contaminant of intensity Q_0, i.e., under the following boundary conditions:

$$c_j(x, z_j, 0) = Q_0\delta(x), \quad j = 1, 2, ..., n, \tag{3.89}$$

where $\delta(x)$ is the delta function.

Under the initial conditions (3.89) the solution (3.88) is simplified in to the form

$$c_j(x, z_j, t) = Q_0 \frac{\exp[-(x-u_j t)^2/(4K_x^j t)}{(4\pi K_x^j t)^{1/2}} +$$

$$+ \frac{2}{h_j} \sum_{l=0}^{\infty} \exp\left[-K_z^j\left(\frac{\partial l}{h_j}\right)^2 t\right] \cos\frac{\pi l}{h_j} z_j \int_{-\infty}^{\infty}\int_0^{\tau} [(-1)^l \mu_j(x, h_j, \tau) -$$

$$- \mu_j(x, 0, \tau)] \exp[K_z^j\left(\frac{\pi l}{h_j}\right)^2 \tau\Big] \times$$

$$\times \frac{\exp\{-(x-x'-u_j(t-\tau))^2/[4K_x^j(t-\tau)]\}}{[4\pi K_x^j(t-\tau)]^{1/2}} \mathrm{d}x'\mathrm{d}\tau. \qquad (3.90)$$

As one can readily see, the first term of this expression is the fundamental solution of one-dimensional diffusion equation in a frame of reference moving at velocity u_j. In each of the layers the contaminant is transported from the source along the direction of the x-axis at different velocities u_j. This leads to a step-structure in the vertical profiles of concentration. At the same time, the second term in (3.90), describing the exchange of contaminant between the layers, gives rise to some 'smoothing' of the vertical profiles.

To estimate quantitatively the contaminant concentration from the solutions (3.90) or (3.88) one has to introduce additional hypotheses on the vertical fluxes of contaminant, μ_j, through the boundaries of the layers. The most natural is to suppose that the fluxes are proportional to the difference of the contaminant concentration in the vicinity of the interface of adjacent layers, i.e.,

$$\begin{aligned} \mu_j(x, 0, t) &= \alpha_{j-1,j}[c_{j-1}(x, h_{j-1}, t) - c_j(x, 0, t)]; \\ \mu_j(x, h_j, t) &= \alpha_{j,j+1}[c_j(x, h_j, t) - c_{j+1}(x, 0, t)], \end{aligned} \qquad (3.91)$$

where parameters α describe a permeability of the layer boundaries.

With the expressions (3.91) the solution (3.90) takes the form

$$c_j(x, z_j, t) = Q_0 \frac{\exp[-(x-u_jt)^2/(4K_x^jt)]}{(4\pi K_x^jt)^{1/2}} +$$

$$+ \frac{2}{h_j} \sum_{i=0}^{\infty} \exp\left[-K_z^j\left(\frac{\pi l}{h_j}\right)^2 t\right] \cos\frac{\pi l}{h_j} z_j \int_{-\infty}^{\infty}\int_0^{t} \Big\{(-1)^l \alpha_{j,j+1} \times$$

$$\times \exp\left[K_z^j\left(\frac{\pi l}{h_j}\right)^2 \tau\right] [c_j(x, h_j, \tau) - c_{j+1}(x, 0, \tau)] - \qquad (3.92)$$

$$- \alpha_{j-1,j}[c_{j-1}(x, h_{j-1}, \tau) - c_j(x, 0, \tau)]\Big\} \times$$

$$\times \frac{\exp\{-(x-x'-u_j(t-\tau))^2/[4K_x^j(t-\tau)]\}}{[4\pi K_x^j(t-\tau)]^{1/2}} \mathrm{d}x'\mathrm{d}\tau.$$

The distribution of contaminant concentration in the plane (x, z) can be estimated from (3.92), if the diffusion coefficients, K_z^j and K_x^j, and the coefficients of permeability, α, are given.

If the velocity differences, Δu, in the layers are high enough, the vertical diffusion of contaminant 'has not time' to smooth significantly the vertical profiles of the contaminant concentration. In this case, to examine qualitatively the shape of these profiles, one can neglect the vertical exchange of substance through the boundaries of layers, i.e., let $\alpha_{j-1,j} = 0$. The vertical profiles of contaminant concentration estimated from the solutions (3.92) under the above suggestion at the distance of 180 m from a linear source are given in Figures 3.19a and b. The vertical profiles of the horizontal-velocity component used in these estimations were measured by Kullenberg (see ref. in [49]) during diffusion experiments in the Baltic Sea. In the first case (Figure 3.19a), the concentration profile is of the inversion type, and in the second case (Figure

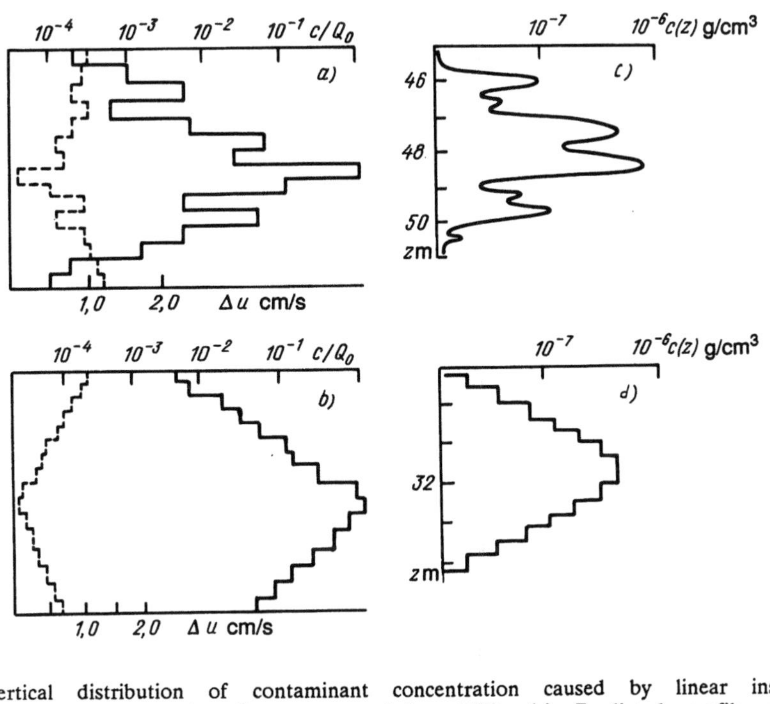

Fig. 3.19. Vertical distribution of contaminant concentration caused by linear instantaneous source in the ocean with fine-structure (after [49]). (a) Predicted profile of inversion form of contaminant concentration at distance $x = 180$ m from a source for diffusion time $t = 1$ h. (b) Predicted profile of step-like form at the same values of x and t. Deviations of the current velocity in the layers from the mean value, 5 cm/s, are shown by the dashed lines. (c) Profile of dye concentration (of inversion form) observed in the sea. (d) Profile of dye concentration (of step-like form) observed in the sea.

3.19b), the profile is step-like. The vertical profiles of dye concentration released to simulate a linear (along the z-axis) source were measured in the sea, and they appeared to be very close in shape to the theoretical profiles (3.19c). As is seen from the figure, the observed profiles are also either of inversion or step shape, though the boundaries between the layers are 'smoothed' some more. This discrepancy between the estimations and the observed data is, naturally, due to neglect of the effect of the contaminant vertical exchange in the solution (3.92).

To determine the vertical fluxes of contaminant in the ocean with fine-structure, Korotenko [50] has considered a problem of contaminant diffusion in two layers that are semi-indefinite in the vertical direction and separated by horizontal interface at $z = 0$. Each of the layers is characterized by its horizontal velocities of the water motion (along the x-axis), u_1 and u_2, and the coefficient of vertical turbulent transport, K_z^1 and K_z^2. The problem is supposed to be a steady one and horizontal diffusion is neglected. Under the above suggestions, the diffusion equations for each of the layers (identified by indexes 1 and 2, respectively) are written in the form

$$\begin{aligned} u_1\ \partial c_1/\partial x &= K_z^1\ \partial^2 c/\partial z^2, \quad x \geqslant 0, \quad z \geqslant 0; \\ u_2\ \partial c_2/\partial x &= K_z^2\ \partial^2 c/\partial z^2, \quad x \geqslant 0, \quad z \leqslant 0. \end{aligned} \tag{3.93}$$

The boundary conditions for equations (3.93) are taken in the form

$$c_1(0,\ z) = f_1(z); \quad c_2(0,\ z) = f_2(z) \quad \text{for } x = 0; \tag{3.94}$$

$$c_1(x,\ +0) = c_2(x,\ -0); \tag{3.95}$$

$$K_z^1\ \partial c_1(x,\ +0)/\partial z = K_z^2\ \partial c_2(x,\ -0)/\partial z = \mu(x) \quad \text{for } z = 0; \tag{3.96}$$

$$c_1(x,\ z) = c_2(x,\ z) = 0 \quad \text{for } x \longrightarrow \infty; \tag{3.97}$$

$$\partial c_1(x,\ z)/\partial z = \partial c_2(x,\ -z)/\partial z = 0 \quad \text{for } z \longrightarrow \infty. \tag{3.98}$$

The conditions (3.95) – (3.96) mean continuity of concentration in the z-direction and continuity of the contaminant flux through the boundary between the layers; the conditions (3.97) – (3.98) mean that both concentration and fluxes of the contaminant tend to zero with increasing of the distance, x and z, from the origin of coordinates to the observation point. $\mu(x)$ denotes the sought-for contaminant flux through the interface between the layers.

The general solution of the problem (3.93) – (3.98) is as follows [50]:

$$c_1(x, z) = \left(4\pi \frac{K_z^1}{u_i}\right)^{-1/2} \int_0^\infty f_1(\zeta) \left\{\exp\left[-\frac{u_1(z + \zeta)^2}{4K_z^1 x}\right] + \right.$$

$$\left. + \exp\left[-\frac{u_1(z - \zeta)^2}{4K_z^1 x}\right]\right\} d\zeta - (K_z^1 u_1)^{-1/2} \int_0^x [\pi(x-\xi)]^{-1/2} \mu(\xi) \times$$

$$\times \exp\left[-\frac{z^2 u_1}{4K_z^1(x - \xi)}\right] d\xi \quad \text{for } x \geqslant 0,\ z \geqslant 0;$$

$$c_2(x, z) = \left(4\pi \frac{K_z^2}{u^2}\right)^{-1/2} \int_0^\infty f_2(\zeta) \left\{\exp\left[-\frac{u_2(z + \zeta)^2}{4K_z^2 x}\right] + \right.$$

$$\left. + \exp\left[-\frac{u_2(z - \zeta)^2}{4K_z^2 x}\right]\right\} d\zeta + (K_z^2 u_2)^{-1/2} \int_0^x [\pi(x-\xi)]^{-1/2} \mu(\xi) \times$$

$$\times \exp\left[-\frac{z^2 u_2}{4K_z^2(x - \xi)}\right] d\xi \quad \text{for } x \geqslant 0,\ z \leqslant 0, \tag{3.99}$$

where ζ and ξ are the integration variables.

Using the boundary condition (3.95) we obtain the Volterra integral equation of the first kind for the desired function

$$F(x) = \int_0^x [\pi(x-\xi)]^{-1/2} \mu(\xi)\, d\xi, \tag{3.100}$$

where

$$F(x) = \frac{\int_0^\infty \left\{\left[f_1(\xi) \exp\left(-\frac{\xi^2 u_1}{4K_z^1 x}\right)\right] \Big/ \left(\pi \frac{K_z^1}{u_1} x\right)^{1/2}\right\} d\zeta}{(K_z^1 u_1)^{-1/2} + (K_z^2 u_2)^{-1/2}} - $$

$$- \frac{\int_0^\infty \left\{\left[f_2(\zeta) \exp\left(-\frac{\zeta^2 u_2}{4K_z^2 x}\right)\right] \Big/ \left(\pi \frac{K_z^2}{u_2} x\right)^{1/2}\right\} d\zeta}{(K_z^1 u_1)^{-1/2} + (K_z^2 u_2)^{-1/2}} \tag{3.101}$$

As an example, a case of continuous point source of intensity Q_0 placed in the upper layer at the point $(0, z_0)$ was considered in [50]. In this case equation (3.100) takes the form

$$Q_0(\pi x)^{-1/2} \exp\left(-\frac{z_0^2 u_1}{4K_z^1 x}\right) = \frac{[u_1^2 (K_z^2 u_2)]^{-1/2}}{[(K_z^1 u_1)^{1/2} + (K_z^2 u_2)^{1/2}]^{-1}} \times$$

$$\times \int_0^x [\pi(x-\zeta)]^{-1/2} \mu(\xi)\, d\xi. \tag{3.102}$$

Equation (3.102) was solved in [50] by the operational method. As a result, the following expression for $\mu(x)$ was obtained:

$$\mu(x) = \frac{Q_0[(K_z^1 u_1)^{1/2} + (K_z^2 u_2)^{1/2}]^{-1} z_0 u_1^{3/2}}{2(K_z^1)^{1/2}(K_z^2 u_2 \pi)^{-1/2} x^{3/2}} \exp\left(-\frac{z_0^2 u_1}{4K_z^1 x}\right). \tag{3.103}$$

This expression shows that the differences of flow velocity in the individual ocean layers can affect significantly the vertical fluxes of the contaminant. Note that if the contaminant source is placed in the upper layer the parameters K_z^1 and u_1 have a stronger influence on the value of contaminant flux than the parameters of the lower layer, K_z^2 and u_2. The distance from the source of contaminant to the interface is, naturally, also rather important. If a great many of the layers with different hydrophysical properties exist in the ocean, the expression for the vertical contaminant flux become more complicated, but general regularities of transformation of the fluxes at the boundaries between the layers remain similar to the above-considered case of a two-layer ocean.

The problem becomes more complicated if some interlayer of thickness h, with significant gradients of both flow velocity and water density, rather than a sharp interface, exists between the homogeneous half-spaces. Apparently, the vertical diffusion coefficient, K_z, in this interlayer depends noticeably on the above-mentioned gradients, so the coefficient K_z must, naturally, be considered as some function of the Richardson number, Ri. Korotenko [50] suggested for the present problem the following expression for K_z:

$$K_z = [(K_0 - \nu)(1 + aRi)^b + \nu), \tag{3.104}$$

where K_0 is the coefficient of vertical turbulent diffusion at $Ri = 0$; ν is the coefficient of molecular diffusion; and a and b are some constants.

For constant contaminant concentration at the upper boundary of the interlayer ($z = 0$) we shall use the following set of two non-stationary equations for the description of contaminant diffusion in the interlayer and the lower half-space:

$$\partial c_1(z, t)/\partial t = K_z \partial^2 c_1(z, t)/\partial z^2 \quad \text{at} \quad t > 0,\ 0 \leqslant z \leqslant h; \tag{3.105}$$

$$\partial c_2(z, t)/\partial t = K_0 \partial^2 c_2(z, t)/\partial z^2 \quad \text{at} \quad t > 0,\ h \leqslant z < \infty; \tag{3.106}$$

under the initial conditions

$$c_1(z, 0) = c_2(z, 0) = 0 \tag{3.107}$$

and the boundary conditions

$$c_1(h, t) = c_2(h, t); \quad K_z \partial c_1(h, t)/\partial z = K_0 \partial c_2(h, t)/\partial z; \tag{3.108}$$

$$c_1(0, t) = c_0 = \text{const}; \quad c_2(\infty, t) = 0, \tag{3.109}$$

where index "1" refers to the interlayer and index "2" to the lower half-space.

Solution of the problem (3.105) – (3.109) was obtained by Korotenko with the help of the operational method and written in the form

$$c_1(z, t) = c_0 \sum_{n=1}^{\infty} d^{n-1} \left[\text{erf}\left(\frac{2(n-1)h + z}{2\sqrt{K_z t}} \right) - d\, \text{erf}\left(\frac{2nh - z}{2\sqrt{K_z t}} \right) \right]; \tag{3.110}$$

$$c_2(z, t) = c_0(1-d) \sum_{n=1}^{\infty} d^{n-1} \left[\text{erf}\left(\frac{z-h}{\sqrt{K_0}} + \frac{(2n-1)h}{\sqrt{K_z}} \right) \frac{1}{2\sqrt{t}} \right], \tag{3.111}$$

where

$$d = (1 - \sqrt{K_z/K_0})/(1 + \sqrt{K_z/K_0}).$$

The expression for the contaminant flux through the lower boundary of the interlayer can be derived from the solutions (3.110) – (3.111) in the form

$$q(h, t) = c_0 \left(\frac{K_0}{\pi t} \right)^{1/2} \frac{[(K_0 - \nu)(1 + aRi)^b + \nu]^{1/2}}{K_0^{1/2} + [(K_0 - \nu)(1 + aRi)^b + \nu]^{1/2}} \times$$

$$\times \sum_{n=1}^{\infty} \left\{ \frac{K_0^{1/2} - [(K_0 - \nu)(1 + aRi)^b + \nu]^{1/2}}{K_0^{1/2} + [(K_0 - \nu)(1 + aRi)^b + \nu]^{1/2}} \right\}^{n-1} \times \tag{3.112}$$

$$\times \exp\left[- \frac{(2n-1)^2 h^2}{4t[(K_0 - \nu)(1 + aRi)^b + \nu]} \right].$$

At $Ri \to 0$ the solutions (3.110) – (3.112) are reduced to the following simple formulas for the concentration and flux of the contaminant diffusing in homogeneous half-space:

$$c(z, t) = c_0\, \text{erf}\left(\frac{z}{2\sqrt{K_0 t}} \right);$$

$$q(h, t) = c_0 \left(\frac{K_0}{\pi t} \right)^{1/2} \exp\left(- \frac{h^2}{4K_0 t} \right). \tag{3.113}$$

In the case of $Ri \to \infty$, the contaminant transport through the lower boundary of the interlayer approaches the molecular one. The most interesting case is

$aRi \gg 1$ but $K_0(aRi)^b \gg \nu$. In this case we arrive at the approximate expression for the contaminant flux:

$$q(h,\ t) \approx c_0\left(\frac{K_0 a^b Ri^b}{\pi t}\right)^{1/2} \exp\left[-\frac{h^2}{4K_0 t\,(aRi)^b}\right] =$$

$$= c_0\left(\frac{K_0 a^b Ri^b}{\pi t}\right)^{1/2} \exp\left[-\frac{h^{2-b}}{4K_0 t\left(\dfrac{\rho a\ \ \Delta\rho}{\rho_0(\Delta u)^2}\right)^b}\right], \tag{3.114}$$

where the Richardson number, Ri, is expressed in terms of the differences of velocity, Δu, and water density, $\Delta\rho$, in the interlayer. As follows from the last equation for $q(t)$, a typical time of contaminant diffusion through the interlayer,

$$t_0 = \frac{h^{2-b}}{4K_0}\left[\frac{\rho_0(\Delta u)^2}{gd\Delta\rho}\right]^b,$$

grows with the increasing of $\Delta\rho$ and h but falls with the increasing of Δu. Contaminant flux through the lower interlayer boundary increases by a factor of $2^{-b}\exp 2^{-b}$ with a twofold growth of Δu, and the flux increases by a factor of $2^{-b/2}\exp(1-2^b)$ with a twofold fall of $\Delta\rho$. A twofold reduction of interlayer thickness h causes the flux to increase by a factor of $[2^{-b/2}\exp(2^{2-b}+1)$. These examples clearly show that layering of the ocean waters significantly affects redistribution of a contaminant. A change of layering parameters can result in noticeable variations in contaminant-concentration distribution and contaminant fluxes.

In practice, when employing the above-given expressions to estimate the concentration and flux of a contaminant in the ocean with fine-structure, the difficulty arises in choosing the diffusion coefficients for each of the layers under consideration. For noticeable differences in velocity and density in the layers, the formula (3.104) can be used to determine K_z. But if the values of Δu in the layers are low enough or, as is often the case, there is no information on Δu at all, to estimate the values of K_z one can use the formulas (3.82) – (3.83), which require as input parameters only the thickness of the interlayer and the density gradient therein. Certainly, one has to bear in mind that the formulas (3.82) – (3.83) were derived under the suggestion that the turbulence is stationary in each of the layers and its features are not varied in time. Yet, in fact, the events of turbulence generation in the ocean body occur in some localized areas of space and in some time instants, with turbulence decaying within time intervals separating those instants. Consequently, in any fixed instant there can exist in the ocean the layers (or patches) of fluid with either developed or decaying turbulence inside them, as well as layers with nearly laminar flow. In such cases, naturally, the formulas (3.82) – (3.83) or (3.104) are quite appropriate for the estimation of turbulence parameters in each of the layers. In addition to that, sometimes no information is available on the parameters h and $\partial\rho/\partial z$ for each of the layers. In this context one faces a problem in determining some effective vertical-diffusion coefficient that would provide a mean transport of contaminant through the whole set of the layers with different hydrophysical features and turbulence levels.

The vertical contaminant diffusion process in the ocean with intermittent turbulized layers turns out to be a rather specific one. If steady interlayers with low turbulence levels are presented, then, as was shown above, the so-called 'blocking' effect can arise because in those interlayers vertical diffusion reduces to the extremely low molecular level. Nevertheless, there exists, on the average, a more intensive vertical exchange in the ocean which can only be accounted for by chaotic arising of the turbulent layers at different depths. Turbulence provides contaminant transport, rapid enough, at distance h in the vertical direction at each of these layers. Afterwards, the turbulence decays in this layer but, in some other instant, another turbulent layer arises. The latter can partly overlag the segment of z-axis that was formerly occupied by the preceding turbulent layer. As a result, the contaminant will be transported in the vertical direction at some additional distance, Δz. Finally, these spontaneously arising turbulent layers, in spite of being separated from each other by pseudo-laminar interlayers in each instant, can provide, on the average, during certain time intervals a rather intensive vertical flux of contaminant.

In the above scheme of vertical diffusion in the presence of intermittent turbulence, it is natural to assume the depth of individual turbulent layer, z_J, the thickness of the layer, h_J, and the root-mean-square level of velocity turbulent fluctuations in the layer, σ_J, to be random quantities [5]. The vertical diffusion coefficient in the J-th turbulent layer should depend on the thickness of the layer, h_J, and on the turbulence intensity in the layer, σ_J, i.e., $K_z^J = f(h_J, \sigma_J)$. The turbulent layers of fluid with diffusion coefficients K_z^J in any time moment are alternated with zones of nearly laminar flow where the diffusion coefficient is close to the molecular one. Variation of the diffusion coefficient with depth is presented schematically in Figure 3.20, both for a continuously turbulized ocean (Figure 3.20a, a 'classical' scheme) and for an ocean layered with respect to the turbulence features (Figure 3.20b).

According to the scheme shown in Figure 3.20b, the distribution of the vertical turbulent diffusion coefficient along the vertical direction can be presented by the random 'pulse' function

$$K_z(z) = \sum_{J=1}^{n} f(\sigma_J, h_J)\, F\left(\frac{z - z_J}{h_J}\right), \tag{3.115}$$

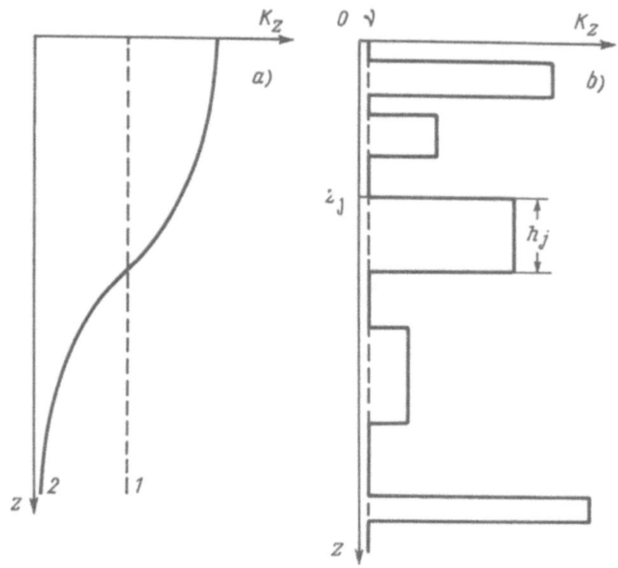

Fig. 3.20. Schematic vertical profiles of vertical-diffusion coefficient, K_z, for the totally turbulent ocean (a) and in the case of intermittent turbulence (b). Dotted line 1 corresponds to the model with diffusion coefficient independent of depth; curve 2 corresponds to the model with diffusion coefficient varying with depth; ν is the molecular-diffusion coefficient; h_J are the thicknesses of the turbulized water layers in the ocean.

where

$$F(x) = \begin{cases} 1 & \text{at} \quad 0 \leqslant x \leqslant 1; \\ 0 & \text{for the other } x. \end{cases}$$

Provided that the values of σ_J and h_J are supposed to be statistically independent of z_J and the probability density of these values to be independent of the number of J, the conditional mean value of function $K_z(z)$ for the ocean of depth H and in the presence of n turbulent layers has the following form:

$$< K_z | _n > = \sum_{J=1}^{n} \int_{\sigma_J} \int_{h_J} \Big[f(\sigma_J, h_J) \; P_{\sigma h}(\sigma_J, h_J) \; d\sigma_J \; dh_J \times$$

$$\times \int_{z_J} F\left(\frac{z - z_J}{h_J}\right) \; P_z(z_J) \; dz_J \Big], \tag{3.116}$$

where P_z is the probability density for z_J, $P_{\sigma h}$ is the joint probability density for σ_J and h_J, and the angular brackets denote statistical averaging.

Assuming the probability density for position of the individual turbulent layers is uniform within the depth interval H, i.e., if $P_z(z_J)\,dz_J = dz_J/H$, we obtain

$$I = \int_{z_J} F\left(\frac{z - z_J}{h_J}\right) P_z(z_J)\,dz_J = \frac{1}{H}\int_{-H/2}^{H/2} F\left(\frac{z - z_J}{h_J}\right) dz_J. \tag{3.117}$$

Provided the thickness of the turbulent layers is much smaller than the total ocean depth, then $I \simeq h_J/H$ and

$$<K_z|_n> = \sum_{J=1}^{n} \int_{\sigma_J}\int_{h_J} \frac{h_J}{H} f(\sigma_J, h_J)\, P_{\sigma h}(\sigma_J, h_J)\, d\sigma_J\, dh_J = \frac{n}{H} <hf>. \tag{3.118}$$

Averaging (3.118) over n we arrive at the unconditional mean value $<K_z> = (<n>/H)<hf>$. Under the natural assumption that the distribution of the number of turbulent layers within interval H obeys the Poisson law, $P_H(n) = [(n_1H)^n/n!]\exp(-n_1H)$ (where $n_1 = H/<\bar{u}>$ is the mean number of turbulent layers per unit depth interval), we arrive at

$$<K_z> = n_1<hf>. \tag{3.119}$$

To specify the function $K_z^J = f(\sigma_J, h_J)$ we will employ dimensional considerations assuming $K_z^J = c\sigma_J h_J$, where c is some non-dimensional universal constant. Then we obtain

$$<K_z> = cn_1 \int_0^\infty\int_0^\infty h^2\sigma\, P_{\sigma h}(\sigma, h)\, d\sigma\, dh, \tag{3.120}$$

and if σ and h are statistically independent then

$$<K_z> = cn_1 \int_0^\infty \sigma P_\sigma(\sigma)\, d\sigma \int_0^\infty h^2\, P_h(h)\, dh. \tag{3.121}$$

Only a few empirical estimates of the probability distribution functions in equations (3.120) – (3.121) have been made till now (see references in [5]). Thus, the probability distribution $P_h(h)$ obtained from observations in the Tunisia Strait was approximated by the log-normal law. On the other hand, according to the measurements carried out in the main thermocline in the northern part of the Pacific, a better approximation of the function $P_h(h)$ was achieved by means of an exponential law, $P_h(h) = \mu_1\exp(-\mu_1 h)$, at $\mu_1 = 0.46\ m^{-1}$, so that $<h^2> = 2/\mu_1^2 = 9.4\ m^2$ and a value of n_1 appeared to be equal to $0.21\ m^{-1}$ for the layer extending from 800 to 1200 m. As measured in the Indian Ocean to the south of Sumatra, a value of n_1 for the upper 250 m layer was found to be equal to

0.021 m^{-1}, $<\sigma>$ = 1.0 cm/s and $<h^2>$ = 21 m^2. For these values of parameters and at c = 0.1 the estimates of $<K_z>$ from equation (3.121) range from 1.6 to 16 cm^2/s.

Empirical two-dimensional probability distributions of $P_{\sigma h}(\sigma, h)$ were obtained from the measurements in the Tunisia Strait. The estimates of $<K_z>$ from equation (3.120) were found to be equal to 16 and 40 cm^2/s from measurements carried out in the eastern and western parts of the strait, respectively. As estimated from data on $P_{\sigma h}(\sigma, h)$ obtained using measurements from the Indian Ocean, the value of $<K_z>$ was found to be equal to 6.8 cm^2/s. The observed scatter of values of $<K_z>$ may be caused by the difference in mean environmental conditions in the individual parts of the World Ocean where these measurements were conducted.

To estimate a possible statistical scatter of the effective vertical turbulent transport coefficient, K_z, we calculate the variance of K_z using the formula $D(K_z) = <K_z^2> - <K_z>^2$. Under the above suggestions on the statistical independence of the parameters σ_J and h_J of z_J in equation (3.115) as well as on the independence of the probability distribution laws for these parameters of number J, we arrive at the expression for the conditional (for n turbulent layers within the depth interval H) mean square value of the vertical turbulent transport coefficient

$$<K_z^2|_n> = \frac{n}{N} <hf^2> + \frac{n^2 - n}{H^2} <hf>^2. \tag{3.122}$$

If the probability distribution for the turbulent layers with depth is the Poisson one, the variance of the vertical turbulent transport coefficient is as follows:

$$D(K_z) = n_1^2 <nf>^2. \tag{3.123}$$

Setting $K_z^J = c\sigma_J h_J$ in equation (3.123), as above, we obtain in the general case

$$D(K_z) = c^2 n_1 \int_0^\infty \int_0^\infty h^3 \sigma^2 P_{\sigma h}(\sigma, h)\, d\sigma\, dh, \tag{3.124}$$

and if σ and h are statistically independent then we obtain

$$D(K_z) = c^2 n_1 \int_0^\infty \sigma^2 P_\sigma(\sigma)\, d\sigma \int_0^\infty h^3 P_h(h)\, dh. \tag{3.125}$$

The estimates of root-mean-square scatter of the values K_z, $\sqrt{D(Kz)}$, obtained from the equations (3.124) and (3.125) using the experimental data mentioned above are of the order of value $<K_z>$.

Let us consider a possible manifestation of variations of the vertical turbulent diffusion coefficient, K_z, in the concentration field of contaminant fallen out from the atmosphere. To do this, we employ the simplest model of vertical diffusion considered in Section 3.1 (equations (3.2) – (3.7)). Theoretical curves for contaminant concentration estimated from equation (3.7) are plotted in Figure 3.21 as functions of time elapsed from some initial time moment. The curves are plotted at several diffusion coefficients and at two values of contaminant flux through the ocean surface: 10^{-13} and 10^{-14} g/(cm^2/s) (the left and the right vertical scales in Figure 3.21, respectively). As one can see from the figure, the range of variation of contaminant concentration at the ocean surface for different values of K_z is spread with time. Thus, for instance, at t = 1 hour the value of c ranges from $0.02 \cdot 10^{-10}$ to $1.81 \cdot 10^{-10}$ g/cm^3 while at t = 10 hours it ranges from $0.07 \cdot 10^{-10}$ to $5.72 \cdot 10^{-10}$ g/cm^3.

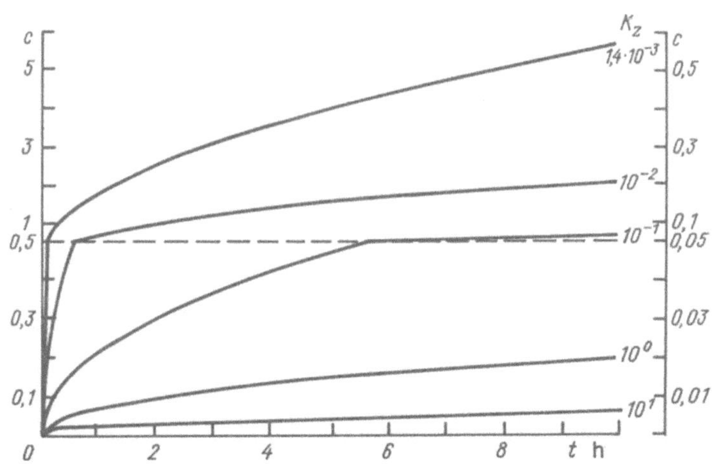

Fig. 3.21. Contaminant concentration, c, at the ocean surface for different values of contaminant fluxes, q, the diffusion coefficients, K_z, and diffusion time t (after [6]).

Thus, the scatter of the values of the vertical turbulent diffusion coefficient caused by intermittency of ocean turbulence can lead to significant variations of contaminant concentration at the ocean surface. The layers with high and low turbulence intensity can arise alternately near the ocean surface in different ocean areas. As a result, the contaminant field at the ocean surface can acquire a 'mosaic' pattern. The differences in concentrations between neighboring patches of this structure depend on the differences in turbulence levels in subsurface ocean layers as well as on the time of existence of the turbulent layers. Typical time of existence of approximately constant level of turbulence in the individual layers of the ocean can be apparently taken equal, to the order of magnitude, to 1 hour. Then, if molecular vertical diffusion takes place in one of the ocean zones while the value of K_z = 10 cm^2/s is prescribed in the neighboring zone, a ratio of surface concentrations in these two zones can be as high as 1:100. The mentioned features of contaminant spread are, naturally, to be taken into account while estimating contaminant concentration in the different areas of the World Ocean.

CHAPTER 4

Practical Methods of Calculating the Contaminant Diffusion in the Ocean

4.1. Numerical Studies of Contaminant Concentration Carried Out Using the Semi-Empirical Diffusion Equation

In the previous chapters we have given a great many examples of analytical solutions to various problems on contaminant diffusion, obtained by means of diffusion equations like (1.22). In a number of cases we managed to simplify this equation, taking into account specific features of the problem. Analytical solutions make it possible to ascertain the influence of different factors upon the contaminant-concentration distribution, as well as to evaluate to influence of boundary and initial conditions on the concentration-field variability in space and time. So when possible we should always aspire to obtain the analytical solution of the diffusion problem with an explicit dependence of the desired contaminant-field characteristics on coordinates, time and parameters governing the process. However, such attempts often fail. It is rather difficult to obtain analytical solutions even for a somewhat complicated form of the domain in which the contaminant diffusion is studied; a complicated structure of the flow-velocity fields in the ocean area under study also cumbers analytical investigation. Analytical solution is often hindered by the exchange coefficients given by complicated functions of coordinates and time. In these cases it is often useful to try to simplify somehow the formulation of the problem so that it can be reduced to a form admitting analytical solutions. For example, sometimes it is possible to replace an integration domain with a complicated boundary by a more simple domain, e.g. by a square or round one. In quite a number of cases analytical solutions can be found more easily if we replace a finite domain by an infinite plane or half-plane. Certainly we have to evaluate the errors appearing in the results of calculations due to such simplifications and assumptions.

As an example, we consider the evaluation of the error caused by the assumption of infinite depth of the ocean in the problem of vertical diffusion of the contaminant fallen out of the atmosphere. The analytical expression obtained for the contaminant-concentration distribution with depth is given in Section 3.1 (see formula (3.6)). Now we define the required relative accuracy of the calculation of the concentration $c(z, t)$ at some depth z_0 as a ratio of the concentration at this depth to that at the ocean surface, i.e.,

$$c(z_0, t)/c(0, t) = a, \tag{4.1}$$

where a is the prescribed accuracy of the solution.

The ratio of concentration at the depth z_0 to that at the depth $z = 0$ is obtained from the analytical solutions (3.6) and (3.7):

$$c(z_0, t)/c(0, t) = \exp(-\xi_0^2) - \sqrt{\pi}\, \xi_0 \,(1 - erf\ \xi_0), \tag{4.2}$$

where $\xi_0 = z_0/(2\sqrt{K_z t})$.

Equating now the right-hand sides of (4.1) and (4.2) we can find the depth z_0 at which the desired accuracy is reached. When this depth is exceeded, we can consider the ocean to be 'infinitely' deep. Obviously, the depth z_0 depends upon the value of the coefficient K_z and the time of diffusion t. For example, for $a = 0.01$ the value of z_0 is described approximately by the following formula:

$$z_0 \approx 3{,}21 \sqrt{K_z t} \tag{4.3}$$

For the case of the molecular-diffusion coefficient the depth z_0 does not exceed 23 cm of 10 hours after the start of diffusion, while for $K_z = 10\ \mathrm{cm^2/s}$, z_0 increases up to 19 m. If we increase the diffusion time up to 10 days, then for $K_z = 10\ \mathrm{cm^2/s}$, z_0 appears to be close to 100 m. For large diffusion times it is usually necessary to take into account the fact that the contaminant is non-conservative and, as a result, we are forced to apply another formula to evaluate the depth of the contaminant penetration (see Section 3.1). In most cases we also can consider the ocean to be infinitely deep. This assumption greatly simplifies the search for an analytical solution of the problem.

If all the attempts to reduce the problem to a form convenient for analytical study fail, we have to apply numerical methods. Modern computers provide sufficiently fast numerical solution of such problems. To solve the differential diffusion equation numerically it is necessary to construct its finite-difference analog. The domain of integration is replaced by a mesh one, usually with a rectangular system of delimiting planes or lines. The outer curvilinear boundary of the domain must also be replaced by a system of broken segments of the surfaces or lines; boundary conditions at the new boundary are 'transferred' from the original contour of the domain. For numerical solution of non-stationary problems of diffusion, discrete time intervals must also be used. Sometimes it is more convenient to use meshes with non-uniform spatial and time steps.

Let us use indices i, j and k to denote the mesh node numbers on the areas x, y and z, respectively; then, for example, a finite-difference approximation of the advection term $u\ \partial c/\partial x$ in the diffusion equation can be written in the form:

$$u\ \partial c/\partial x \approx u_{i,j,k}(c_{i,j,k} - c_{i-1,j,k})/\Delta x \tag{4.4}$$

Here $u_{i,j,k}$ is the velocity component $u(x, y, z)$ in the node i, j, k; $c_{i,j,k}$ and $c_{i-1,j,k}$ are the contaminant concentrations in the corresponding nodes of the meshes; Δx is the mesh spacing along the x-axis.

In a non-stationary case the diffusion equation can be written in a general form:

$$\partial c/\partial t + AC = f, \tag{4.5}$$

where A is some operator and f is a contaminant-source function.

In a finite-difference approximation with respect to time equation (4.5) can be written in the following form:

$$(c_{l+1} - c_l))/\Delta t + Ac_l = f_l, \tag{4.6}$$

or in the form

$$(c_{l+1} - c_l)/\Delta t + Ac_{l+1} = f_l, \tag{4.7}$$

where the indices l and $l+1$ denote the numbers of discrete moments of time counted from some initial moment.

The finite-difference approximation of the form (4.6) is called the explicit approximation and that of the form (4.7) the implicit one. If equation (4.5) is solved with the initial condition $c = \varphi$ for $t = 0$, φ being a given function, then for both types of approximation the initial condition will be $c_0 = \varphi$. If a solution of the initial value problem for differential equation (4.5) is smooth enough, then the accuracy of the finite-difference approximation of differential equation (4.5) can be evaluated by expanding the desired function in Taylor's series and substituting the obtained series in the formulas (4.6) and (4.7). This procedure enables us to make sure that approximations (4.4), (4.6) and (4.7) are of the first order of accuracy. But the following type of finite-difference approximations of equation (4.5) will have the second order of accuracy with respect to t:

$$(c_{l+1} - c_l)/\Delta t + A(c_{l+1} + c_l)/2 = f_l, \tag{4.8}$$

where

$$f_l = f_l(t_{l+1/2}) \quad \text{or} \quad f_l = [f(t_l) + f(t_{l+1})]/2. \tag{4.9}$$

If some parameters of the diffusion equation, for example, the diffusion coefficient, change rapidly in some space-time domain, we are forced to make the steps smaller or to use another method guaranteeing the desired accuracy of approximation. In particular, if we study numerically diffusion problems with instantaneous contaminant sources, when contaminant concentration at the initial moment of time is described by the δ-function, it is necessary, generally speaking, to use infinitesimal steps. In this case we can use the analytical solution of the problem for small distances from the source and short diffusion times supposing that diffusion takes place in unbounded space. After that we can solve the problem numerically, assuming the contaminant source 'smeared' over the certain area to be the initial concentration field.

When numerical methods are applied it is important to make sure that the chosen finite-difference scheme is stable and convergent. The schemes will have such properties if certain relations between the space and time steps and parameters of the problem hold. For example, if an explicit finite-difference scheme for the non-stationary diffusion equation with sole vertical diffusion term is used, then the following relation must hold:

$$\Delta t < (\Delta z)^2/(2K_z), \tag{4.10}$$

where K_z is the constant vertical diffusion coefficient.

In the case of the stationary diffusion equation with the terms describing advection along the axes x and z and with the vertical diffusion term, the stability and convergency condition can be written in the form (see ref. [8])

$$\Delta x < \frac{u(\Delta z)^2}{2K_z} \frac{1}{1 + w\Delta z/(2K_z)}, \tag{4.11}$$

where u and w are the constant velocity components along the axes x and z, respectively.

More detailed consideration of stability and convergency criteria for finite-difference schemes is beyond the scope of this monograph. A great deal of works are devoted to this question, and those who are interested in such problems in more detail are referred to the monograph by G.I. Marchuk [61] and collection of papers [39]. Here we shall dwell only upon some results of the application of numerical methods to the problems of contaminant diffusion in the ocean.

Apparently, a numerical method of calculating the contaminant diffusion in the sea was applied for the first time in [76]. This paper presented the results of calculations carried out successively for the following problems on the horizontal contaminant diffusion: diffusion with constant diffusion coefficients, diffusion with coefficients given by the "4/3 power law" and diffusion with coefficients depending on the relative depth of the basin. The possibility of contaminant decay was also taken into account by the diffusion equation. In the last paragraph of [76] the explicit scheme was used to study numerically the contaminant diffusion from the instantaneous distributed source in the coastwise area with a complicated configuration of the coastline and irregular depth in the presence of a stationary but spatially varying field of the horizontal current velocity components. At the coastline the condition of non-penetration of the contaminant through the boundary of the water-land interface was set. Mesh spacing in this problem was equal to 100 m and the time step was equal to 1000 sec. The calculation was performed for 17 successive locations of the diffusing contaminant spot. Figure 4.1 shows the plan of the coastwise sea area, the initial distribution of the contaminant spot with circular distribution of concentration isolines and the shape of the spot 16,000 sec after the beginning of diffusion. The analysis of the obtained pictures of diffusion revealed that the spot was first diffusing almost symmetrically and was transported by current from the right to the left. Approaching the cape outstanding far into the sea, the spot acquired more and more complicated configuration; it was drawn near the cape by the current flowing along it, and extended along the current. The center of the spot (a point with a maximal contaminant concentration) moved along a smooth curve, going round the cape, and the isoline of unit concentration value was 'tangential' to the coastline near the cape, creating a pollution zone of the coastwise area. Having passed the cape, the spot began to spread into the bay situated beyond the cape, where a stagnant contamination zone could appear due to the presence of closed water circulations.

Complicating successively the diffusion problem which is to be solved by numerical methods, we managed to evaluate in [76] the role of different factors influencing the formation of the contaminant-concentration field. Thus, taking into account the dependence of the horizontal diffusion coefficient upon the scale of the phenomenon, we obtained already for the very first steps notable differences in concentration distribution inside the diffusing spot in comparison with the case for constant K_L. For a distributed instantaneous contaminant source shown in Figure 4.1, a numerical investigation with the constant diffusion coefficient equal to 10^4 cm^2/s gave a more rapid decrease of the contaminant concentration in the central (100 – 150 m) zone of the spot in comparison with the results of the calculation carried out under the assumption K_L depending on the phenomenon scale (radius vector of the observation point). In the spot zones more distant from the center the concentration obtained in the latter case appeared to be less than that calculated from the diffusion equation with the constant coefficient. Obviously, the fact that the diffusing contaminant is non-conservative could influence the concentration distribution for the diffusion times of the order of several hours only in the case of a contaminant with a commensurable half-life. In contrast, the influence of bottom topography taken into account by a formula like (2.80) appeared to be substantial in a number of

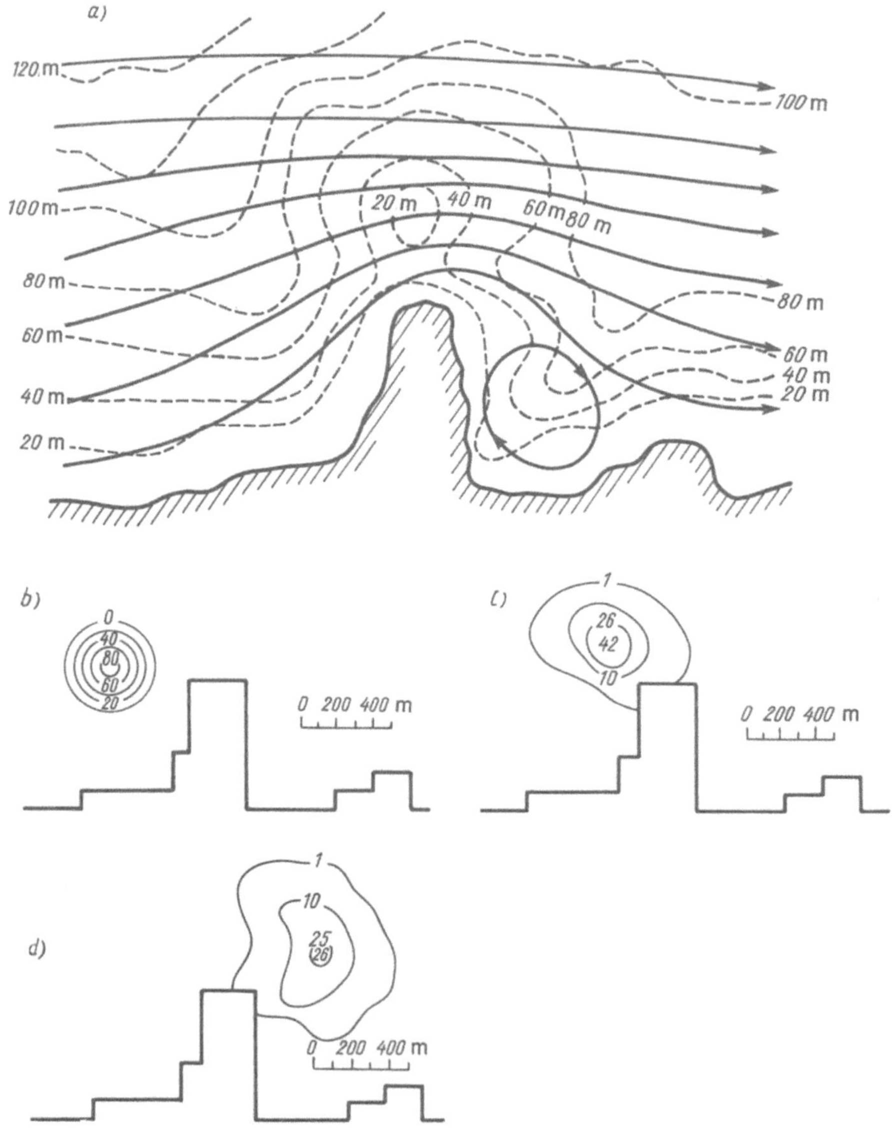

Fig. 4.1. Spreading of the contaminant spot in a coastwise sea area, calculated numerically. a – coastwise sea area with a scheme of currents and bottom contours; b – approximation of the coastline by a broken function and initial distribution of the contaminant concentration in the diffusing spot; c – position of the contaminant spot 8000 s after the diffusion started; d – position of the contaminant spot 16,000 s after the diffusion started.

cases. Thus, the rate of the contaminant-concentration decrease during the diffusion of the spot considered in a basin with a depth equal to 100 m was considerably higher than that obtained for the case of diffusion in a deep sea. Multiplication of the horizontal diffusion coefficient by the factor $f(l/H)$ results, naturally, in a more rapid growth of the diffusing contaminant spot. Thus, 10,000 s after the diffusion started, the radius of the spot was equal to 600 m (within the limits of the unit concentration isoline) for $f(l/H) = 1$ and 700 m for $f(l/H)$ given by the graph show in Figure 2.62.

Numerical calculation of the contaminant diffusion in the coastwise sea area was performed in [2]. In this numerical study in the current-velocity field was not given, but calculated according to the wind field over sea. Both the wind and the current were considered to be steady and not to vary with time. As the diffusing contaminant was supposed to have no influence upon the water dynamics, the diffusion equation was solved separately, making use of the horizontal and vertical velocity components (u, directed normal to the coastline (x-axis), and w, respectively) obtained in the dynamic problem. Diffusion in the direction parallel to the coastline was neglected. The diffusion process was considered to be essentially non-stationary; at the initial moment concentration c was set equal to zero throughout the domain under study. There was no contaminant flux through the free surface and the bottom of the basin; at the coastline the concentration was considered to be a known function of the vertical coordinate and time, while at the distance of 50 km from the coast $c(x, z, t)$ was set equal to zero. To imitate the contaminant source similar to an instantaneous point source, or, vice versa, to a linear source acting for a long time, the function $c(0, z, t)$ was taken in the form:

$$c(0,\ z,\ t) \ = \ c_0 \ \exp[-k(k_1 - z)^2 - k_2(k_3 - t)^2]. \tag{4.12}$$

Varying the coefficients k_1, k_2 and k_3, we can change the contaminant-source characteristics.

For numerical experiments the diffusion equation was replaced by its finite-difference analog solved by the Thomas method. Boundary conditions were approximated to a second order of accuracy. Mesh spacing along the x-axis was non-uniform: $\Delta x = 25$ m in a region between the coastal line and the distance 100 m, $\Delta x = 100$ m for 100 m $< x \leqslant 5$ km and $\Delta x = 5$ km for $5 < x \leqslant 50$ km. For the depth equal to 40 m the vertical step was equal to 1 m, and the time step Δt was equal to 12 hours. The values of the diffusion coefficients were chosen within the limits $K_z = (0, 1, ..., 1)$ cm^2/s and $K_x = (10^3 ... 10^4)$ cm^2/s. Different wind types in the coastwise sea area causing different types of water circulations were considered. Naturally, substantially different types of the contaminant-concentration distribution, forming contaminant 'plumes' of one or another type, were obtained (Figure 4.2). Thus, in the case of set-up wind, the position of the contaminant source appeared to be of the most importance.

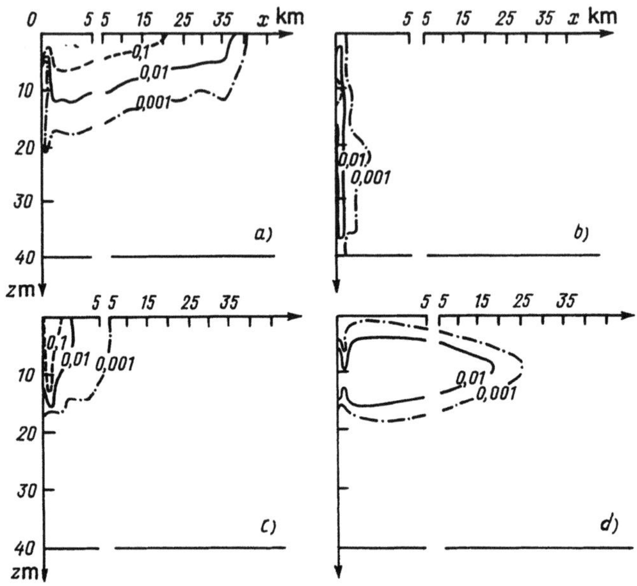

Fig. 4.2. Distribution of the contaminant concentration on a vertical plane for the set-down (a, b) and set-up (c, d) wind, calculated numerically for the case of a surface (a, c) and submerged (b, d) source location (after [2]).

When being discharged onto the surface sea layer, the contaminant began to spread fast at large distances from the coast. Thus, 24 hours after the source began to act, the contaminant (i.e., the isoline corresponding to 0.1% from the concentration of contaminant discharged) was already observed at distances up to 40 km from the coast. At the same time, when the source was located at the 10 m depth, the contaminant almost ceased spreading in the horizontal direction but began to move close to the coast, forming a zone of strong water pollution there. It is clear that for the set-up wind the picture changed. In the case of the source situated at the surface, the contaminant was spreading along the x-axis only for several km; on the contrary, in the case of a deep location of the source, noticeable concentrations of the contaminant could be found at large distances from the source, although not in the surface layers, but in intermediate ones. Naturally, different contaminant-concentration distribution resulted for various values of diffusion coefficients within the limits indicated above; however, these distinctions were quantitative, but not qualitative, because the basic configuration of the contamination zones was formed under the influence of the current-velocity field, which is strongly dependent on the wind structure.

The results of some numerical studies of the contaminant diffusion in the Caspian Sea and in the Baltic Sea are given in [39]. The calculations were based on the finite-difference analog of the non-stationary diffusion equation with the two terms describing horizontal advection, the term describing horizontal contaminant diffusion and one term taking into account the possibility of the contaminant being non-conservative. The velocity field was, in turn, found by solving a simplified system of water dynamic equations with the terms describing horizontal viscosity, Coriolis force and pressure gradients. The following boundary conditions were imposed: tangential wind friction was given at the surface and zero velocities at the bottom. The run-off of rivers into the sea was also taken into account. The initial condition for the diffusion equation was zero contaminant concentration in the sea waters. At some segments of the coastline the contaminant flux was prescribed and at the others the absence of the flux or a certain value of concentration was assumed. The horizontal diffusion coefficient in the calculations performed for the Caspian Sea was equal to $1.5 \cdot 10^6$ cm^2/s. The diffusion equation was approximated by a finite-difference formula with the first order of accuracy with respect to time and the second order of accuracy with respect to spacial variables. Numerical experiments were carried out for the northern part of the Caspian Sea for the case of north-east winds blowing over the water area and for continuous and instantaneous contaminant sources. The results of calculations for the diffusion time from 1 hour to 18 days are presented in Figure 4.3. The analysis of the results showed that contaminants in the sea are transferred mainly under the influence of advective factors. However, if a contaminant point source is located in a zone of horizon-

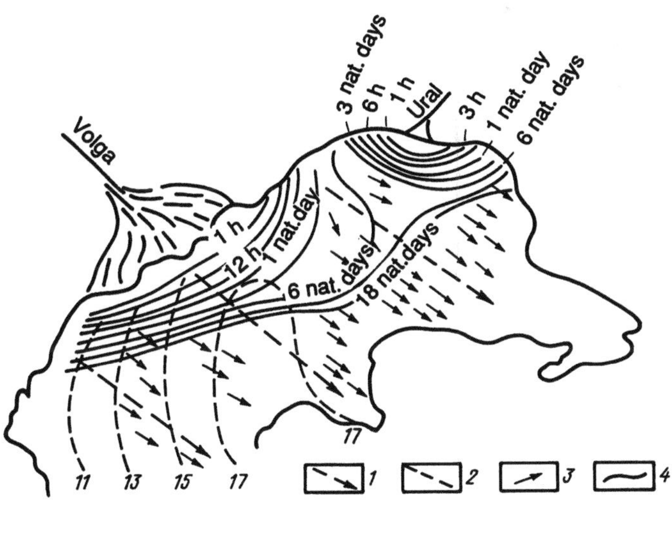

Fig. 4.3. Position of the 0.1%-concentration isoline after different times elapsed since the beginning of sources' action, obtained numerically for the case of contaminant diffusing from the mouth areas of the Volga and Ural rivers, the wind being north-west (after the paper by Shkudova from [39]). 1 – wind direction; 2 – isolines of wind velocity; 3 – current direction; 4 – 0.1%-isoline of contaminant concentration.

tal eddy structures of the current-velocity field, then the diffusion and advection terms can provide a contribution of the same order into the contaminant spreading.

It is clear that the contaminant-concentration field depends strongly on the time of the source action. We see from Figure 4.3 that in the case of imitating the runoffs of the Volga and the Ural rivers by the contaminant sources with a certain extension along the coastline, the concentration isolines 'move' with time seawards. One day after the sources are 'switched on', the isoline with the contaminant concentration of 0.1% (with respect to the initial concentration) recedes from the coastline to the distance of up to 70 km. For larger diffusion time t, the velocity of the contaminant front advance slows down and the picture becomes stationary for t ranging from 15 to 20 days. Naturally, the form of such a stationary contaminant field depends on the type of water circulation, i.e., eventually on the accepted field of wind blowing over the sea. Changes in the wind field will cause (with some delay) changes in the contaminant-concentration field. For shallow sea areas the characteristic time of such a changing will, apparently, be of the same order as the time necessary for the concentration field to become stationary, since the current-velocity field has a smaller relaxation time.

Numerical methods were used to calculate the spreading of contaminant from stationary point sources located in the Bay of Gdánsk and the port of Klaipeda region. The current-velocity field was considered to be given and stationary, and the horizontal turbulent transport coefficient was equal to $5 \cdot 10^5$ cm^2/s. Computation was carried out via the Thomas method and the special method according to which the desired stationary contaminant-concentration field is sought as the limit of the solution to the corresponding non-stationary problem when the diffusion time tends to infinity. Numerical study of the problem revealed that when the north-east currents are prevailing, the contaminant is transported from the Bay of Gdánsk in the same direction, reaching the Kurshskaya spit with still noticeable concentration values. The contaminant is transported from the port of Klaipeda to the north, not receding far from the coastline. The collection of works [39] also contains the results of numerical computation of water circulations and contaminant spreading in the Black Sea. This problem was also discussed earlier in [106].

Numerical computation of the spreading of a contaminant on a global scale was performed in the already mentioned paper [96]. The problem of oxygen distribution in the waters of the World Ocean was solved on the latitude-longitude meshes with a 5-degree spacing. The finite-difference scheme was accurate to the second order with respect to time and horizontal coordinates. A time step was equal to 5 days, and the coefficient of horizontal diffusion was assumed to be of the order of 10^8 cm^2/s. Solution of the problem was obtained as a limit of evolution of the solution to a corresponding non-stationary problem when the diffusion time tends to infinity. The regime was considered to become stationary when the concentration values for two successive moments of time coincided with 1% accuracy. For such a criterion used in the global problem under study, the time necessary for the regime to become stationary appeared to be about 43 years.

While the numerical computation of the global oxygen spreading in the ocean waters was carried out for a distributed source, the paper [40] presented the results of numerical solution of the global problem for an instantaneous contaminant point source. The problem simulated the diffusion of radioactive contamination from a hypothetical nuclear explosion in the Pacific Ocean. A two-dimensional approach was used when solving the problem, i.e., vertical processes were not taken into account. The fluxes of the radioactive contaminant (strontium-90) through the coastline were assumed to be zero, and the absence of isotopes in the ocean waters was taken as an initial condition. The non-stationary diffusion equation took into account the terms describing horizontal advection, horizontal diffusion and isotope decay. The horizontal turbulent diffusion coefficient was taken as 10^7 cm^2/s. The current-velocity field was borrowed from another paper, which presented the results of a calculation of the integral stream function in the baroclinic World Ocean.

To overcome the difficulty occurring when solving numerically the diffusion problem with an instantaneous contaminant point source, the following trick was used in [40]. First, the fundamental solution of the two-dimensional diffusion equation was constructed; the obtained solution was the approximated by the following broken function: the contaminant value was constant on the segment between the explosion point and the point removed from it at the distance r = 900 km; in the region between 900 km and 1800 km the dependence of concentration on r was described by a fifth-order polynomial; for r = 1800 km the concentration o was assumed to be zero. Thus, the radius of 'smearing' of the initial contaminant distribution was accepted as equal to 1800 km, guaranteeing, for a 5-degree spacial spacing, a small error of approximation at the first stages of calculation for the time step equal to 5 days.

The numerical finite-difference scheme used in [40] to solve the global diffusion problem was first-order accurate with respect to time and horizontal coordinates. Figure 4.4 shows the results of the calculation of the contaminant spreading for a number of time moments after the moment of explosion. The isotope source was located in the Pacific Ocean at the intersection of the equator with the meridian 175° W. The figure shows the position of the isotope-concentration isoline $C = 0.3\ C_0$ at 1, 3, 5, 10, 20, 35 and 70 years after the moment of explosion, where C_0 is the isotope concentration provided that all the radionuclides released during the explosion were distributed uniformly in the waters of the World Ocean. As we see from the figure, distribution of the contaminant concentration in the ocean is conditioned mainly by the type of global

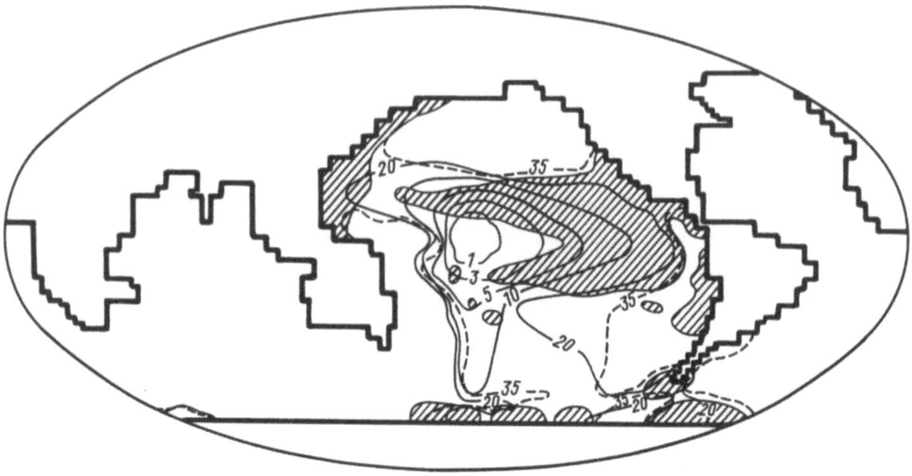

Fig. 4.4. Numerical calculation of the positions of 0.3 $\bar{c}_0$-strontium-90 concentration isolines after 1, 3, 5, 10, 20, 35, and 70 years elapsed since the nuclear explosion (after [40]). The area with isotope concentration, exceeding 0.3 $\bar{c}_0$ 70 years after the explosion, is shaded.

ocean water circulations. The contaminant spot is elongated by the Equatorial Counter-Current in the zonal direction and reaches the coasts of North America only 25 – 30 years after the explosion. As time passes, an area with noticeable isotope concentration takes a more complicated form. The contaminant diffuses into the South Trade Current and is transported into the subtropical water circular motion, also getting into the southern part of the Pacific Ocean. The contaminant is transported further into the Atlantic and Indian oceans. Seventy years after the explosion the isotope can be found in the form of separate spots at large distances from the place of explosion. Localization of these spots is determined by the joint effect of advection and diffusion, as well as by the isotope decay, resulting in a decrease of the general level of ocean contamination.

Obviously, the numerical calculation of the global contamination transport is rather approximate and can be considered only as an example of possible estimations of the averaged contaminant distribution in the World Ocean for the case of a point source. The obtained vertically averaged velocity of spreading of the contaminant concentrations must also be regarded as roughly tentative and mean, since we know for a fact that the radioactive products of American nuclear tests in the Pacific Ocean spread in individual horizontal water layers at large distances from the place of explosion in short times. Thus, 4 months after the nuclear explosions near the atolls of Eniwetok and Bikini, the radioactive contamination was observed at the distance of 2000 km from the place of explosion, and 8 months after the explosion at the distance of 7000 km (see the paper by Nelepo in [91]).

Such a discrepancy between the experimental isotope transport velocities and those obtained by solving the problem numerically is caused, first of all, by neglecting the dependence of the transport equation parameters (e.g. the current velocities and transport coefficients) on the vertical coordinate. The values of these parameters averaged over the depth of the ocean are much smaller than their actual values and differ essentially from the values of u, v and K_L in the surface layers of the ocean. So, to evaluate the actual velocities of the contaminant transport in the ocean, we need data on three-dimensional fields of the current velocities and the transport coefficients, and on their variability with time.

Numerical calculation of the contaminant spreading in a coastwise ocean area for a three-dimensional field of current velocities was performed by Kochergin and Bokovikov [52]. The variability of the diffusion coefficients was taken into account. The authors used dynamic equations with the terms describing the Coriolis force and vertical turbulent friction to describe the field of horizontal current velocity components. The non-stationary water-temperature field was, in turn, described by the heat-conduction equation with the term of vertical heat conduction. The dynamic and heat conduction equations were supplemented by the equations of turbulent-energy balance b and turbulent-energy dissipation ε. The obtained set of equations was closed by the relation between the turbulent viscosity coefficient and the quantities b and ε. The non-stationary contaminant diffusion equation took into account all three advection terms and diffusion terms with the diffusion coefficients K_x, K_y and K_z along the x-, y- and z-axes, respectively. The coefficient K_z was considered to be a function of the vertical coordinate z. Possible decay of the diffusing contaminant was also taken into account. At the surface tangential wind friction and water temperature were prescribed, while the turbulent energy, dissipation and contaminant fluxes through the free surface were set equal to zero. At the bottom (the depth $H(x, y)$) the current velocity, the turbulent energy, the turbulent-energy dissipation and the vertical contaminant flux were assumed to be zero, while the water temperature was prescribed. At the rigid lateral boundary (supposed to be vertical) the normal velocity component and normal contaminant flux component were equal to zero. At the liquid lateral boundary the values of the horizontal current velocity components were given, while the contaminant concentration was supposed to be zero. At the initial moment there were no currents in the sea, the water temperature was distributed as a known linear function, and the contaminant concentration was described by a smooth function localized in a neighborhood of the contaminant source (i.e., the point with the coordinates x^*, y^* and z^*). Since the turbulent transport coefficients were the desired functions, a certain value of the vertical turbulent viscosity coefficient was set at the initial moment of time. The viscosity coefficient K, heat-condition coefficient K^T and diffusion coefficient K^C were assumed to be proportional to each other, i.e., related by the formulas $K^T = \alpha_1 K$, $K^C = \alpha_2 K$, where α_1 and α_2 are constants.

Since the authors supposed that the diffusing contaminant did not influence the hydrodynamics of the basin, the problem was split into two (similarly to the papers mentioned above in this section). First, a solution for the hydrothermodynamical characteristics of the basin was sought, and after that the obtained values of the current-velocity components and vertical-diffusion coefficient (as functions of the coordinate and time) were substituted into the diffusion equation, which remained linear with respect to the contaminant concentration $c(x, y, z, t)$. When studying these problems numerically they reduced the equations and the boundary conditions to the dimensionless form and, since all the external factors in this problem were stationary, the resulting hydrothermodynamical equations were solved as non-stationary with the limit of $t \longrightarrow \infty$ to obtain stationary solution. A solution was considered to become stationary when the difference between the values of the vertical turbulent viscosity coefficients K_z, obtained at the two successive iterations, became small enough.

When solving the diffusion equation in [52], the vertical current velocity was calculated from the continuity equation using horizontal components, while the values of the horizontal diffusion coefficients were chosen according to the scale of the phenomenon under consideration ($K_x = 10^3$ cm^2/s, $K_y = 10^2$ cm^2/s, the x-axis is aligned with the coastline, the y-axis is normal to the x-axis). To approximate better the contaminant-concentration field in the vicinity of the source, the method of 'embedded' meshes was applied.

Besides the basic meshes, the method uses additional meshes with a double spacial step. In this case, the first order of approximation on the basic meshes can be reached together with the second order of approximation on the second meshes. Since the contaminant source in the diffusion problem was supposed to act continuously (beginning from the time moment $t = 0$), the diffusion problem was solved by the above-mentioned numerical method in the limit of $t \longrightarrow \infty$ in non-stationary equations.

Numerical study of the diffusion equation was carried out in [52] for an offshore area of the hypothetical sea with a complicated bottom topography and coastline geometry. The domain under consideration extended for 9.35 km along the x-axis and for 5.1 km along the y-axis, normal to the coast. The horizontal basic mesh spacing was equal to 425 m, a time step was equal to $1.8 \cdot 10^3$ s, and 21 layers were used in vertical direction. The contaminant source was located at the depth of 27 m. Calculations were performed for different wind fields, different types of temperature stratification, bottom topography and coastline geometry. The increase of the wind force (causing the increase of the current velocities) resulted in the reduction of the contaminant concentration due to the more rapid advective transport of the contaminant from the discharge area. The main factor determining the contaminant concentration in proximity to the coast were three-dimensional surging circulations in the off-shore sea area. A complicated coastline could cause the emergence of local circulations and local zones with increased contaminant concentrations. Figure 4.5 shows the example of the calculated contaminant consideration on the vertical plane (x, z), going through the contaminant source ($y = 2.55$ km). In this figure, the current is directed from the left to the right, causing the elongation of the contaminant spot in this direction. Analyzing the obtained patterns of the contaminant concentration, the authors of [52] reached a grounded conclusion that when numerically studying diffusion problems in ocean areas with sharp gradients of the velocity field or the diffusion coefficients (for example, in a coastwise area) it is necessary to use the schemes with increased order of accuracy. In the areas with smoothly varying hydrodynamical characteristics satisfactory results can be obtained using first-order schemes.

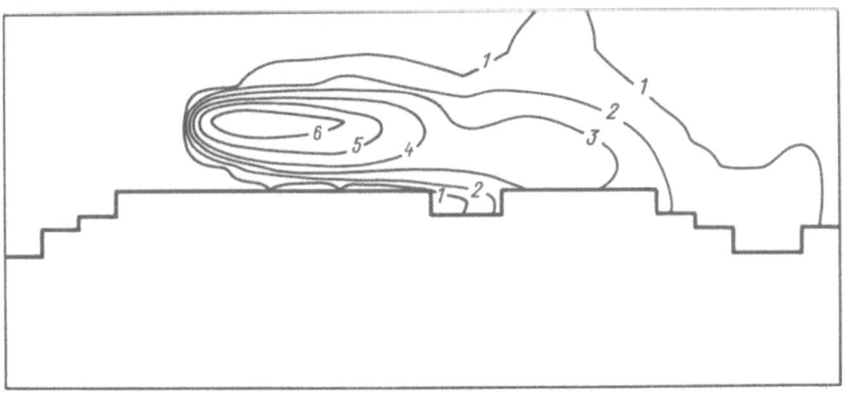

Fig. 4.5. Numerical calculation of the contaminant-concentration distribution on a vertical plane (x, z), passing through the source (after [52]). Contaminant-concentration values to the isolines (given in percentage with respect to the contaminant concentration in the source): 1 – 0.1%; 2 – 0.5%; 3 – 1%; 4 – 2%; 5 – 5%; 6 – 10%.

Kochergin and Bokovikov carried out the calculations for an even more complicated system of hydrothermodynamic equations and diffusion equation [53]. Besides the vertical turbulent terms, the dynamic equations for the horizontal components of the velocity involved the terms with horizontal viscosity. The heat-conduction equation took into account the advective and horizontal turbulent heat-conduction terms. Besides this, the system was supplemented by the salt-diffusion equation with the terms similar to those of the heat-conduction equation; the sea-water density was considered as dependent both on temperature and salinity. The equations of turbulent-energy balance and turbulent-energy dissipation were similar to those used in [52]. The system of hydrothermodynamic equations was supplemented by the appropriate boundary conditions for the desired functions. All the terms were kept in the diffusion equation, and the boundary condition at the bottom took into account the possibility of gravitational sedimentation of the contaminant. The diffusion equation was solved, as before, independently from the system of hydrodynamic equations, for the contaminant was supposed not to influence the hydrodynamic and thermodynamic characteristics of the basin. In this case the current velocity and vertical-diffusion coefficient obtained when solving the hydrothermodynamic equations were treated as input parameters for the diffusion problem. The authors sought the solution by the same numerical method and in the same domain as in [52], using the smaller vertical spacing in the upper layer of the sea. The analysis of the patterns of the contaminant-concentration distribution obtained for different values of the parameters showed that, besides the current velocity and the diffusion coefficient, the bottom topography and the coastline form are important factors.

Gravitational sedimentation of the contaminant becomes essential only when the sedimentation velocity becomes comparable, in order of magnitude, with the typical vertical velocity of the current in the area under consideration. Taking into account these conclusions we can state that to calculate properly the contamination behavior in the ocean, we need certain preliminary data on the wind field, bottom topography, contaminant properties and contaminant sources' characteristics. Figure 4.6 shows the example of the horizontal distribution of the contaminant concentration calculated in [53] for the case of a localized continuous source situated at the depth of 27 m.

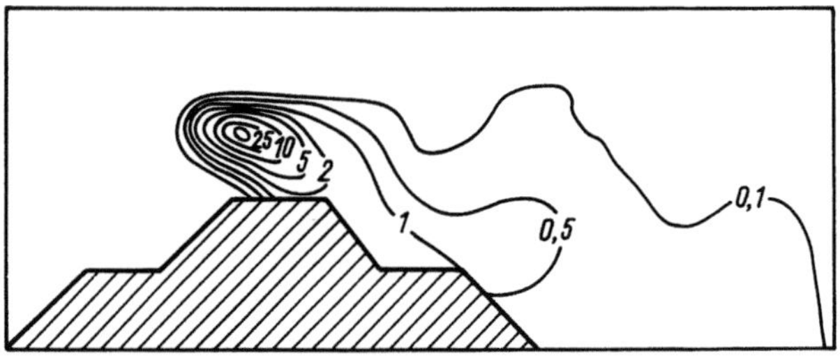

Fig. 4.6. Numerical calculation of the contaminant-concentration distribution on a horizontal plane (x, y) at the depth of 27 m, where the continuously acting contaminant point source is positioned (after [53]). Figures on the contaminant-concentration isolines are given in percentage with respect to the contaminant concentration in the source.

The coupled system of hydrothermodynamic and contaminant-diffusion equations with more grounded closure hypotheses was proposed and analyzed in [57] and [86].

The dynamic equations for the horizontal velocity components u and v, and the heat- and salt-transport equations were written in the form

$$\frac{\partial u}{\partial t} - \Omega v = \frac{\partial}{\partial z}\left(K_z \frac{\partial u}{\partial z}\right); \tag{4.13}$$

$$\frac{\partial v}{\partial t} + \Omega u = \frac{\partial}{\partial z}\left(K_z \frac{\partial v}{\partial z}\right); \tag{4.14}$$

$$\frac{\partial T}{\partial t} = \frac{\partial}{\partial z}\left(K_z^T \frac{\partial T}{\partial z}\right) - u\frac{\partial T}{\partial x} - v\frac{\partial T}{\partial y}; \tag{4.15}$$

$$\frac{\partial S}{\partial t} = \frac{\partial}{\partial z}\left(K_z^S \frac{\partial S}{\partial z}\right) - u\frac{\partial S}{\partial x} - v\frac{\partial S}{\partial y}, \tag{4.16}$$

and the water density ρ was supposed to be a linear function of T and S:

$$\rho = \rho_0[1 + \alpha(T - 27) + \beta(S - 36)]. \tag{4.17}$$

Here T is the water temperature measured in the Celsius scale, S is the salinity measured in promilles, Ω is the Coriolis parameter, ρ is the water density, ρ_0 is the standard density, K_z is the coefficient of the vertical turbulent viscosity, K_z^T and K_z^S are the vertical turbulent diffusivities for heat and salt, respectively, and α and β are the thermal and salinity coefficients of sea-water expansion.

Equations (4.13) – (4.17) were supplemented by the equation for the balance of turbulent energy b and that for the temperature-inhomogeneities measure θ:

$$\frac{\partial b}{\partial t} = K_z\left[\left(\frac{\partial u}{\partial z}\right)^2 + \left(\frac{\partial v}{\partial z}\right)^2\right] + \frac{\partial}{\partial z}\left(K_z^b \frac{\partial b}{\partial z}\right) - \varepsilon - B; \tag{4.18}$$

$$\frac{\partial \theta}{\partial t} = K_z^T\left(\frac{\partial T}{\partial z}\right)^2 + \frac{\partial}{\partial z}\left(K_z^\theta \frac{\partial \theta}{\partial z}\right) - \varepsilon_T, \tag{4.19}$$

where K_z^b and K_z^θ are the coefficients for vertical transport of turbulent energy and temperature-inhomogeneities measure, respectively, ε is the rate of turbulent-energy dissipation, ε_T is the rate of the temperature-inhomogeneities dissipation, and B is the buoyancy flux.

Further, to close equations (4.13) – (4.19) the following expressions were used:

$$K_z = l\sqrt{b}; \tag{4.20}$$

$$\varepsilon = a_\varepsilon\, b^{3/2}/l; \tag{4.21}$$

$$\varepsilon_T = \lambda(Ri) \cdot 6\chi \left(\frac{\partial T}{\partial z}\right)^2; \tag{4.22}$$

$$b = g\alpha\, K_z^T\, \partial T/\partial z + g\beta\, K_z^S\, \partial S/\partial z; \tag{4.23}$$

$$l^2 = b/[\varphi(Ri)N^2]; \tag{4.24}$$

$$Ri = N^2/[(\partial u/\partial z)^2 + (\partial v/\partial z)^2]; \tag{4.25}$$

$$N^2 = \frac{g}{\rho}\frac{\partial \rho}{\partial z}, \tag{4.26}$$

where l is the turbulent scale, Ri is the Richardson number, N is the Väisälä–Brunt frequency, χ is the molecular thermal diffusivity, and $\lambda(Ri)$, $\varphi(Ri)$ and the coefficient α_ε are some functions of Ri. It is also supposed that ratios of the coefficients for turbulent transport of different properties are functions of the Richardson number:

$$K_z^T/K_z = \alpha_T(Ri); \quad K_z^S/K_z = \alpha_S(Ri); \quad K_z^b/K_z = \alpha_b(Ri);$$
$$K_z^\theta/K_z = \alpha_\theta(Ri); \quad K_z^c/K_z = \alpha_C(Ri), \tag{4.27}$$

where K_z^c is the coefficient of vertical turbulent contaminant diffusion.

Apparently, all the functions introduced in the system (4.13) – (4.27) and depending on Ri can be assumed, with a high degree of generality, to be power functions, but, generally speaking, with different exponents. After defining the functions of Ri and prescribing the physical constants, the system (4.13) – (4.27) becomes closed relative to 19 unknowns: u, v, T, s, ρ, b, θ, K_z, K_z^T, K_z^S, K_z^b, K_z^θ, K_z^c, ε, ε_T, B, l, Ri and N. Under appropriate boundary and initial conditions the system can be solved by one or another method.

If we suppose again that the contaminant discharged into the ocean does not influence essentially its hydrothermodynamic characteristics, then a contaminant-diffusion equation like (1.26) will be independent of the system (4.13) – (4.27) and can be solved separately after solving the system (4.13) – (4.27) and finding the velocity components and the coefficient of vertical turbulent diffusion. It is clear that the system of hydrothermodynamic equations (4.13) – (4.27) can be generalized for a three-dimensional case, the terms of vertical advection and horizontal transport of different properties being taken into account. However, the horizontal turbulent transport coefficients are not obtained from the system, but are given *a priori* from auxiliary considerations. The horizontal-diffusion coefficients in the diffusion equation should be set in a similar way. To do this, most often the scale considerations are used in which one determines the coefficients K_x^c and K_y^c using a characteristic scale of the diffusion phenomenon. A functional form of dependence of the diffusion coefficients on the phenomenon scale L should be found according to the recommendations given in Section 1.4. If we use the "4/3 power law" to evaluate the coefficients K_x^c and K_x^c, then it is natural to take the energy-dissipation rate from the solution of the system (4.13) – (4.27).

For any way of evaluating the turbulent diffusion coefficients it is necessary to bear in mind that the finite-difference approximation of the diffusion equation causes the emergence of certain fictitious contaminant fluxes related with variations of the current-velocity and contaminant-concentration fields with scales less than the mesh spacing. These fictitious ('computational') contaminant fluxes along each of the coordinate axes can be written in the form $\overline{u'c'}$, $\overline{v'c'}$, $\overline{w'c'}$, where the primes denote time variations of corresponding fields, and the overbar means averaging over the space-time interval of the meshes. Using the customary semi-empirical expression for fluxes in terms of the field gradients and transport coefficients, we obtain $\overline{u'c'} = -a_x \Delta c/\Delta x$, $\overline{v'c'} = -a_y \Delta c/\Delta y$ and $\overline{w'c'} = -a_z \Delta c/\Delta z$, where the quantities a_x, a_y and a_z are called the computational diffusion coefficients. In a number of cases these coefficients can cause rather intensive diffusion fluxes, which is why it is always desirable to evaluate previously the computational diffusion coefficients and to compare then with the diffusion coefficients in the original differential equation. The latter, generally speaking, must exceed the values of the quantities a_i, because otherwise the contaminant transport will be caused, mainly, by the 'undermeshes' phenomena depending on the parameters of the meshes and not on the physical properties of the process modeled.

In a number of diffusion problems the contaminant concentration can appear to be high enough (especially in the neighborhood of the source) so that it will be impossible to neglect the influence of the contaminant upon the water density. In such a case we would add to the sea-water state equation a term like γc, where γ is a coefficient characterizing the water density variations corresponding to the contaminant-concentration variations. Now the whole system of equations (4.19) – (4.27), (1.26) becomes intercoupled and should be solved simultaneously. We note that in this case the diffusion equation can become essentially nonlinear, since the coefficient of vertical turbulent diffusion K_z^c now becomes a function of concentration c (due to the dependence of K_z on the vertical density stratification of the water). Such a type of nonlinearity can cause a number of interesting effects in the process of contaminant spreading, in particular, the emergence of sharp fronts at the boundaries of diffusing spots or contaminant jets. Investigation of such effects can be of great interest for oceanology, although so far nobody has used nonlinear diffusion equations to solve the problems of the contaminant spreading.

4.2. The Walking-Particle Method

The examples of computations of contaminant spread in the ocean considered in the previous section show the rather great possibilities of the above methods for the solution of various diffusion problems for natural sea conditions. But these methods also have some weak points. The difficult problem is still the determination or choice of turbulent diffusion coefficients. Even if we use for the calculations complicated coupled equations like (4.13) – (4.27), in which current velocities and diffusion coefficients are determined in the course of the solution, it is necessary to prescribe a great number of empirical functions and constants that are not completely known. Implicit schemes of computation of the semi-empirical diffusion equation and analytical methods of its solution both give non-zero values of contaminant concentration in the entire area considered immediately after the beginning of the source action, and, of course, it is unreal. Explicit approximation permits the obtaining of a sharply defined boundary of the diffusion contaminant spot, but the location of this boundary and its spacial velocity are determined mainly by the choice of spacial and time steps of the computational scheme and not by the actual mechanics of the diffusion process. Due to these disadvantages of the computational methods based on the solution of the semi-empirical diffusion equation, the methods of direct numerical analysis of discrete-particle (tracer) motion in turbulent flows became widespread. It is evident that the trajectory of the tracer, whose velocity does not differ from the environmental fluid velocity, is determined by the following equations:

$$\mathrm{d}X/\mathrm{d}t = U(X, Y, Z, t) = \bar{u}(X, Y, Z, t) + u'(X, Y, Z, t); \tag{4.28}$$

$$\mathrm{d}Y/\mathrm{d}t = V(X, Y, Z, t) = \bar{v}(X, Y, Z, t) + v'(X, Y, Z, t); \tag{4.29}$$

$$\mathrm{d}W/\mathrm{d}t = W(X, Y, Z, t) = \bar{w}(X, Y, Z, t) + w'(X, Y, Z, t), \tag{4.30}$$

where X, Y, Z are current-tracer coordinates along the x, y and z axes; U, V and W are instantaneous values of water (tracer) velocity components along corresponding axes in the point X, Y, Z; $\bar{u}$, $\bar{v}$, $\bar{w}$ are mean-velocity components; and u', v' and w' are fluctuation (turbulent) 'additions' to the mean velocities.

Computing equations (4.28) – (4.30) numerically we can use schemes of various kinds. For example, in the scheme of second order approximation in time and with linear interpolation of current velocity between grid nodes the solution of equation (4.28) (and, similarly, of equations (4.29) and (4.30)) is obtained at the $(n+1)$-th time step using the formula [39]

$$X^{n+1} = X^n + U^{n+1}\,\Delta t/\Delta x, \tag{4.31}$$

where $U^{n+1} = (U_1 + U_2)/2$ and

$$U_1 = U^n[L, M](1-S)(1-S_1) + U^n[L+1, M]\, S(1-S_1) + \\ + U^n[L, M+1](1-S)S_1 + U^n[L+1, M+1]SS_1;$$

$$U_2 = U^{n+1}[P, M](1-Q)(1-S_1) + U^{n+1}[P+1, M]\, Q[1-S_1] + \\ + U^{n+1}[P, M+1](1-Q)S_1 + U^{n+1}[P+1, M+1]QS_1;$$

$$L = \text{ent}(X^n); \quad M = \text{ent}(Y^n); \quad S = X^n - L; \quad S_1 = Y^n - M;$$

$$P_1 = X^n - U^n[L, M]\, \Delta t/\Delta x; \quad P = \text{ent}(P_1); \quad Q = P_1 - P_2;$$

where superscripts denote time-step number; Δt is the value of the time step; Δx is spacial step of the grid in the nodes of which the current-velocity components are given, and ent means the entire part of a number.

The mean-velocity components $\bar{u}$, $\bar{v}$, $\bar{w}$ in (4.28) – (4.30), which are considered as given in the nodes of the computational grid, determine the advective transport of a tracer by the current. As to the fluctuation-velocity components u', $v'$and w', due to which the tracer diffusion 'walking' occurs, their assignment is a more complicated problem. As fluctuation-velocity components are by definition, random, they should be described by some distribution laws rather than by deterministic expressions. The most commonly used for this purpose is the normal law, which describes the fluctuation-velocity component distribution with fairly good accuracy. As the expectation of fluctuation-velocity components is equal to zero their normal distribution laws are determined by their variances only (in general, they are different for each velocity component). Under the assumption that the distribution laws for the fluctuation-velocity components are normal and with given values of variances, the numerical calculation of the tracer trajectories may be performed with the aid of a random-number generator. The values of u', v' and w' obtained with the aid of this generator for the moment $t=0$ are added to the values of $\bar{u}$, $\bar{v}$ and $\bar{w}$, and then, according to equation (4.31) (and analogous equations for Y and Z coordinates), we may calculate the tracer location for the moment Δt. Then, with the new set of values of random fluctuation velocities, one may calculate the next step of tracer trajectory, etc. Of course, for simulation of the action of an instantaneous or continuous point or spacially distributed contaminant source we must suppose that the source releases a great number of tracers per unit time. In this case, the number of tracers in unit fluid volume and in some time moment is assumed to be proportional to the concentration of continuously distributed contaminant that would be diffused in the same time moment and in the same velocity field from a source with the same intensity and space-time characteristics. If the averaged velocity field is stationary the calculation for an instantaneous contaminant point source can be made in a different way. Let us suppose that a source releases one tracer in some time moment. Then we calculate its trajectory until the given moment t. After this, let us repeat the calculation until the given moment for the same tracer but with other random values of velocity components u', v' and w' for every node of computational grid. After many such calculations we can obtain the repetition of the tracer hit in every elementary computational space element. When the number of experiments tends to infinity this repetition will tend to the probability of the hit that we may identify with the contaminant concentration in the solution of the customary diffusion problem, as was previously explained in more detail in Section 1.7.

If the tracer walking area is limited, then, of course, it is necessary to assign the boundary conditions for tracer motions on the area boundaries. If the boundary is impenetrable for the contaminant whose diffusion is simulated by tracer motions, then the boundary conditions are conditions of the mirror reflection of tracers from the boundary. If the tracer trajectory crosses the area boundary at the next computational step then, in this case, in the next time moment we must assign to the tracer the coordinates of the point in the mirror that is the reflection of the point in which the tracer would be at this moment in the absence of the boundary. If in the case of continuously distributed contaminant concentration in the customary diffusion problem its value on the area boundary is zero we must assume that the boundary absorbs the tracers. In the case of such a boundary condition every tracer that crosses the area boundary at some computational step is assumed to be absorbed and does not take part in the following computations. The condition of the constant contaminant concentration on the area boundary may be simulated by an arrangement on the of boundary some tracer sources, which cause the required concentration to form near the boundary. In this case the tracers coming from the outer area must be absorbed on the boundary. A more complicated boundary condition occurs in the case of a 'semi-penetrable' basin coast. For example, such a condition may occur in problems on the diffusion of oil that is partly absorbed and partly reflected by the coast and again spreads seaward. Certainly, in the case of such a boundary condition we must assign the 'reflection coefficient', showing which part of the tracers is reflected from the coast and which is absorbed. In the computer program a block must be present that 'kills' every λ-th part of the tracers arrived at the unit length of coastline or unit surface of bottom.

In the algorithm of numerical calculation of tracer walking described above, at every time step fluctuation velocities are assigned by a random-number generator independently from their values on the previous step. Such random walking of particles that do not 'remember' their previous step is called Markovian. In a turbulent flow the walking process may be considered Markovian when the Lagrangian time-scale (1.157) is less than time step Δt of the computational scheme. Physically this means that the 'lifetime' of turbulent eddies in the flow must be less than the interval Δt. For large eddies in the ocean this condition, naturally, may be rather severe, but one does not forget that in the numerical calculation of tracer walking the spacial scale of turbulent 'additions' to the velocity field is prescribed by the spacial step of the computational grid. When the spacial scale of fluctuation velocities is small one can, naturally, expect small values of their 'lifetime'. So using walking tracer methods for the solution of diffusion problems the Markov hypothesis may be applied under condition of a fairly detailed prescription of the field of mean current velocity components. If this field is available only with a large averaging scale then for using the Markov hypothesis we must increase the computational steps Δt, thereby decreasing the accuracy and resolution of the computation. If, for some reasons, we cannot consider the process as Markovian, then more complicated methods with dependent-particle random walking may be used.

Variance of tracer location in the direction of the x-axes for diffusion times less than the Lagrangian scale is described, as shown in Section 1.7, by the formula (1.153) or, correspondingly, by the formulas (1.154) and (1.155). These formulas involve the Lagrangian velocity correlation function, which must be prescribed for the calculation of tracer motion in the case of dependent walking. Increments in the coordinates of the tracer, ΔX, ΔY and ΔZ, after N time steps can be obtained, according to formulas (4.28) – (4.30), from the following expressions:

$$\begin{aligned}
\Delta_N X &= \sum_{i=1}^{N} \bar{u}_i \Delta t + \sum_{i=1}^{N} u_i' \Delta t = \sum_{i=1}^{N} \Delta \bar{X}_i + \sum_{i=1}^{N} \Delta x_i'; \\
\Delta_N Y &= \sum_{i=1}^{N} \bar{v}_i \Delta t + \sum_{i=1}^{N} v_i' \Delta t = \sum_{i=1}^{N} \Delta \bar{Y}_i + \sum_{i=1}^{N} \Delta y_i'; \\
\Delta_N Z &= \sum_{i=1}^{N} \bar{w}_i \Delta t + \sum_{i=1}^{N} w_i' \Delta t = \sum_{i=1}^{N} \Delta \bar{Z}_i + \sum_{i=1}^{N} \Delta z_i';
\end{aligned} \tag{4.32}$$

or after squaring and averaging

$$\begin{aligned}
\Delta_N X^2 &= \overline{\left(\sum_{i=1}^{N} \Delta \bar{X}_i\right)^2} + \overline{\left(\sum_{i=1}^{N} \Delta x_i'\right)^2}; \\
\Delta_N Y^2 &= \overline{\left(\sum_{i=1}^{N} \Delta \bar{Y}_i\right)^2} + \overline{\left(\sum_{i=1}^{N} \Delta y_i'\right)^2}; \\
\Delta_N Z^2 &= \overline{\left(\sum_{i=1}^{N} \Delta \bar{Z}_i\right)^2} + \overline{\left(\sum_{i=1}^{N} \Delta z_i'\right)^2},
\end{aligned} \tag{4.33}$$

where the overbar means statistical averaging, and ΔX_i and Δx_i (as well as ΔY_i and Δy_i, ΔZ_i and Δz_i) are assumed to be uncorrelated with one another.

In the case of independent particle walking the last term in the right-hand side of equations (4.33) transform to a simple expression like (1.159), and in the case of dependent walking these terms must be calculated according formulas such as (1.153) – (1.155). It should be remembered that here for the sake of simplicity we assume that the particles do not interact with one another and also that medium statistical characteristics do not depend on coordinates. Moreover, the sizes of the tracers are assumed to be small and their density is close to the fluid density, so particle-motion characteristics do not, practically, differ from characteristics of motion of fluid moles in the correspondent space points. It is evident that the last assumption is valid for simulating the diffusion of distributed passive indicators (for example, dyes) in the ocean by walking particles. On the other hand, the laws of motion of large solid markers (blocks of ice, for example) are, of course, different, as has been said in Section 2.4.

The walking-particle method permits us to take into account in numerical calculations a possible decay (or 'multiplication') of the diffusing contaminant. This phenomenon may be simulated by the particle 'loss' in the diffusion process. The number of 'lost' particles is commonly assumed to be proportional to their quantity. In this case the distribution of probability of particle 'survival' has the form

$$p = e^{-\alpha t}, \tag{4.34}$$

where the value of α determines the lifetime of particles.

In general α may be, in turn, a function of time and coordinates. In this case the probability of tracer survival depends on α values in all points of the tracer trajectory, so for two particles coming from the same point but moving after that along different trajectories the probability of survival during a given time interval is different.

So in the case the survival probability for each particle must be calculated separately according to the formula

$$p = \exp\left\{-\int_0^t \alpha(x, y, z, \tau)\, \alpha\tau\right\}, \tag{4.35}$$

where in numerical calculation the integration is replaced by summation according to the relationship

$$\int_0^t \alpha(x, y, z, \tau)\, \alpha\tau = \sum_{i=1}^{N} \alpha(x_i, y_i, z_i, i\Delta t)\, \Delta t. \tag{4.36}$$

When the probability reaches a threshold value p the particle is assumed to be lost in the numerical calculation and is not considered after that.

When the parameter α is constant the probability of particle survival does not depend on its trajectory and is determined by the lifetime of the particle in the ocean. So for particles leaving a source simultaneously the survival probability will change with time in the same way independently of their trajectories. As a consequence, in the case of a constant (or depending only on time but not on coordinates) value of α it is unnecessary to calculate the survival probability for each particle, but it is sufficient to know the functional dependence of the probability on diffusion time. So if n_0 tracers were released in the ocean at an initial time moment, then after time t there must remain

$$n = n_0 \exp\left\{-\int_0^t \alpha(\tau)\, \alpha\tau\right\} \tag{4.37}$$

particles in the ocean. If a contaminant source is continuous, the n_0 in the formula (4.37) means the number of particles leaving the source during the calculational time interval Δt. For tracers leaving the source in the next intervals the number of survival particles may also be calculated according to formula (4.37) but in this case the integration must be performed from another initial moment. All lost tracers are not considered in the following calculations.

The calculation of contaminant diffusion by the walking-particle method for Lake Baikal was performed by Gadibirova and Galkin [99] in 1970. The behavior of contaminant released by a plant in a coastal area of the lake was simulated. The problem was solved in the Cartesian coordinate system with the x-axis directed along the coast line, the y-axis directed normally to the coast and the z-axis directed downward from the water surface. Since there was no information about current velocity in the whole area investigated the authors proposed giving the velocity field in the form $u_i = f(t)\varphi_i(x, y, z)$, where $f(t)$ is a random function of time and $\varphi_i(x, y, z)$ is some non-random function of coordinates, $i = 1, 2, 3$.

The form of the function $\varphi(x, y, z)$ was specified with the aid of the assumption that current velocity was independent of the coordinates and that the total water flux through the planes parallel to the planes (z, x) was equal to zero. To prescribe the random function $f(t)$ observational data on current velocity, obtained with hydrological current meter installed at 10 m depth near the contaminant source, were used. The current velocity was measured in steps of 10 min, the total duration of observation run up to 5 days. Because of such discreteness of observational data on current velocity the time step in the numerical calculation of contaminant diffusion was also chosen equal to 10 min. Figure 4.7 shows the horizontal distribution of contaminant concentration (in percents of the concentration of released contaminant) 5 days after the beginning of source action. It is seen from the figure that the jet of discharge waters stretches along the coast under action of the mean current directed coastwise. At the same time, since the amplitude of the current-velocity component along the axis had variations there were pseudoperiodical structures in the jet. It is

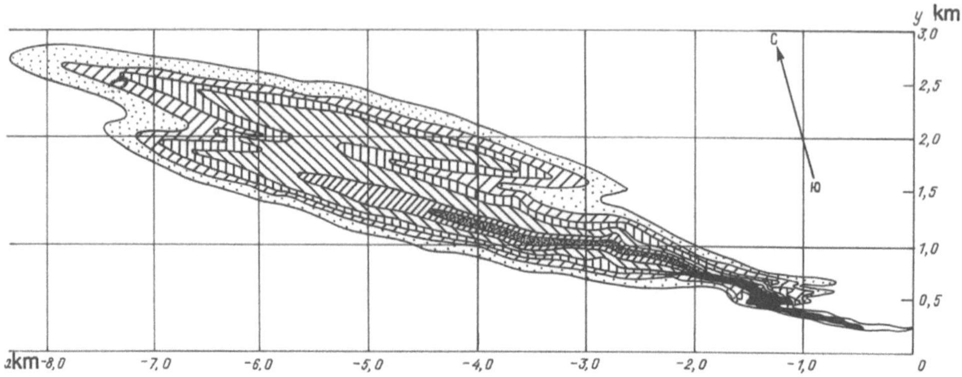

Fig. 4.7. Distribution of contaminant concentration in horizontal plane 5 days from the beginning of action of continuous point source (in percentages of the initial concentration). The calculation was performed for the coastal area of Lake Baikal by the walking-particle method. Isolines contour successively zones with the following values of concentration: > 50, 10–50, 5–10, 1–5, 0.5–1, 0.1–0,5, 0.01–0.1% (after the paper by Gadibirova and Galkin from [100]).

interesting that the contaminant diffusion in the jet was anisotropic. For example, the determination of effective diffusion coefficients using the final values of particle coordinates gives the following values: $K_x = 2.75\ 10^6\ \text{cm}^2/\text{s}$, $K_y = 3.7\ 10^3\ \text{cm}^2/\text{s}$, $K_z = 46\ \text{cm}^2/\text{s}$. More intensive horizontal exchange in the x-axis direction may be due to both anisotropy or large-scale turbulence in the sea coastal zone and to the influence of the vertical velocity shear on the effective coefficient in horizontal exchange, as was considered in detail in Section 2.2.

The results of the solution of some other diffusion problems for Lake Baikal performed by the walking-particle method are given in the monograph [18]. In one of the problems the seasonal variability of the distribution of a contaminant discharged into the sea is considered. In this problem the mean current velocities and the fluctuation velocities were prescribed as sine-shaped functions with the maximum in November and the minimum in June. In the second problem the formation of a cloud of slowly decaying contaminant was investigated. It was found that the contaminant area reached a steady state only after two year from the beginning of action of the contaminant source. In both problems the diffusion was found to be anisotropic and, as a result, the contaminant spot was retained against the coast, polluting the lake coastal regions.

Zhurbas, Mamedov and Tataraeve [47] used the walking particle method to calculate the contaminant diffusion in the Caspian Sea. The behavior of a contaminant released by a continuous point source located near the water's edge in the south-western part of the Caspian Sea was investigated. The field of mean current velocities was assumed to be known and the coast line was fitted by a broken line. The components of turbulent current velocities were produced by a generator of random normally distributed numbers with prescribed variances. To take into account an anisotropy of turbulent exchange in the sea coastal area, the root-mean-square deviation for fluctuation velocity along the coastline σ_x was assigned as to 5 cm/s and for fluctuation velocity normal to the coastal line as equal to σ_y = 10 cm/s. The choice of these values for the root-mean-square deviation was based on the known data on the coefficients of horizontal diffusion in this part of the sea, K_x and K_y, which in turn were related to the σ_x

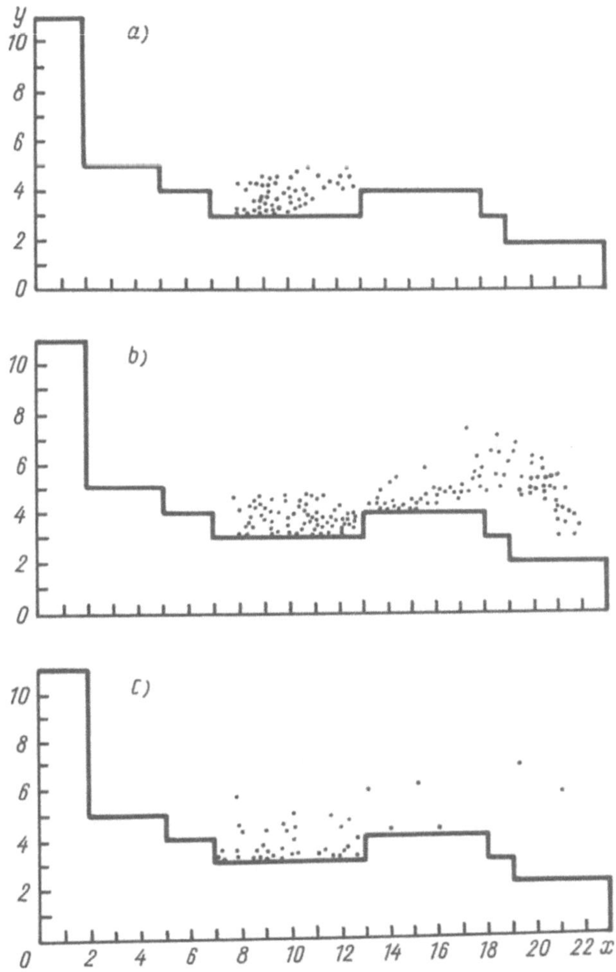

Fig. 4.8. Results of the computations of motion of tracers released by a point source located near the coast: (a) 60 hours from the beginning of the source action and under the condition that on the coastline 1/400 part of all tracers was absorbed, (b) 240 hours from the beginning (1/400 part was absorbed), (c) 240 hours from the beginning (all particles were absorbed) (after the paper by Zhurbas, Mamedov and Tatarayev from [47]).

and σ_y by the relationships $K_x = l\sigma_x$ and $K_y = l\sigma_y$, where l is the characteristic scale of the phenomenon. In the numerical calculation the value of l was taken to be equal to the spacial step of the grid, which was equal to 2 miles. Figure 4.8 shows the pattern of tracer location 50 and 240 hours after the beginning of a source action. The two former figures were obtained under the assumption that at the coastline 1/400 part of the tracers was absorbed when tracers touched the coastline in the diffusion process. The last figure shows the results of the calculation obtained under the assumption that all tracers arriving at the coastline during the 240 hours of diffusion are absorbed at this line. These figures show that when a contaminant source is located near the water's edge, the coastal areas of the sea are highly polluted, and the size of this area and pollution intensity are strongly dependent on the boundary conditions. Because the tracers are absorbed on the coastline the contaminant concentration in the sea decreases, but, of course, in this case the beach areas become heavily polluted from the contaminant that, naturally, accumulates there.

Ivanov and Filippov [36] computed contaminant diffusion in the western part of the Black Sea by the walking particle method. The second-order approximation of time derivatives and linear interpolation of current-velocity values between the grid nodes were used in these computations. The possible decay of contaminant in the diffusion process was also taken into account in the computational scheme. The field of mean-velocity components is rather complicated since it is formed under the action of several currents and run-off of the Danube river. Fluctuation velocities are again assumed to be distributed according to the normal laws with prescribed variances. The variances were obtained using experimental data on current velocities recorded by hydrological current meters. The data sets were smoothed by the filter with a time-scale equal to 30 min, that is in accordance with the space step of computational grid equal to 5 km. As a result of this analysis the values of root-mean-square fluctuation velocities were accepted as σ_x = 2.4 cm/s, σ_y = 3.3 cm/s for coastal regions, and σ_x = 15 cm/s, σ_y = 15.1 cm/s for seaward regions. The numerical computations of tracer motions were performed for four continuous contaminant sources (which started at the moment $t=0$), three of which were located in the coastal region of the sea and one of which was in seaward region. For every source the computations were extended in time up to 6 days, and after that time steady-state patterns of contaminant distributions were formed. The intensity of every source was equal to 1 tracer per 30 min. At the coastline the condition of slip of tracer along the coast was accepted.

Figure 4.9 shows the result of the computation of contaminant diffusion from two sources for diffusion times 1, 3 and 6 days. The first source was located at 10 km from the coastline in the zone of current, directed along the coast, and the second source was located in close proximity to the coastline. From the first source the contaminant was spread mainly along the current, and from the

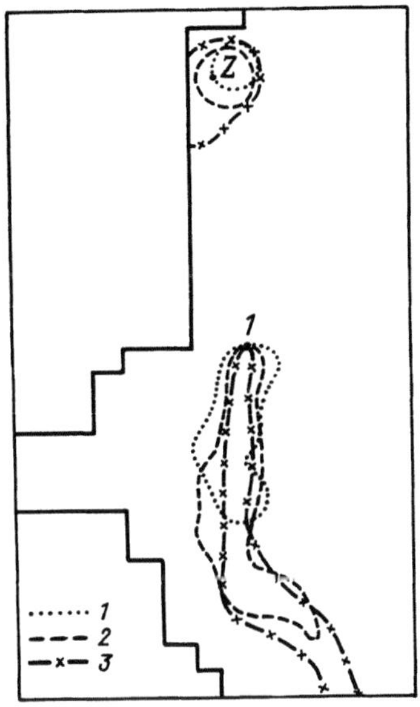

Fig. 4.9. The areas of the contaminant spread formed by two point sources and calculated by the walking particle method: (1) 1 day after the beginning of the source action (2) 3 days after, (3) 6 days after (after [36]).

second source, in the vicinity of which the velocity of constant current was small, the contaminant spot was formed mainly under the action of turbulent velocity fluctuations. From this source the contaminant reached the coastline, polluting off-shore regions. An analysis of the computation results, obtained for all four contaminant sources, shows that significant contaminant concentrations reach the shore if the contaminant sources are located not further 5 km from the coastline. If the sources are located further seawards the contaminant is transported by current, being quickly diluted and usually not reaching the shore. Of course, these conclusions refer, strictly speaking, only to the sea region considered above, but it is likely that they are valid for all off-shore regions with hydrological conditions similar to the ones considered above.

The results of the contaminant-diffusion computations considered above, performed by the walking particle method, were based on the model of turbulent velocity field described by the Gaussian distribution law with prescribed second moments. But such a field may be described also by its spectral characteristics. Such an approach to the numerical simulation of contaminant diffusion in a random turbulent velocity field was proposed in [65] and [132]. In this approach it is assumed that the field has a spectral expansion. All the space considered is divided in N parts, and the realization of the velocity field, at the assumption of its isotropy, homogeneity and stationarity, is presented in the form

$$\mathbf{u}(\mathbf{x}, t) = \sum_{k=1}^{N} P_k^{1/2}[\xi_k \sin(\lambda_k \mathbf{x} + \omega_k t) + \eta_k \cos(\lambda_k \mathbf{x} + \omega_k t)], \tag{4.38}$$

where ξ_k and η_k are vectors whose components are random independent Gaussian quantities; $\mathbf{x}$ and λ_k are points in three-dimensional space; ω_k are random quantities, chosen independently of λ_k, in given segments of a division of time axis; P_k is the total density distribution of random points in space; k is number of subspace; and t is time.

For every realization of the velocity field the trajectory of contaminant particle is then calculated according to equation (4.38). If the trajectories of two independent particles are simulated one may obtain the correlation function of the field. And if the correlation function or the spectrum of the field are known it is possible to obtain using them the statistical characteristics of the contaminant-concentration field and the coefficients of turbulent diffusion by the random walking method.

The correlation function of the velocity field is necessary also for computation of the contaminant diffusion by the method used by Demyshev, Eremeev, Ivanov and Knysh [22] and proposed previously by Saffman and Thiebach (see references in [22]). In this method it is supposed that in the expression for the turbulent contaminant flux $\overline{u'c'}$ the quantity c' is a function of fluctuation velocity $u_i'(\mathbf{x}, t)$. In general, in this case the contaminant flux is determined by the random velocity field moments of all orders. But under assumption that the field is quasi-Gaussian the situation becomes simpler and the expression for $\overline{u'c'}$ takes the form

$$\overline{u_i'(\mathbf{x}, t)\, c'[u_i'(\mathbf{x}, t)]} = -\int_0^t \overline{u_i'(\mathbf{x}, t)\, u_j'(\mathbf{x}, t)}\, \overline{\frac{\partial c(\mathbf{x}, t')}{\partial x_j}}\, dt'. \tag{4.39}$$

If $t \gg T_u$, where T_u is the Lagrangian velocity time-scale, the expression (4.39) transforms into customary semi-empirical expression for contaminant flux with the turbulent diffusion coefficient determined by equation (1.156) or (1.160). The calculation of contaminant horizontal diffusion using the formula (4.39) was performed in [22] for the ocean areas with synoptic eddy formations in the velocity field. In the calculation the spacial step was equal to 6.8 miles, and the time step was equal to 12 hours. For the different initial locations of contaminant spots relative to synoptic eddies the spot motion and diffusion character were different. The spots located on the circumference of the eddy were often drawn into the central part of the eddy, and the spots were stretched in the direction of the motion fluid in the eddy. If the contaminant source was located initially in the core of the eddy the contaminant diffusion is inhibited, and the contaminant remained in the core of the eddy for the long time. If the contaminant spot was located initially at the boundary of two eddies, it could be divided into two parts, which then moved and diffused independently.

Demyschev, Eremeev, Ivanov and Knysh [23] calculated the three-dimensional diffusion in the field of synoptic eddies in the ocean by the walking particle method. The calculation was performed using expressions similar to (4.28) – (4.30) but written in some other form:

$$\frac{dX_i}{dt} = \eta_i(\mathbf{x}, t) + g_{ij}(\mathbf{x}, t) f_j(t); \; X_i\big|_{t=t_0} = X_i^0 \tag{4.40}$$

$$i, j = 1, 2, 3,$$

where X_i are coordinates of random trajectory of contaminant particle; $\eta_i(\mathbf{x}, t)$ and $g_{ij}(\mathbf{x}, t)$ are some non-random functions; $f(t)$ is δ-correlated Gaussian process with zero mean value; and X_i^0 are initial coordinates of the particles.

If we suppose that the functions η_i and g_{ij} are related with the field of mean velocities u_i and with the turbulent diffusion coefficients by the relationships

$$\eta_i(\mathbf{x}, t) = \bar{u}_i(\mathbf{x}, t) + \tfrac{1}{2}\, g_{ij}(\mathbf{x}, t)\; \partial g_{kl}(\mathbf{x}, t)/\partial x_k; \tag{4.41}$$

$$g_{il}(\mathbf{x}, t)\; g_{i,j}(\mathbf{x}, t) = K_{ij}(\mathbf{x}, t), \tag{4.42}$$

then the formula (4.40) will correspond to turbulent diffusion equation of the form

$$\frac{\partial \bar{c}}{\partial t} + \bar{u}_i \frac{\partial c}{\partial \mathbf{x}_i} = \frac{1}{2} \frac{\partial}{\partial \mathbf{x}_i} K_{ij} \frac{\partial \bar{c}}{\partial \mathbf{x}_j}\,. \tag{4.43}$$

To solve equation (4.40) – (4.42) it is necessary to know the field of mean velocity and the elements of the matrix g_{ij}. If the diffusion coefficient matrix K_{ij} is known, then the values of g may be obtained from the relationship

$$g = \sum_{k=1}^{3} \frac{\sqrt{\lambda_k}\; \prod_{i \neq k} (K - \lambda_k E)}{\prod_{i = k} (\lambda_i - \lambda_k)}, \quad i = 1, 2, 3, \tag{4.44}$$

where E is the unit matrix, and λ are the eigenvalues of the matrix K.

For the quasi-Gaussian field of the current-fluctuation velocities and for small values of the correlation scale the diffusion coefficients may be estimated according to the relationship

$$K_{ij} = \tfrac{1}{2}\, \overline{u_i'(\mathbf{x}, t)\, u_j'(\mathbf{x}, t)}\; \Delta t. \tag{4.45}$$

Numerical calculations using the above-given relationships were performed in [23] with the time step equal to 1 day and with the number of particles in the ensemble equal to 5000. The velocities of particles located between the grid nodes (the horizontal steps were equal to 4.25 miles, and the vertical step was equal to 10 m) were obtained by the parabolic interpolation. The particles that reached the boundary of the area considered were excluded from further calculations. To determine the contaminant-concentration field the mass of every particle was distributed between the eight nearest nodes of the grid with the aid of three-linear interpolation. Then the sum of the contributions from all the particles in the particular node was divided by the volume of the grid spacial cell. The quantity thus derived was then accepted as the contaminant concentration in the cell considered.

In [23], the numerical calculations of contaminant-spot diffusion were again performed for different initial locations of the spots relative to synoptic-eddy formations in the ocean. It came out that the spot was trapped by the eddy under the specific relationship among the distance from the spot to the center of the eddy, the intensity of horizontal diffusion and the hydrodynamical parameters of the eddy. In most cases, when the effective horizontal-diffusion coefficient was equal to 10^5 cm^2/s, the entrapment of the contaminant spot occurs for distances of no more than 65 – 70 miles from the center of the eddy. For larger distances from the center of the eddy or lesser values of the diffusion coefficient the contaminant spot was not trapped by the eddy but began to escape from it. At the initial stages of the entrapment the contaminant spot is curved, as a rule, embracing the central part of the eddy. Further, the contaminant can fill in the center of the eddy and remain there a rather long time, weakly diffusing. Figure 4.10 shows an example of the initial stages of the entrapment of the contaminant spot in the synoptic eddy. In this example the initial distribution of the contaminant was given in the form of a disk located at the depth of 400 m; the effective coefficient of the horizontal diffusion was equal to 10^4 cm^2/s; the coefficient of the vertical diffusion was equal to 1 cm^2/s. Ten days after the beginning of the diffusion the spot revealed a characteristic bending in the direction of the water motion in the synoptic eddy. In the next stages of the diffusion some localized maxima of the contaminant concentration occurred due to the inhomogeneity of the current-velocity field that began to be entrapped in the central part of the eddy. In the three-dimensional case as well as in the two-dimensional one, if the contaminant spot is initially located between two synoptic eddies either the complete breakup of the spot or the formation of two maxima in the spot, which nevertheless remains unified and not divided into two parts, can be observed.

In using the walking particle method for the solution of diffusion problems questions often arise that must be answered before the numerical computations are begun. Specifically, based on the required accuracy of the evaluation of the

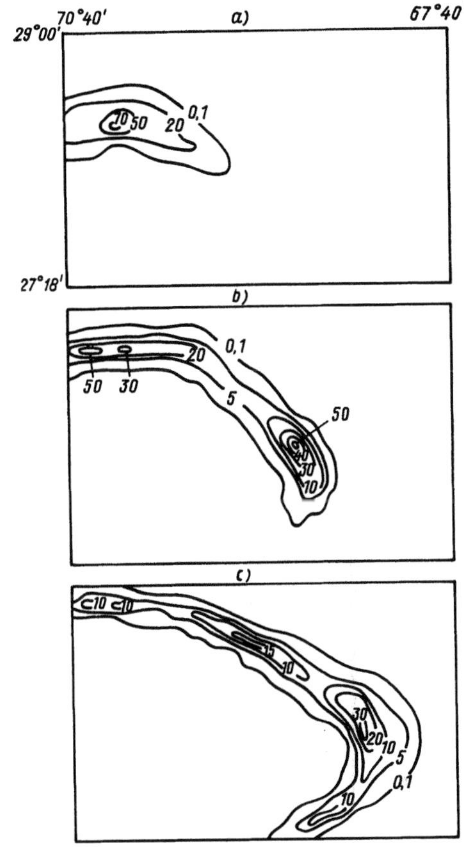

Fig. 4.10. The results of the numerical computations of the transformation of the contaminant spot, located at the circumference of the synoptic eddy: (a) 10 days after the beginning of the diffusion, (b) 20 days after, (c) 30 days after (after [23]).

contaminant concentration in different sea areas it should be necessary to estimate how many tracers should be released by the source per unit time. The choice of spacial-time steps of the computational grid is also related to this question. Indeed, if in every elementary cell of the grid a small number of tracers is present then the accuracy of the estimation of the contaminant concentration is also not great. As a first approximation, let us assume that the relative accuracy of estimation of the concentration is equal to the ratio of the square root of the tracer number in the cell to this number. Thus, with 10% accuracy of estimation of the concentration about 100 tracers must be in a cell, and with 1% accuracy there must bu up to 10,000 tracers. It is also very important to be sure, before the beginning of the calculation, that the hypotheses used, for example, the hypothesis on the independence of the statistical characteristics of particle motions (the moments of the distribution of their velocities) on the spacial coordinates in the area considered and also on the diffusion time, are held. It is also necessary to compare the Lagrangian time-scale of the current investigated with the chosen space-time step of the computational grid. Without such comparison it is difficult to draw a conclusion about the dependence or independence of the consequent walking of the tracer in the current.

In the walking-particle method it is usually supposed that the turbulent 'additions' to the particle-coordinate increments in the formulas (4.32) have a random character and are determined by their statistical characteristics. But recently (see [147]) it was shown that the random (chaotic) walking of particles can also be simulated by special deterministic functions; e.g. we may assign

$$X_{t+1} = X_t + \mu \sin X_t, \tag{4.46}$$

where X_t is the particle coordinate in the time moment t; X_{t+1} is the coordinate in the next moment $t+1$; and t takes the values 0, 1, 2, ...

When the value of the parameter μ is about 4.6 the model (4.46) provides particle motion of a random-like type, and the variance of particle coordinate increases linearly with the diffusion time t. It was found that some expressions can be chosen for X_{t+1} that also give a power growth of the variance like t^m with $1 < m < \infty$. Thus, a more general expression for X_{t+1} was proposed in [147]:

$$X_{t+1} = X_t + a|X_t|^p \sin(X_t), \tag{4.47}$$

where a and p are some parameters with $0 \leqslant p \leqslant 1$. If $p=0$ the formula (4.47) transforms into (4.46), and if $p > 0$ the multiplier before the sine increases the 'steps' of the particle in its walking. This causes the variance of the particle coordinate to increase according to a power law of the form t^m, where the power m can take values from 1 at $p=0$ to infinity when p tends to 1. In this case the values of the parameters a must lie in some interval beyond which the diffusion character of the particle motion is disturbed.

As we have seen, the walking-particle method is based on the Lagrangian approach to the investigation of diffusing-particle motion. As known, in this method the characteristics of moving particles of fluid are investigated, while in the Eulerian method the characteristics of the fluid medium are investigated in fixed points of the space. In the Lagrangian method every fluid particle must be 'marked', and that gives us the opportunity to trace the evolution of its motion in time. Usually, the Cartesian coordinates of the particle $x_0 = a$, $y_0 = b$, $z_0 = c$ in some initial time moment are used as much 'markers' (tracers). In this case the coordinates of the following positions of the particle x, y, z in an arbitrary moment t will be functions of a, b, c and t:

$$x = f_1(a, b, c, t); \; y = f_2(a, b, c, t); \; z = f_3(a, b, c, t), \tag{4.48}$$

and it is evident that the functions f_1, f_2, f_3 at $t = t_0$ transform identically to a, b, c.

Transition from the Lagrangian to the Eulerian variables can be made by solution of equation (4.48) relative a, b, c. Such solution must be unique and may be written in the form

$$a = \varphi_1(x, y, z, t); \; b = \varphi_2(x, y, z, t); \; c = \varphi_3(x, y, z, t), \tag{4.49}$$

where φ_1, φ_2, φ_3 are some functions.

From the mutual solvability of equations (4.48) and (4.49) it follows that functional determinants

$$J = \begin{vmatrix} \frac{\partial x}{\partial a} & \frac{\partial y}{\partial a} & \frac{\partial z}{\partial a} \\ \frac{\partial x}{\partial b} & \frac{\partial y}{\partial b} & \frac{\partial z}{\partial b} \\ \frac{\partial x}{\partial c} & \frac{\partial y}{\partial c} & \frac{\partial z}{\partial c} \end{vmatrix} \tag{4.50}$$

$$J_1 = \begin{vmatrix} \frac{\partial a}{\partial x} & \frac{\partial b}{\partial x} & \frac{\partial c}{\partial x} \\ \frac{\partial a}{\partial y} & \frac{\partial b}{\partial y} & \frac{\partial c}{\partial y} \\ \frac{\partial a}{\partial z} & \frac{\partial b}{\partial z} & \frac{\partial c}{\partial z} \end{vmatrix} \tag{4.51}$$

do not identically become zero or infinity.

The basic hydrodynamic equations, in particular, the continuity equation and the equations of motion, may be written in the Lagrangian variables. For example, the continuity equation in the Lagrangian variables takes the form

$$J = 1. \tag{4.52}$$

It is also possible to write down the equation for contaminant diffusion in a moving fluid in the Lagrangian variables. In this case the advective terms in the Eulerian diffusion equation vanish in the equation written in the Lagrangian form, so in the latter case the frame moves as if with the migrating particles. But as payment for this simplification of the diffusion equation the diffusion terms that occur in this equation are more complicated than in the Eulerian diffusion equation. The diffusion equation in the Lagrangian form was proposed by Corrsin [120] for the instantaneous values of the concentration and the fluid velocities in the two-dimensional flow. A more general derivation of the diffusion equation in the Lagrangian form for the two-dimensional flow was given by Okubo and his colleagues [149]. This derivation is based on the two-dimensional equation for the molecular diffusion in the commonplace Eulerian form:

$$\frac{\partial c}{\partial t} + \frac{\partial}{\partial x_i}(u_i c) = D_{ij}\frac{\partial^2 c}{\partial x \partial y}; \quad i, j = 1, 2, \tag{4.53}$$

where c is the contaminant concentration depending on two coordinates and time; u_i are the components of the instantaneous flow velocity; D_{ij} is the tensor of the molecular-diffusion coefficients; and the repeated indexes mean the summation.

Since only a two-dimensional flow is considered, the velocity field in equation (4.53) may be both divergent and convergent, and thus, in general, $\partial u_i/\partial x_i \neq 0$. It is evident that equation (4.53) can be rewritten in the form

$$\frac{\partial c}{\partial t} + u_i\frac{\partial c}{\partial x_i} + \left(\frac{\partial u_i}{\partial x_i}\right) c = D_{ij}\frac{\partial^2 c}{\partial x_i \partial x_j}. \tag{4.54}$$

To transform equation (4.54) to the Lagrangian form let us write the identity for the contaminant concentration in the Eulerian (c) and Lagrangian (Γ) forms:

$$c(\mathbf{x}, t) \equiv \Gamma(\mathbf{x}_0, t), \tag{4.55}$$

where $\mathbf{x}$ is the two-dimensional vector of the Eulerian variables (with x and y components), and $\mathbf{x}_0$ is the two-dimensional vector of the Lagrangian variables (with x_0 and y_0 components; for further convenience they will be denoted as a and b); t is the time.

Since the Lagrangian frame follows the moving fluid particles (contaminant particles) then

$$\frac{\partial \mathbf{x}(\mathbf{x}_0, t)}{\partial t} = \mathbf{u}(\mathbf{x}(\mathbf{x}_0, t), t) \tag{4.56}$$

under the initial condition

$$\mathbf{x}(\mathbf{x}_0, 0) = \mathbf{x}_0. \tag{4.57}$$

From these equations it follows that

$$\left(\frac{\partial \Gamma}{\partial t}\right)_{\text{Lag}} = \left(\frac{\partial c}{\partial t} + u_i\frac{\partial c}{\partial x_i}\right)_{\text{Euler}} \tag{4.58}$$

Then using the formulas for coordinate transformations

$$\frac{\partial}{\partial a} = \frac{\partial x}{\partial a}\frac{\partial}{\partial x} + \frac{\partial y}{\partial a}\frac{\partial}{\partial y}; \tag{4.59}$$

$$\frac{\partial}{\partial b} = \frac{\partial x}{\partial b}\frac{\partial}{\partial x} + \frac{\partial y}{\partial b}\frac{\partial}{\partial y}$$

and solving (4.59) relative to $\partial/\partial x$ and $\partial/\partial y$ we obtain

$$\begin{aligned} \frac{\partial}{\partial x} &= \left(\frac{\partial x}{\partial a}\frac{\partial y}{\partial b} - \frac{\partial x}{\partial b}\frac{\partial y}{\partial a}\right)^{-1}\left(\frac{\partial y}{\partial b}\frac{\partial}{\partial a} - \frac{\partial y}{\partial a}\frac{\partial}{\partial b}\right); \\ \frac{\partial}{\partial x} &= \left(\frac{\partial x}{\partial a}\frac{\partial y}{\partial b} - \frac{\partial x}{\partial b}\frac{\partial y}{\partial a}\right)^{-1}\left(\frac{\partial x}{\partial a}\frac{\partial}{\partial b} - \frac{\partial x}{\partial b}\frac{\partial}{\partial a}\right). \end{aligned} \tag{4.60}$$

For further convenience we will use the following notations:

$$J = \left(\frac{\partial x}{\partial a}\frac{\partial y}{\partial b} - \frac{\partial x}{\partial b}\frac{\partial y}{\partial a}\right) = \begin{vmatrix} \frac{\partial x}{\partial a} & \frac{\partial y}{\partial a} \\ \frac{\partial x}{\partial b} & \frac{\partial y}{\partial b} \end{vmatrix} = [x, y] \tag{4.61}$$

$$\left(\frac{\partial x}{\partial a}\frac{\partial}{\partial b} - \frac{\partial x}{\partial b}\frac{\partial}{\partial a}\right) = [x, \bullet]; \tag{4.62}$$

$$\left(\frac{\partial y}{\partial a}\frac{\partial}{\partial b} - \frac{\partial y}{\partial b}\frac{\partial}{\partial a}\right) = [y, \bullet]. \tag{4.63}$$

Using these notations we rewrite equations (4.60) in the form

$$\begin{aligned} \partial/\partial x &= -J^{-1}[y, \bullet]; \\ \partial/\partial y &= J^{-1}[x, \bullet]. \end{aligned} \tag{4.64}$$

Since

$$\frac{\partial u_i}{\partial x_i} = \frac{\partial}{\partial x}\left(\frac{\partial x}{\partial t}\right) + \frac{\partial}{\partial x}\left(\frac{\partial x}{\partial t}\right), \tag{4.65}$$

using the expressions (4.64), we then obtain

$$\begin{aligned} \frac{\partial u_i}{\partial x_i} &= J^{-1}\left[\frac{\partial y}{\partial b}\frac{\partial}{\partial a}\left(\frac{\partial x}{\partial t}\right) - \frac{\partial y}{\partial a}\frac{\partial}{\partial b}\left(\frac{\partial x}{\partial t}\right)\right] + J^{-1}\left[\frac{\partial x}{\partial a}\frac{\partial}{\partial b}\left(\frac{\partial y}{\partial t}\right) - \frac{\partial x}{\partial b}\frac{\partial}{\partial a}\left(\frac{\partial y}{\partial t}\right)\right] \\ &= J^{-1}\frac{\partial}{\partial t}\left(\frac{\partial x}{\partial a}\frac{\partial y}{\partial b} - \frac{\partial x}{\partial b}\frac{\partial y}{\partial a}\right) = J^{-1}\frac{\partial J}{\partial t} \end{aligned} \tag{4.66}$$

or

$$\frac{\partial u_i}{\partial x_i} = \frac{\partial}{\partial t}(\ln J). \tag{4.67}$$

The repeated use of the operation that leads to the formulas (4.64) gives the expressions for the second derivatives in the form

$$\frac{\partial^2}{\partial x^2} = J^{-1}[y, \bullet]\,\{J^{-1}[y, \bullet]\} = J^{-1}\{y, J^{-1}[y, \bullet]\};$$

$$\frac{\partial^2}{\partial x \partial y} = -J^{-1}[x, \bullet]\,\{J^{-1}[y, \bullet]\} = -J^{-1}\{x, J^{-1}[y, \bullet]\};$$

$$\frac{\partial^2}{\partial y \partial x} = -J^{-1}[y, \bullet]\,\{J^{-1}[x, \bullet]\} = -J^{-1}\{y, J^{-1}[x, \bullet]\};$$

$$\frac{\partial^2}{\partial y^2} = J^{-1}[x, \bullet]\,\{J^{-1}[x, \bullet]\} = J^{-1}\{x, J^{-1}[x, \bullet]\}; \tag{4.68}$$

where the square brackets have the meaning determined by the formulas (4.61)–(4.63).

If we now substitute the expressions (4.55), (4.58), (4.66) and (4.68) into the Eulerian diffusion equation (4.54) we obtain the diffusion equation in the Lagrangian form

$$\frac{\partial \Gamma}{\partial t} + J^{-1}\frac{\partial J}{\partial t}\Gamma = J^{-1}\{D_{11}[y, J^{-1}[y, \bullet]] - D_{12}[x, J^{-1}[y, \bullet]] -$$
$$- D_{21}[y, J^{-1}[x, \bullet]] + D_{22}[x, J^{-1}[x, \bullet]]\}\,\Gamma. \tag{4.69}$$

If we multiply (4.69) by J, then equation (4.69) takes the more compact form

$$\frac{\partial (J \cdot \Gamma)}{\partial t} = \{D_{11}[y, J^{-1}[y, \bullet]] - D_{12}[x, J^{-1}[y, \bullet]] -$$
$$- D_{21}[y, J^{-1}[x, \bullet]] + D_{22}[x, J^{-1}[x, \bullet]]\}\,\Gamma. \tag{4.70}$$

Equation (4.70) describes the evolution of the contaminant-concentration distribution in the Lagrangian variables for the case of molecular diffusion. To take into account the turbulent character of the ocean water motion it is necessary, generally speaking, to average equation (4.70) and to accept some hypotheses to close the expression thus obtained. But since this procedure is very complicated we may use a non-so-rigorous but more simple method, namely, we can assume that the transformations given above refer to a transfer equation of the type (4.54) already averaged, where the quantities c and u_i mean averaged quantities, and instead of the exchange coefficient tensor D_{ij} we must use in this case the tensor of the turbulent-diffusion coefficients K_{ij}. In this case the derivation of equation (4.70) remains valid, but now Γ means the averaged concentration Γ, and J means the averaged functional determinant J, i.e., the diffusion equation (4.70) takes in this case the form [149]

$$\partial(\bar{J}\cdot\bar{\Gamma})/\partial t = \{K_{11}[\bar{y}, \bar{J}^{-1}[\bar{y}, \bullet]] - K_{12}[\bar{x}, \bar{J}^{-1}[\bar{y}, \bullet]] - - K_{21}[\bar{y}, \bar{J}^{-1}[\bar{x}, \bullet]] + K_{22}[\bar{x}, \bar{J}^{-1}[\bar{x}, \bullet]]\}\ \bar{\Gamma}. \tag{4.71}$$

Certainly, to solve equation (4.71) it is necessary to supply it with initial and boundary conditions for the function $\bar{\Gamma}$.

It is easily seen that in the absence of diffusion ($K_{ij} = 0$) equation (4.71) has the simple solution

$$\bar{\Gamma}(\bar{a}, \bar{b}, t) = \bar{J}^{-1}(\bar{a}, \bar{b}, t)\ \bar{\Gamma}(\bar{a}, \bar{b}, 0). \tag{4.72}$$

Though the diffusion equation (4.71) is linear relative to the function I, in the case of an arbitrary deformation of the velocity field the equation probably does not have an analytical solution. But under some simplifying assumptions on velocity field such a solution may be obtained. In particular, if one supposes, as was done in [149], that the tracer motions in the flow are combined from large-scale motions and small-scale fluctuations that do not correlate with one another, then the deformations of the velocity field $\partial x/\partial a$, $\partial x/\partial b$, $\partial y/\partial a$ and $\partial y/\partial b$ may be considered as functions of the time only and do not depend on a and b. In this case the diffusion coefficients K_{ij} also do not depend on a and b. Under this assumption equation (4.71) may be rewritten in the form (the overbars in J and Γ will be omitted from now on)

$$\frac{\partial(J\Gamma)}{\partial t} = J^{-2}\Big[\{K_{11}e_{22}^2 - (K_{12} + K_{21})e_{12}e_{22} + K_{22}e_{12}^2\}\ \frac{\partial^2\Gamma}{\partial a^2} + + \{-2K_{11}e_{21}e_{22} + (K_{12} + K_{21})(e_{11}e_{22} + e_{21}e_{21}\} - 2K_{22}e_{11}e_{12}\}\ \frac{\partial^2\Gamma}{\partial a\,\partial b} + \{K_{11}e_{21}^2 - (K_{12} + K_{21})e_{11}e_{21} + K_{22}e_{11}^2\}\ \frac{\partial^2\Gamma}{\partial b^2}\Big] \tag{4.73}$$

where the following notations are introduced: $\partial x/\partial a = e_{11}$, $\partial x/\partial b = e_{12}$, $\partial y/\partial a = e_{21}$ and $\partial y/\partial b = e_{22}$.

If we further perform the Fourier transform of the function $\Gamma(a, b, t)$ according to the formula

$$\tilde{\Gamma}(\lambda, \mu, t) \equiv \iint_{-\infty}^{\infty} \Gamma(a, b, t)\ \exp[i(a\lambda + b\mu)]\ da db \tag{4.74}$$

and apply this transform to equation (4.73), then the solution of the equation transformed in this way (under the condition of zero contaminant concentration at infinity) takes the form

$$\tilde{\Gamma}(\lambda, \mu, t) = \tilde{\Gamma}(\lambda, \mu, 0)\, J^{-1}(t) \exp[-\{B_1(t)\lambda^2 + B_2(t)\lambda\mu + B_3(t)\mu^2\}], \quad (4.75)$$

where

$$B_1(t) = \int_0^t \{K_{11}e_{22}^2 - (K_{12} + K_{21})e_{12}e_{22} + K_{22}e_{12}^2\} J^{-2} dt';$$

$$B_2(t) = -\int_0^t \{2K_{11}e_{21}e_{22} - (K_{12} + K_{21})(e_{11}e_{22} + e_{12}e_{21}) + $$
$$+ 2K_{22}e_{11}e_{12}\} J^{-2}\, dt';$$

$$B_3(t) = \int_0^t \{K_{11}e_{21}^2 - (K_{12} + K_{21})e_{11}e_{21} + K_{22}e_{11}^2\} J^{-2} dt'.$$

In the formula (4.75) the quantity $J^{-1}\exp[-\{B_1\lambda^2 + B_2\lambda\mu + B_3\mu^2\}]$ is the Fourier transform of the Green function $\Gamma_g(a, b, t)$, i.e., of the solution of equation (4.73) for the instantaneous point source. The inverse Fourier transform of the function $\tilde{\Gamma}(a, b, t)$ can be performed according to the formula

$$\Gamma(a, b, t) \equiv (4\pi^2)^{-1} \iint_{-\infty}^{\infty} \tilde{\Gamma}(\lambda, \mu, t) \exp[-i(a\lambda + b\mu)]\, d\lambda d\mu. \quad (4.76)$$

If we apply the inverse Fourier transform to the Green function we obtain

$$\Gamma_g(a, b, t) = \frac{1}{4\pi J(B_1B_3 - B_2^2/4)} \exp\left\{-\frac{B_3a^2 - B_2ab + B_1b^2}{4(B_1B_3 - B_2^2/4)}\right\}. \quad (4.77)$$

With an arbitrary form of the initial distribution of the contaminant concentration $\Gamma(a, b, 0)$ the solution of equation (4.73) may be written in the form

$$\Gamma(a, b, t) = \iint_{-\infty}^{\infty} \Gamma(a', b', 0')\, \Gamma g(a - a', b - b', t)\, da'db'. \quad (4.78)$$

We can give to the expression (4.78) a specific form if, for example, the initial distribution of the concentration is Gaussian,

$$\Gamma(a, b, 0) = \frac{N}{2\pi\sigma_a\sigma_b(1-\rho_0^2)^{1/2}} \exp\left[-\frac{1}{2(1-\rho_0^2)}\left\{\frac{a^2}{\sigma_a^2} - 2\rho_0\frac{ab}{\sigma_a\sigma_b} + \frac{b^2}{\sigma_b^2}\right\}\right], \quad (4.79)$$

where N is the total number of diffusing tracers, σ_a and σ_b are the initial variances of tracer locations in the directions a and b, correspondingly; and ρ_0 is the correlation coefficient of the initial tracer distribution.

After substituting (4.79) in (4.78) we now obtain the solution in the form

$$\Gamma(a, b, t) = \frac{N}{4\pi J(R_1R_3 - R_2^2/4)} \exp\left\{-\frac{R_3a^2 - R_2ab + R_1b^2}{4(R_1R_3 - R_4^2/4)}\right\}. \tag{4.80}$$

where

$$R_1(t) = \sigma_a^2/2 + B_1(t); \quad R_2(t) = \rho_0\sigma_a\sigma_b + B_2(t);$$
$$R_3(t) = \sigma_b^2/2 + B_3(t). \tag{4.81}$$

The quantities $2R_1$ and $2R_3$ represent the total variances of tracer displacements in the directions a and b, $2R_2$ is the total variance, and $2B_i$ $(i = 1, 2, 3)$ are the contributions to the variances due to the simultaneous actions of the velocity-field deformation and the turbulent diffusion.

The solution (4.80) may be transformed from the Lagrangian to the Eulerian form with the aid of inverse coordinate transforms of the form

$$x = e_{11}a + e_{12}b; \qquad y = e_{21}a + e_{22}b \tag{4.82}$$

or

$$a = e_{11}^{-1}x + e_{12}^{-1}y; \quad b = e_{21}^{-1}x + e_{22}^{-1}y. \tag{4.83}$$

Applying these transforms to (4.80) we obtain the expression for the contaminant concentration in the Eulerian representation,

$$c(x, y, t) = \frac{N}{4\pi J(t)(R_1R_3 - R_2^2/4)} \exp\left\{-\frac{Q_1x^2 - Q_2xy + Q_3y^2}{4J^2(R_1R_3 - R_4^2/4)}\right\}, \tag{4.84}$$

where

$$Q_1 = e_{21}^2R_1 + e_{21}e_{22}R_2 + e_{22}^2R_3;$$
$$Q_2 = 2e_{11}e_{21}R_1 + (e_{11}e_{22} + e_{12}e_{21})R_2 + 2e_{12}e_{22}R_3;$$
$$Q_3 = e_{11}^2R_1 + e_{11}e_{12}R_2 + e_{12}^2R_3, \tag{4.85}$$

and the functions $R_i(t)$ are given by the expressions (4.81).

The analysis of the solution (4.85) shows that if the initial distribution of the contaminant has the Gaussian form then, in the case considered, it also remains Gaussian in the next time moments for any values of the diffusion coefficients and the quantities $e_{ij}(t)$. The isolines of the contaminant concentration have the form of ellipses whose axes can turn in time.

It is more difficult to obtain the solution of the diffusion equation in the Lagrangian representation in analytical form for more complicated velocity fields and initial distributions of diffusing tracers. But investigations of the diffusion processes in the Lagrangian variables will of course be developed intensively in spite of their complexity, since they permit the obtaining of many interesting and important regularities of contaminant behavior in moving fluids.

4.3. Estimations of Contaminant-Concentration Fields from Discharge Facilities

The regularities of contaminant behavior in the ocean presented in the above sections, as well as the methods used to estimate the diffusion of the contaminant patches or jets, can be employed directly to determine the concentration of contaminant from discharge of industrial and domestic wastes in the coastal areas of the seas and the oceans. However, estimation of the concentration of contaminants caused by the discharge facilities has a number of characteristic features, some of which will be discussed in the present section.

Before estimating the concentration of the pollutants discharged into the sea medium one has to establish beforehand the amount of a given substance and its concentration that can be considered as safe either for the water users or for the sea inhabitants. Answering this question, it is worthwhile to use the concept of the assimilation capacity, A_{mi}, of the ocean or its individual areas. By this concept, presented in paper [37], one means the amount of pollutant i (either in the whole area considered or in the unit volume of sea water) that can ba accumulated, decomposed, transformed (under influence of the biological and chemical factors) and removed (due to sedimentation, diffusion or another process) out of the given sea volume in a unit time without disturbing the normal functioning of the ecological system, m, existing in the volume. The assimilation capacity depends on many factors, e.g., the dynamical state of the water mass, the structure of the biological system and the characteristic properties of its functioning and the properties of pollutants. Among the factors that provide 'self-purification' of the water mass, the following are mentioned in [37]: the hydrodynamical transport of a contaminant, its microbiological oxidation, biological transformation, biochemical oxidation, chemical and physical transformations, sedimentation of the abiogenic particles and biosedimentation. Of those processes the hydrodynamical phenomena, microbiological oxidation and biosedimentation are, seemingly, the most important ones.

To estimate the assimilation capacity of the water mass, A_{mi}, it is important to know those species in an ecological system that are the most sensitive to the pollution of the medium by a given pollutant. Using the most sensitive elements of the ecological system one has to determine the maximum permissible concentration, c_{oi}, of a given pollutant, i (or of a number of pollutants), in the medium, as well as the 'multiplicity coefficient' K_i of pollutant amount (in the unit volume) that can be processed by a given ecological system, m, or removed from the volume during the assumed averaging time due to different processes, i.e.,

$$A_{mi} = C_{oi} K_i V, \tag{4.86}$$

where V is the volume of the water mass under consideration.

There can be two approaches to working out the requirements for pollution control of a basin. Within the framework of the first approach, the arising of even limited zones in which the pollutants' concentration exceeds the permissible levels is prohibited. In the lines of the second approach, the existence of limited zones of pollution is admitted close to sites of waste-water discharge where the quality of the water is not supposed to satisfy all of the operational requirements. In the first case, naturally, the building of sanitation works that can bring the waste-waters features to levels providing for the safety of the inhabitants of the given basin and the water users is required. In the second case, purification of the water is permitted to be not so careful which, naturally, leads to a significant reduction in the cost of building and operating the sanitation works.

The choice of the permissible concentrations of harmful substances in a basin is a rather complicated problem. As mentioned above, definition of these concentrations has to be based on the organisms of a biocenosis existing in a basin that are the most sensitive to the substance in question. One has to take into account the fact that the harmful influence of a given substance can manifest itself long after its contact with an organism. The substance can affect the genetic system of the organism as well, so the consequences will be revealed only in the following generations of the organisms. Sea-medium inhabitants can sometimes endure short-duration effects of harmful substances without given rise to serious consequences; however, pathological changes in the organisms are possible during prolonged contact, even if the contaminant concentration is low. The effect of one of the harmful substances can increase the sensitivity of a given organism to the influence of another substance so, in the general case, the individual reactions of an organism cannot be treated as independent factors when considering the sum of the dangerous ingredients. All the phenomena mentioned above make it necessary to take into account not only the permissible concentration of each of the harmful substances present in the waste water (as is mostly the case) but the time of existence of that or other concentrations of a given substance in the water mass as well in order to elaborate the permissible discharge levels. If the quantitative characteristic of change of the physiological processes in the organism caused by substance i is denoted as ΔB_i, the water-quality criterion can be written as follows (see article by Aitsam *et al.* in [62]):

$$\triangle B_i < \triangle B_i^0 = f_1(c_i, t_i), \tag{4.87}$$

where c_i is the concentration of substance i, t_i is the time of influence of the given concentration of the substance on an organism, $\triangle B_i^0$ is the permissible level of physiological change in the organism, and f_1 is some function.

If two harmful substances produce simultaneous and dependent influence on an organism, equation (4.87) has to be written as follows:

$$\triangle B_{ij} < \triangle B_{ij}^0 = f_2(c_i, t_i), \tag{4.88}$$

where $i, j = 1, 2, 3, ..., k$; k is the amount of the harmful ingredients in the waste water; $\triangle B_{ij}^0$ are the permissible levels of the physiological changes caused by simultaneous effect of both of the harmful substances; and f_2 is some function. Equation (4.88) can be readily generalized to the case of simultaneous and dependent influence produced on an organism by a larger number of the harmful substances as well.

Using formulas of the form (4.87) or (4.88), one has to bear in mind that each of the organisms reacts to the dangerous substance in its individual way. Consequently, one has to consider the given formulas as stochastic rather than deterministic ones, if there is a great number of the organisms of the same species (and of the same age and sex). In this case a concept of the probability of corresponding physiological changes in organisms, F, is to be used, and to define water-quality criterion one has to impose permissible probabilities of unfavorable changes, F_0, i.e., to suppose [62]

$$F_0 < F\{\triangle B_{ij} < \triangle B_{ij}^0\}, \quad i, j = 1, 2, ..., k. \tag{4.89}$$

That probabilistic criterion of the permissible water quality in the given water-usage area can, evidently, be represented as follows:

$$F_0 > F\{\triangle B_{ij} > \triangle B_{ij}^0\}. \tag{4.90}$$

Since the values of $\triangle B_{ij}$ are functions of c_i and t_i, equation (4.90) is reduced to a form

$$F_0 > F\{c_i > c_{0i};\ t_i > t_{0i}\} = \int_{c_{0i}}^{\infty}\int_{t_{0i}}^{\infty} p(c_i, t_i)\ dc_i dt_i, \tag{4.91}$$

where $p(c_i, t_i)$ is the two-dimensional probability density of the concentration values and the durations of existence of these concentrations in the point (x_0, y_0, z_0) under consideration.

If the point (x_0, y_0, z_0) is at some distance from the waste-water discharge site, the contaminant concentration at (x_0, y_0, z_0) depends substantially on both the direction and the magnitude of the current velocity in the sea area considered. If the current is directed from the point of discharge to that of water usage, then the concentration of pollutants will be, evidently, maximal at (x_0, y_0, z_0), but for reverse current direction, the water medium in the water-usage zone will be rather weakly polluted, if at all, due to the waste-water source considered. If the modulus of the current velocity, $|u|$, is small enough, the pollution area cannot spread up to the point (x_0, y_0, z_0) as well. Thus, the argument in the function $p(c_i, t_i)$ can be replaced by the values of $|u|$ and φ, the latter being the direction of the current-velocity vector relative to the geographical coordinates or to the direction from the waste-water discharge site to the water-usage point. In this case the condition (4.91) has the following form:

$$F_0 > \int p(|u|, \varphi, t_i) \, \mathrm{d}|u| \, \mathrm{d}\varphi \, \mathrm{d}t_i. \tag{4.92}$$

Integration in equation (4.92) has to be carried out over that part of three-dimensional space $(|u|, \varphi, t_i)$ where the contaminant concentrations determined from the values of $|u|$, φ and t_i exceed the permissible levels. Either the domain of integration or the form of the probability density function, $p(|u|, \varphi, t_i)$, are complicated ones and, consequently, that integration can be carried out, in most of the cases, by the numerical methods only. The probability density function can be determined by the processing of a large body of observational data on the direction and magnitude of the current velocities in the area of the expected waste-water discharge. If a researcher has this 'current-rose' at his disposal, the contaminant concentrations at the water-usage site can be estimated afterwards, by this or another method, for each of the selected gradings of the modulus and the direction of the current velocities, and then the value of t_i can be found.

The time of existence of a certain contaminant concentration in a given water-mass element depends either on the discharge facility operation regime and the current field or on the temporal variability of the above-mentioned factors of self-purification of the water. Random, chaotic variations of contaminant concentration caused by turbulent motions of the waters take place in the ocean (and in freshwater reservoirs), as well as regular changes, either temporal or spacial, of the intensity of the latter factors. If the effect of a harmful substance on an organism is long-term enough, the latter can become more sensitive or, on the contrary, adapt itself to a given substance; in the latter case, the values of c_{0i} and t_{0i} in equation (4.91) can be, generally speaking, functions of time. The harmful substance flow rate, Q_i, in the waste-water source cannot always be imposed by a deterministic function as well. Thus, for instance, as shown by Aitsam [1], the concentration of a number of contaminants in the waste-water sewage collectors of Moscow and Tallinn is described by the random functions of time with the Gaussian probability distributions.

Naturally, the permissible physiological changes can be different for plankton, neuston, nekton and benthos, and the reactions of the various groups of organisms to the harmful substances are not the same ones. The harmful substances assimilated by organisms of the lower trophic levels afterwards can be transferred through the feeding chain to the higher-level organisms. Thus, some individual species can be affected by pollutants, even if there were no direct contacts with the waste water, due to their consumption of the organisms that have accumulated a harmful substance earlier (or in another area).

Consequently, in order to make a well-grounded choice of the permissible harmful-contaminant concentrations and the existence times of the pollution zones it is necessary to take into account all of the possible trophic relations between the individual components of a biocenosis rather than just considering the reaction of a given organism to the pollutants. To answer all these questions one evidently has to carry out extensive studies on the physiology of the organisms making up the biocenosis, the mechanisms of the assimilation of pollutants by the organisms and the release of the pollutants, the quantitative features of the transfer of pollutants through the trophic chains and the chemical and biochemical transformations of these pollutants, which take place either in the environment or in the organisms.

Of no less difficulty is the problem of water self-purification from the harmful contaminants. To consider this problem it is necessary to have knowledge of the mechanism of pollutant input into the ocean waters (including a direct input from the atmosphere), the mechanisms of the transport of pollutants by the water motions, the dependence of the rates of chemical and physical processes of the transformation of the pollutants on the temperature and salinity of the waters and their acidity, content of oxygen and other factors, the mechanisms of contaminant sedimentation, the processes of biochemical decomposing of the pollutants either in the water mass or in the floor sediments, etc. To solve all these questions a good deal of cooperative effort has to be undertaken by researchers from the different fields of science in order to improve the investigation methods, carry out laboratory and field observations and develop model descriptions of the phenomena. Significant advances in all the mentioned areas have been obtained until now (see, for instance, collections of papers [28, 33, 34, 62]), yet a rather large body of studies is still to be fulfilled in the future.

The water-quality levels approved presently for pollutants of different forms are still not well-grounded enough. Nevertheless, following these standards, one can prevent to a considerable extent the undesirable changes in the conditions of either the water medium of the oceans or their biocenosis. Consequently, the design work on discharge facilities has to include, as a main requirement, a warranty that the concentration of discharge pollutants in the sea waters will not exceed a certain value in a somewhat large volume of the water mass. To meet this requirement the designer has to take into account all the forms of pollutant transformation in the sea medium, as well as to design the purifiers with operational performances providing a necessary level of waste-water purification.

The coastal purifiers can be considered as a certain filter-rectifier [1]; its effect on the time-varying concentration of contaminant is described by the following relation:

$$\bar{c}_i(t) = \int_{-\infty}^{\infty} h(t-t')\, c_i(t')\, \mathrm{d}t', \tag{4.93}$$

where $c_i(t)$ is the substance concentration in the waters input into the purifier; $\bar{c}_i(t)$ is the concentration of the same substance in the waters passed through the purifier; $h(t-t')$ is the purifier weighting function; and t' is the integration variable. The form of the weighting function depends either on the purifier type or on its operation. The mean contaminant concentration must be reduced substantially as a result of purifier operation, and high-frequency variations of $c_i(t)$ have to be smoothed as well. In particular, the filtrating effect of the purifier has to manifest itself in a sharp drop in the levels of the short-lived yet abrupt rise ('the volley-like discharges') of pollutant concentration that are the most dangerous ones either for the water-body inhabitants or the water users. In most cases it is rather difficult to determine the form of the purifier weighting function theoretically, so direct measurements of the functions $c_i(t)$ and $\bar{c}_i(t)$ taken in the different conditions of purifier operation can be used. If the form of the weighting function $h(t)$ is known, the quality of the discharged water can be predicted for the varying loads on the purification facility. The weighting function form can be readily determined through theoretical estimations only for the particular purifier designs known as "averaging facilities". The effect of the latter is similar to that produced by a sliding averaging procedure on random function, i.e., in this case the weighting function can be written as follows:

$$h(t) = \begin{cases} a/T_0, & -T_0/2 \leqslant t \leqslant T_0/2; \\ 0, & |t| > T_0/2, \end{cases} \tag{4.94}$$

where T_0 is the averaging time, and a is some constant depending on the structural features of the averaging facility.

If the Fourier transform is taken of the expression (4.93), we obtain a spectral representation of the mathematical model describing the effect of the purifier:

$$\overline{E}_i(\omega) = H(\omega)\ E_i(\omega), \tag{4.95}$$

where $E_i(\omega)$ is the spectral density of the concentration fluctuations of a substance i in the waters entering the purifier; $\overline{E}_i(\omega)$ is the spectral density of the fluctuations of the same substance in the waters after purification; and $H(\omega)$ is the weighting function spectral density or the purifier transfer function which indicates transformation of the spectrum $E_i(\omega)$ due to the effect of the purifier.

Thus, for instance, the transfer function $H(\omega)$ for the weighting function of the form (4.94) is as follows:

$$H(\omega) = a^2 \frac{\sin^2 \omega t/2}{(\omega t/2)^2}. \tag{4.96}$$

In more complicated cases the purifier transfer function is to be determined through experiments. Thus, Aitsam [62] studied operation of the Kuryanovskaya aeration station (the city of Moscow) and showed that the variances of the pollutant-concentration fluctuations in the purified waters, $\overline{\sigma}_i^2$, are reduced sharply as compared to the variances σ_i^2 in the waters entering the station. At the same time the shapes of spectral functions $E_i(\omega)$ are only slightly changed due to the effect of the station. If the functions $E_i(\omega)$ and $\overline{E}_i(\omega)$ are normalized on the variances, they are reduced to nearly the same form. Consequently, in this case the transfer function H is independent of ω and leads only to a reduction of the level of the spectral function $E_i(\omega)$, while its shape is not modified. As a result, the function H can be written as follows:

$$H = \sigma_i^2/\overline{\sigma_i^2} = \text{const}, \tag{4.97}$$

and the weighting function $h(t)$ has the form

$$H(t) = \sigma_i/\overline{\sigma}_i = \text{const}. \tag{4.98}$$

A value of H found by Aitsam from different observation runs was close to $10^{-2} - 10^{-3}$. The formulas (4.97) – (4.98), naturally, cannot be treated as universal ones because different purifier designs or different operation conditions can result in different transfer functions.

The spectral functions $E_i(\omega)$ of the waters entering the purifiers, as well as the functions $c_i(t)$, are unknown a priori in most of the cases. In this relation the model representations of different kinds can be used for $E_i(\omega)$ and $c_i(t)$. Thus, for instance, Aitsam [62] proposed to present the functions $c_i(t)$ in the form

$$c_i(t) = \overline{c_i} + \sum_{n=1}^{N} A_{n,i} \sin(n \frac{2\pi}{T_i} t + \varphi_{n,i}) + c_i'(t), \tag{4.99}$$

where $\bar{c}_i$ is the contaminant concentration steady component; $c_i'(t)$ is the random fluctuation of concentration; and a term

$$\sum_{n=1}^{N} A_{n,i} \sin(n \frac{2\pi}{T_i} t + \varphi_{n,i})$$

describes periodical variations of concentration caused by cyclical operation of the water-using facilities (with the amplitudes $A_{n,i}$ and the phase shifts $\varphi_{n,i}$).

Some suggestions can be made for the random component, $c_i'(t)$, as well. In particular, one can suppose $c_i'(t)$ to be a stationary and ergodic random function with the normal probability distribution and the correlation function of the form

$$R_i(\tau) = \sigma_i^2 \exp(-\alpha\tau^2) \tag{4.100}$$

where σ_i^2 is the normal distribution variance, and α is a constant.

The spectral density of random process with the correlation function (4.100) has the form

$$E_i(\omega) = \sigma_i^2 \sqrt{\frac{\pi}{\alpha}} \exp\left(-\frac{\omega^2}{4\alpha}\right). \tag{4.101}$$

Provided the transfer function of a given purifier, $H(\omega)$, is known, then, using the known form of $E_i(\omega)$ the spectral density of the contaminant-concentration fluctuations in the conditionally purified water discharge into the sea is estimated by the formula (4.95). The further transformation of the contaminant-concentration field in the sea can be analyzed either by employing the spectral analysis methods (see Section 1.8) with $E_i(\omega)$ as the initial condition or, as is mostly the case, using semi-empirical models of contaminant transport by the ocean waters.

The behavior of pollutants in the zone close to discharge facilities depends, in many respects, on the facility design and the outflow regime of the polluted waters. In practice the discharge facilities usually have used heads with one or several escape orifices or heads with nozzles (either with a confuser or without it). If facilities with a great number of nozzles (or escape orifices) separated from each other are used, a high degree of the polluted water's original dilution is attained, yet, in this case, the original pollution zone grows in size. To increase the original dilution, the outflow velocities of the jets of polluted water at the nozzles outlets can be increased as well. Some recommendations on the design of the confuser of the discharge facilities are presented, for instance, in the works by Lapshev, Shilov and Grabovkii [62]. In particular, they proposed using nozzles of the injector type, which provide a high level of turbulization of the outflowing fluid jet. Naturally, the rate at which the jet is mixed with ambient fluid increases due to this artificial turbulization and, as a result, the contaminant concentration decreases more sharply with distance from the discharge site. The degree of the original dilution of the waste water in the jets depends as well on the density difference between the jets and ambient fluid. Besides, the above-mentioned authors also considered nonuniformity of the outflow of waste-water jets from the nozzles positioned at the head section of the discharge collector. Nonuniformity of fluid-flow rates from such a nozzle system leads to nonuniformity of the original dilution of the jets and, thereby, to decreasing of the efficiency of the total original mixing between the waste waters and ambient water mass. To rule out the possibility of this undesirable effect, the nonuniformity of fluid-flow rates at the nozzles outlets apparently has to be less than $10-20\%$.

If the discharge facility is positioned deep in the water, the contaminant concentration can be reduced by a factor of 10^3-10^4 at distances from 100 to 1000 m from the discharge site due to both original and diffusion dilution. In most cases this decrease turns out to be sufficient enough to meet the existing sanitary requirements as to the sea-water quality. Certainly, the pollution zone close to the discharge facilities can be diminished due to the higher degree of waste-water purification at the coastal aeration stations. The main oceanological factors that either encourage or prevent dilution of the waste waters and their purification are to be taken into account as well. Of those factors the following are of the most significance: the sea water's density stratification, the magnitudes and directions of current velocity, turbulent mixing intensity, oxygen content in the water body and the development of phytoplankton and zooplankton. If the depth of the discharge facility and its distance from the coastline are chosen properly, the cloud of polluted waters can always be submerged and will not reach the coastal zone of the sea. This conclusion has been borne out by the construction experience of the deep-water discharge facility near the city of Yalta, as well as the operation of a number of foreign discharge facilities (see a review [34]).

The methods employed at present to estimate the behavior of contaminant jets injected into the sea medium have been discussed in detail in Section 2.3. As one can readily see, these methods are, in principle, the integral ones because only the averaged (or typical) parameters of the jets are taken into account. The empirical profiles describing distribution of temperature and velocity of the water and of the contaminant concentration across the jet are usually introduced in order to convert the partial differential equations of hydrothermodynamics to ordinary differential equations. In this approach, the turbulence of the jets and ambient medium is taken into account by introducing some empirical relations which, for instance, describe the entrainment effect. The evolution of the free submerged jets either in homogeneous or in stratified fluid can be described satisfactorily by the integral methods. Lateral depth-independent currents of ambient fluid in the discharge-facility area are also taken into account by these methods sufficiently well. But it is difficult to take into account a velocity gradient in such a flow within the framework of the integral approaches to the problem. The effects of contaminant-source non-steadiness cannot be studied by the integral methods, nor can the interactions of the jets with the free surface of the water basin and its floor.

Owing to the mentioned limitations of the integral methods, a more complicated technique that employs partial differential equations of hydrothermodynamics was used in recent years to estimate the behavior of the contaminant jets in the ocean. In these equations, the terms describing the turbulent transport must be expressed in terms of other unknowns to obtain closed equations. The simplest method, closing the equations by introducing the turbulent-transport coefficients gives, as usual, no marked advantages as compared to the integral methods because adequate choosing of the turbulent-transport coefficients appears to be a difficult problem. At the same time the extremely complicated closure models (for instance, in terms of the higher-order statistical moments of fluctuation characteristics) are not very useful as well due either to uncertainty in imposing boundary conditions for the desired characteristics of the flow or to the significant errors arising in numerical computations (analytical solutions in the complicated models cannot be obtained in most cases). Because of it, the models based on joint consideration of the equations of motion, turbulent-energy balance, b, and turbulent-energy dissipation, ε, balance are presently regarded as being, along with the integral methods, the most suitable ones for resolving the problems of propagation of the contaminant jets. Such a set of equations is presented, in the general form, in Section 4.1 (equations (4.13) – (4.27)). Certainly, this set of equations can be written down in somewhat different forms or simplified to some extent, yet the equations describing balance of the values of b and ε are to be retained without fail. The possible methods of constructing these b, ε-systems used to estimate the behavior of the contaminant jets are discussed in detail in the review by Rody presented in [63]. In recent years a number of problems concerning the behavior of jets, non-interacting with the free or rigid boundaries of the main water basin, have been considered numerically with the help of the b, ε-models. In the fist class of problems jets with plain or round cross-sections having the same density as that of the ambient fluid or with some additional buoyancy were considered. In the second class cases of the outflow of the jets onto the surface of the water basin and cases of the submerged outflow with the fluid in the water basin at rest or in motion were studied. Behavior of the jet in a fluid at rest is determined, to a considerable extent, by the turbulence of the jet itself, while in the moving fluid the jet turbulence is interacting with the ambient medium turbulence, the latter becoming the dominant factor of the jet transformation processes in the far zone.

The solution of the problems on contaminant jets obtained on the basis of the b, ε-models showed that the models yielded satisfactory results for jets in the media with current-velocity shear; the empirical constants in the set of the equations for this case are the same as in the case of the jets propagating in the medium with uniform velocity field. Interactions of the jets with the main currents and with the free or rigid boundaries of the water basin are described satisfactorily by the b, ε-models as well. The b, ε-models also simulate the recirculational flows arising at margins of the jets discharged close to the water-basin boundaries.

As an example, the behavior of the contaminant jet estimated numerically in accordance with the b, ε-model for nonstratified fluid moving at velocity v_c is shown in Figure 4.11. The contaminant density has the same value as that of the ambient fluid, and the jet axis is directed along the medium current-velocity vector. Evolution of the ratio $U_c/\Delta U$ (where ΔU is the excess of fluid velocity at the jet axis above the value of U_c) with distance is shown in Figure 4.12 by

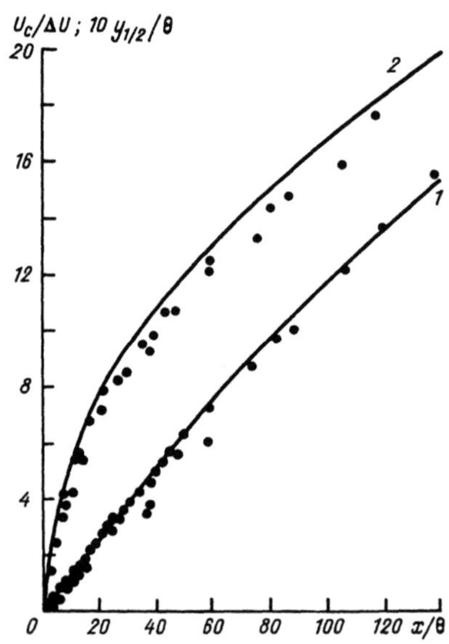

Fig. 4.11. Behavior of a contaminant jet flowing into the non-stratified fluid as estimated from b, ε-model (after the paper by Rody in [63]). 1 – the dependence of non-dimensional ratio $U_c/\Delta U$ on non-dimensional distance x/θ, where ΔU is the excess of velocity at the jet axis over the ambient-fluid velocity U_c; x is the distance from the site of the jet discharge; θ is the momentum excess thickness. 2 – the dependence of non-dimensional half-width of the jet, $y_{1/2}/\theta$, on the non-dimensional distance x/θ.

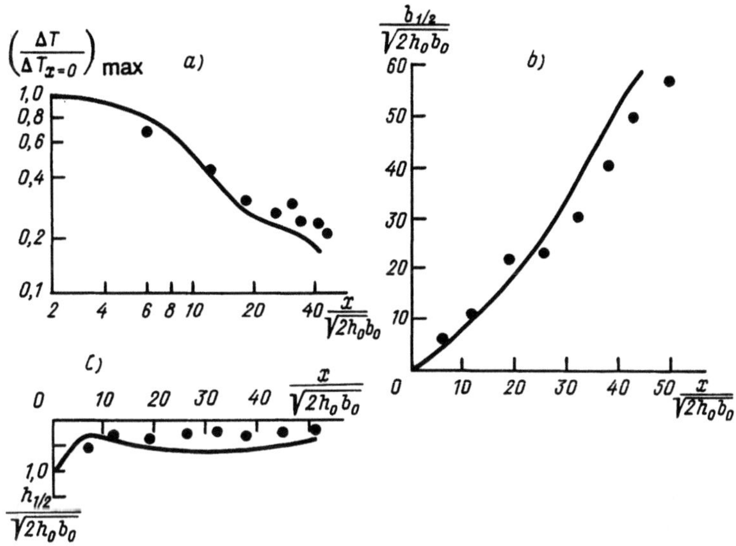

Fig. 4.12. Parameters of the heated contaminant jet flowing out onto the surface of the water body as estimated from b, ε-model. The variations of normalized temperature of the jet, its half-width and half-depth with the distance site (the plots a, b and c, respectively; for notations see the text) are shown after the paper by Rody from [63].

curve 1. The distance x measured from the point of the jet outflow is normalized by the so-called 'momentum excess thickness', θ, which is defined for the round jet by the formula

$$\theta = \left[2\pi \int_0^\infty \frac{U(U-U_c)}{U_c^2} \times r dr\right]^{1/2},$$

where r is the radius vector with its origin at the jet axis, and U is the fluid velocity at distance r from the jet axis. Variation of the normalized half-width of the jet, $y_{1/2}$, with the normalized distance, x/θ, is shown in the figure by curve 2. The half-width of the jet is defined as a distance from the jet axis to the point where $U-U_c = \Delta U/2$. Using the non-dimensional variables permits one to present the results of computations for different values of the ratio of the jet velocity to external current velocity by the universal curves. Thus, for instance, the case of outflow of the jet into the fluid at rest (when $\theta = \infty$) corresponds to the origin of coordinates at the plots. The experimental points given in the figure show good agreement between estimations and observational data over a wide range of the values of x/θ.

The characteristics of a heated rectangular jet of contaminant discharged onto the surface of the water basin are shown in Figure 4.12 as estimated from the b, ε-model. The plot at Figure 4.12a shows a decrease of the temperature at the jet axis; the temperature is normalized by the difference of the jet temperature at the discharge point and the temperature of the ambient fluid T_0. The jet outflow velocity is assumed to be equal to U_0 and the ambient fluid is supposed to be at rest. The distance along the jet axis, x, is normalized by the value of $\sqrt{2h_0b_0}$, where $2b_0$ and h_0 are the 'initial' values of the width and the depth of the jet, respectively. The dependence of half-width of the jet, $b_{1/2}$, on non-dimensional distance is shown in Figure 4.12b and the dependence of "half-depth" of the jet, $h_{1/2}$, on $x/\sqrt{2h_0b_0}$ in Figure 4.12c. Half-width and half-depth of the jet are defined as the distances measured from the jet axis (which is, in the case in question, positioned at the surface of the water basin) to the points where the excess of water temperature in the jet above the ambient-fluid temperature is equal to half of its maximal value for a given cross-section of the jet. The experimental points are in good agreement with the theoretical curves for this case as well; however, some more significant deviations of the experimental data from the theoretical curves are observed for large distances from the discharge site.

Some other examples of application of the b, ε-models to the estimation of propagation of contaminant jets discharged into the water medium can be found in the above-mentioned review by Rody presented in [63]. The results of computations given in this review describe distributions of temperature and velocity of the fluid in vertically directed jets, in jets penetrating into the fluid along the rigid walls and outflowing at some angles relatively to the direction of the main fluid current. In all these cases the available experimental data are in good agreement with theoretical results. In this context one, seemingly, can conclude that now the b, ε-models are verified enough to estimate the behavior of the contaminant jets in the zones close to the discharge sites where the jets' turbulence intensity still exceeds the natural turbulence intensity of the sea medium.

As one can see, estimation of the concentration field of the contaminants discharged into the ocean appears to be a rather complicated problem. To start calculations one has to know, as initial information, the main parameters of the discharged waters, namely, the discharge volumes, the contaminant concentrations and the temporal variability of these characteristics. Besides, it is necessary to know the weighting function of purifiers, either existing or being designed, in order to determine the properties of the waters that are purified and intended for discharge into the ocean. Known structural properties of the heads of the discharge facilities make it possible, in turn, to evaluate the original dilution of the contaminant jets in the sea water. For this purpose the above considered models (b, ε-models or the integral ones) providing estimation of the near zones of the jets flowing out of discharge facilities are to be used. As soon as the intensity of intrinsic jet turbulence falls to that of the natural ocean turbulence, the methods of jet estimation considered in Section 2.3 become valid. Some individual links of this calculation chain can, naturally, be substituted by more simple evaluation procedures of pollution-field transformation. In most cases the results of theoretical estimations have to be verified by experiments that can be fulfilled using fluorescent dyes, isotopes and discrete particles. Such a combination of the theoretical methods with experimental ones must finally assure that the pollution levels will not exceed the permissible normals and that the ocean waters will remain clear and rich in various forms of marine life.

References

1. Aitsam, A. M. 'Turbulent diffusion of the pollutants discharged into the sea with waste waters', Dr. Thesis, Inst. of Oceanology, USSR Academy of Sciences, Moscow, 1968 (in Russian).
2. Belousova, E. J. and Latun, V. S. 'Storm surges circulation of waters and spreading of admixture in the coastal zone of the sea', *Marine Hydrophysical Research*, 1975, **3**, 54–65 (in Russian).
3. Beliayev, V. S., Kolesnikov, A. G., and Nelepo, B. A. 'Estimation of intensity of radioactive contamination in the ocean based on new data on the exchange processes', 3-rd UNO Int. Conf. on Peaceful Employment of Nuclear Power. Geneva, 1964. UNO paper A/conf. 28/P/766 Ukrainian SSR, May 1964 (in Russian).
4. Beliayev, V. S., Kolesnikov, A. G. and Nelepo, B. A. 'Spreading of radioactive contaminants in the ocean', *Atm. Ocean. Phys.*, 1967, **3** (3), 1092–1100 (in Russian).
5. Beliayev, V. S. and Ozmidov, R. V. 'A model of vertical turbulent exchange in the ocean with intermittent turbulence', *Dokladi Ak. Nauk. SSSR*, 1980, **254** (4), 995–998 (in Russian).
6. Beliayev, V. S. and Ozmidov, R. V. 'Influence of the ocean turbulence intermittency on the formation of the ocean surface structure', *Doklady Ak. Nauk. SSSR*, 1982, **263** (5), 1225–1229 (in Russian).
7. Benilov, A. Yu. 'On the generation of turbulence in the ocean by the surface waves', *Atm. Ocean. Phys.*, 1973, **7** (3), 292–303 (in Russian).
8. Berliand, M. E. *Modern Problems of Atmosphere Diffusion and Contaminant of the Atmosphere.* Leningrad, Gidrometeoizdat, 1975 (in Russian).
9. Brosin, H., Kremser, U. and Muraviev, S. S. 'The experimental study of dye diffusion in the Baltic Sea', *Oceanology*, 1977, **17** (2), 221–226 (in Russian).
10. Van Dam, I. 'Horizontal diffusion in the North Sea off the coast of the Netherlands and disposal of industrial waste waters', in *Intern. Ass. on Hydr. Res.*, 11-*th Congr.*, 1965, 1966, **6**, 585–597 (in Russian).
11. Interaction of the Atmosphere, Hydrosphere and Lithosphere in the Near-Shore Zone. Edited by Z. K. Belberov *et al.* Publ. House of the Bulgarian Acad. Of Sci., Sofia, 1980 (in Russian).
12. Interaction of the Atmosphere, Hydrosphere and Lithosphere in the Near-Shore Zone. *Results of the Intern. Experiment 'Kamchiya-78'.* Edited by Z. K. Belberov *et al.* Publ. House of the Bulgarian Acad. Of Sci., Sofia, 1982 (in Russian).
13. Vinogradov, A. S. 'On the vertical diffusion of the radioactive admixture in the ocean', *Marine Hydrophysical Research*, 1976, **3**, 87–93 (in Russian).
14. Vinogradov, A. S. 'Using radioactive isotopes as tracer of hydrophysical processes in the ocean', in *Some Aspects of Meso-Scale Variability of Temperature Field in the Ocean.* Inst. Of Oceanology, USSR Academy of Sciences, Moscow, 1977, 71–110 (in Russian).
15. Vinogradov, A. S. 'Vertical diffusion of cosmogenic isotopes in the ocean' *Oceanology*, 1978, **18**, (1), 50–57 (in Russian).
16. Vinogradov, A. S. 'Vertical diffusion of radioactive fallout in the ocean', *Oceanology*, 1979, **19** (2), 239–245 (in Russian).

17. Vinogradov, A. S. 'The influence of the upper mixed layer on the radioactive fallout diffusion in the ocean', *Oceanology*, 1982, **22** (5), 744–749 (in Russian).
18 Galkin L. M. *Solving of the Diffusion Problems by Monte-Carlo Method*. Moscow, Nauka, 1975 (in Russian).
19. Garger, E. K. On estimation of the horizontal diffusion parameters by turbulent-features', *Trudy Inst. Eksperiment. Meteorologii*, Gidrometeoizdat, Moscow, 1972, **27**, pp. 81–89 (in Russian).
20. Gezentsvei, A. N. 'Large-scale horizontal turbulent exchange in the Black Sea', *Trudy Inst. Okeanologii Akad. Nauk SSSR*, 1961, **52**, 115–132 (in Russian).
21. Goroshko, V. I. 'Turbulent diffusion of contaminant in the near-shore zone of the sea', in *Studies of the variability of Hydrophysical Fields in the Ocean*. Moscow, Nauka, 1974, 145–150 (in Russian).
22. Demyshev, S. G., Eremeev, V. N., Ivanov, L. M. and Knysh V. V. 'Interaction of passive scalar admixture with eddy formations in the open ocean', *Atm. Ocean Phys.*, 1982, **18** (3), 276–285 (in Russian).
23. Demyshev, S. G., Eremeev, V. N., Ivanov, L. M. and Knysh V. V. 'Passive scalar transport in eddy fields of the open ocean', *Atm. Ocean. Phys.*, 1983, **19** (2), 185–194 (in Russian).
24. Zhurbas, V. M. 'Tracks of turbulent jets of contaminant in a steady stratified medium', Vodnye Resursy, 1977, **4**, 165–172 (in Russian).
25. Zhurbas, V. M. 'The main mechanism of oil spreading in the sea', in *Progress in Sciences, Fluid and Gas Mechanics Series*, VINITI Ak. Nauk SSSR, Moscow, 1978, v. **12**, pp. 144–159 (in Russian).
26. Zhurbas, V. M. and Mamedov, R. M. 'Eperimental study of dye plumes diffusion in the sea surface layer', *Oceanology*, 1976, **16** (6), 990–994 (in Russian).
27. Zhurbas, V. M., Muraviev, S. S. and Tataraev, T. M. 'Experimental study of turbulent diffusion of dye plumes in the sea surface layer', *Oceanology*, 1975, **15** (4), 611–615 (in Russian).
28. *Contamination of the Salt Seas*. Edited by G. Okolotovish, 1976, Inst. of Marine Fishing, Gdynia, Poland (in Russian).
29. Zeidler, R. Kremser, U. and Muraviev, S. S. 'Diffusion test using dye tracer close to the shore', in *Interaction of the Atmosphere, Hydrosphere and Lithosphere in the Near-Shore Zone*. Sofia, Publ. House of the Bulg. Acad. Sci., 1980, 92–100 (in Russian).
30. Zalesny, V. B. and Ivanov, Yu. A. 'Numerical experiments on the analysis of a two-dimensional modal of thermohaline and wind circulation in the ocean', *Atm. Ocean. Phys.*, 1978, **14** (2), 184–194 (in Russian).
31. Zats, V. I. 'On stability of submerging of waste waters fields in the intermediate layer of the stratified sea', *Vodnye Resursy*, 1975, **1**, 180–191 (in Russian).
32. Zats, V. I., Nemirovsky, M. S., Shulgina, E. F., Kurakova, L. V., Kuftarkova, E. A., Senichkina, L. G. and Chepurnova, E. A. 'A combined approach to studying deep-water discharges of waste water', *Vodnye Resursy*, 1979, **6**, 181–191 (in Russian).
33. Zats, V. I. and Nemirovsky, M. S. 'Using of fluorescent tracers method in studies of the exchange processes in zones of waste waters discharge in the sea', *Biology of the Sea*, Naukova Dumka, Kiev, 1972, **27**, 31–41 (in Russian).
34. Zats, V. I., Nemirovsky, M. S., Andryuschenko, B. F. *et al. Experimental Studies of the Deep-Water Discharge of Waste Waters in Yalta Area*. Naukova Dumka, Kiev, 1973 (in Russian).
35. Zats, V. I. and Ozmidov, R. V. 'Turbulence and mixing in the Black Sea', in *Studies on the Water Dynamics and Hydrochemistry of the Black Sea*. Pt. 2, 1978, pp. 237–291. Published by COMECON Coordination Centre for Studies of Chemical, Physical, Biological and Other Processes in the Most Important Ocean Areas, Moscow (in Russian).
36. Ivanov, K. and Filippov, Yu. G. 'Propagation of dynamically passive admixture in the surface sea layer', *Oceanology*, 1978, **18** (3), 423–426 (in Russian).
37. Izrael, Iu. A. and Tsyban, A. V. 'On the assimilation capacity of the World Ocean',

Doklady Ak. Nauk SSSR, 1983, **272** (3), 702–705 (in Russian).

38. Isaev, G. V. and Ovsienko, S. N. 'Extension of a viscous film on the sea surface under conditions of a wind drift', *Meteorology and Hydrology*, 1983, **2**, 74–81 (in Russian).
39. *Studies of Turbulence and Solving of the Contaminants Transport Problems*. Trudy Gos. Okeanograph. Inst., No. 141, ed. by E. B. Borishov, Moscow, 1977 (in Russian).
40. Kagan, B. A., Oganesyan, L. A. and Ryabchenko, V. A. 'The distribution of passive admixture from an instantaneous point source in the World Ocean (a numerical experiment)' *Atm. Ocean. Phys.*, 1978. **14** (6), 621–627 (in Russian).
41. Kagan, B. A. and Ryabchenko, V. A. 'Numerical experiments on the seasonal evolution of carbon cycle in the ocean', *Atm. Ocean. Phys.*, 1982, **18** (4), 373–382 (in Russian).
42. Kamenkovich, V. M. 'On the coefficients of eddy diffusion and viscosity in large-scale oceanic and atmospheric motions, *Atm. Ocean. Phys.*, 1967, **3** (12), 1326–1333 (in Russian).
43. Karabashev, G. S. 'An underwater fluorimeter', *Trudy Inst. Okeanologii Akad. Nauk SSSR*, 1965, **77**, 110–115 (in Russian).
44. Kolesnikov, A. G. 'Some results of direct measurements of vertical turbulent exchange intensity in a sea', in *Some Problems and Data on Oceanological Studies*, Akad Nauk SSSR, 1959, v. 1, pp. 20–28 (in Russian).
45. Kolesnikov, A. G. 'Vertical turbulent exchange in a stratified sea', *Izv. Ak. Nauk SSSR, ser. geofiz.*, 1960, No. 11, pp. 1614–1623 (in Russian).
46. Kolmogorov, A. N. 'Local turbulent structure in incompressible fluids at very high Reynolds number', *Doklady Ak. Nauk SSSR*, 1941, **30** (4), 299–303 (in Russian).
47. COMECON Coordination Center for Studies of Chemical, Physical, Biological and Other Processes in the Most Important Ocean Areas. Bull. No. 5, Moscow, 1977 (in Russian).
48. COMECON Coordination Center for Studies of Chemical, Physical, Biological and Other Processes in the Most Important Ocean Areas. Bull. No. 17, Moscow, 1984 (in Russian).
49. Korotenko, K. A. 'The influence of the current velocity field fine structure on mass transport in the ocean', *Oceanology*, 1983, **23** (2), 249–254 (in Russian).
50. Korotenko, K. A. 'On the vertical matter flow in the ocean with stratified structure of the current velocity field', *Oceanology*, 1983, **23** (3), 396–398 (in Russian).
51. Korotenko, K. A. 'On the vertical transfer of substance in the stratified ocean pycnocline', *Atm. Ocean. Phys.*, 1983, **19** (7), 773–776 (in Russian).
52. Kochergin, V. P. and Bokovikov, A. G. 'A three-dimensional numerical model of contaminant dispersion in the near-shore zone of a deep basin', *Atm. Ocean. Phys.*, 1980, **16** (7), 729–737 (in Russian).
53. Kochergin, V. P. and Borovikov, A. G. 'Numerical experiments aimed at investigation of the distribution of pollutants in the sea', *Meteorology and Hydrology*, 1984, **3**, 73–79 (in Russian).
54. Kremser, U. 'Comparison of the experimental data on the maximum concentration of admixture in diffusing dye patches in the sea with theoretical expressions', *Oceanology*, 1981, **21** (2), 244–248 (in Russian).
55. Kremser, U. and Ozmidov, R. V. 'Turbulence, diffusion and fine-structure of the fields in the Baltic Sea', in *Studies on the Waters Dynamics of the Baltic Sea*. COMECON Coordination Center for Studies of Chemical, Physical, Biological and Other Processes in the Most Important Ocean Areas, Moscow, 1977, 152–197 (in Russian).
56. Kuznetsov, V. R. 'Probability of passive contaminant concentration in the turbulent flows with a cross shear', *Izv. Ak. Nauk SSSR, Mekhanika Zhidkosti i Gaza*, 1972, No. 5, 85–91 (in Russian).
57. Lozovatsky, I. D. 'A model of the vertical structure of turbulent layers in the ocean', *Atm. ocean. phys.*, 1982 **18** (1), 70–78 (in Russian).
58. Lozovatsky, I. D. and Ozmidov, R. V. 'Statistical characteristics of the local structure of developed small-scale turbulence in the Kuroshio current', *Oceanology*, 1979, **19** (5), 757–765 (in Russian).
59. Liubimtsev, M. M. and Van'yan, P. L. 'On probability density of the contaminants concentration fluctuations in the turbulent flows', COMECON Coordination Center for Stu-

dies of Chemical, Physical, Biological and Other Processes in the Most Important Ocean Areas. Moscow, 1984, Bull. No. 18, 56–60 (in Russian).

60. Makkaveev, P. N. and Yakushev, E. V. 'Simulation of vertical distributions of total carbon dioxide in the photic layer of the oligotrophic ocean areas'. *Oceanology*, 1983, **23** (5), 801–806 (in Russian).
61. Marchuk, G. I. *Numerical Solving of the Atmosphere and Ocean Dynamics Problems*. Leningrad, Gidrometeoizdat, 1974 (in Russian).
62. Materials from the 3rd Symp. on self-purification of the water-basins and waste-water mixing. Moscow–Tallin, 1969 (in Russian).
63. *Prediction Methods for Turbulent Flows*. Edited by B. Collman, Hemisphere Publishing Corporation, 1980
64. Mirabel, A. P. and Monin, A. S. 'On statistical properties of admixture mixing by two-dimensional turbulence', *Atm. Ocean. Phys.*, 1983, **19** (9), 902–912 (in Russian).
65. Mikhailov, G. A. and Sabelfeld, K. K. 'On numerical simulation of the diffusion process of the admixture in the random velocity fields', *Atm. Ocean. Phys.*, 1980, **16** (3), 229–235 (in Russian).
66. Monin, A. S. 'On the interaction between the vertical and horizontal diffusion of mixture in the sea', *Oceanology*, 1969, **9** (1), 76–81 (in Russian).
67. Monin, A. S. and Ozmidov, R. V. *Turbulence in the Ocean*. D. Reidel Publ. Company, 1985 (in Russian).
68. Monin, A. S. and Yaglom, A. M. *Statistical Hydromechanics*. Part 1, 1965, Part 2, 1967, Moscow, Nauka (in Russian).
69. Nelepo, B. A. 'An investigation of the radioactivity of the water in the Atlantic Ocean'. *Doklady Ak. Nauk SSSR*, 1960, **134** (4), 810–811 (in Russian).
70. Nelepo, B. A. *Nuclear Hydrophysics*. Moscow, Atomizdat, 1970 (in Russian).
71. Novikov, E. A. 'On turbulent diffusion in a flow with cross gradient of velocity', *Prikl. Matem. Mekhan.*, 1958, **22** (3), 576–579 (in Russian).
72. Ozmidov, R. V. 'Experiments on horizontal turbulent diffusion in a sea and in an artificial basin of moderate depth', *Izv. Ak. Nauk SSSR, ser. geofiz.*, 1957, **6**, 756–764 (in Russian).
73. Ozmidov, R. V. 'On the role of turbulent eddies of different scale in the diffusion process', *Izv. Ak. Nauk SSSR, ser. geofiz.*, 1958, **2**, 272–273 (in Russian).
74. Ozmidov, R. V. 'On calculating of the horizontal turbulent diffusion of contaminant spots in a sea', *Dokl. Ak. Nauk SSSR*, **120** (4), 761–763 (in Russian).
75. Ozmidov, R. V. 'Dependence of the horizontal turbulent diffusion coefficient in a wind-generated currents on a relative depth of a basin', *Izv. Ak. Nauk SSSR, ser. geofiz.*, 1959, **8**, 1242–1246 (in Russian).
76. Ozmidov, R. V. 'Horizontal turbulent diffusion of the pollutant patches in the sea', *Trudy Inst. Okeanol. Ak. Nauk*, 1960, **37**, pp. 164–181 (in Russian).
77. Ozmidov, R. V. 'Diffusion of a contaminant in the homogeneous isotropic turbulence field', *Izv. Ak. Nauk SSSR, ser. geofiz.*, 1960, **1**, 174–175 (in Russian).
78. Ozmidov, R. V. 'On the turbulent exchange in the stably stratified ocean', *Atm. Ocean Phys.*, 1965, **1** (8), 853–860 (in Russian).
79. Ozmidov, R. V. 'On some features of energy spectra of oceanic turbulence', *Doklady Ak. Nauk SSSR*, 1965, **161** (4), 828–831 (in Russian).
80. Ozmidov, R. V. 'Dependence of horizontal turbulent exchange in the ocean on a scale of phenomenon', *Atm. Ocean. Phys.*, 1968, **4** (11), 1224–1225 (in Russian).
81. Ozmidov, R. V. *Horizontal Turbulence and Turbulent Exchange in the Ocean*. Moscow, Nauka, 1968 (in Russian).
82. Ozmidov, R. V. 'The international expedition to study the turbulent diffusion of mixtures in the sea', *Oceanology*, 1969, **9** (6), 1128–1130 (in Russian).
83. Ozmidov, R. V. 'International symposium on the studies of turbulence and the diffusion of admixtures in the sea', *Oceanology*, 1976, **16** (4), 726–728 (in Russian).
84. Ozmidov, R. V. 'The international meeting on the models of admixture diffusion in the ocean', *Oceanology*, 1982, **22** (6), 1043–1044 (in Russian).

85. Ozmidov, R. V. 'Small-scale turbulence and fine structure of hydrophysical fields in the ocean', *Oceanology*, 1983, **23** (4), 533–537 (in Russian).
86. Ozmidov, R. V. 'The basic laws of pollutants spreading in the World Ocean', *Meteorology and Hydrology*, 1984, **8**, 51–58 (in Russian).
87. Ozmidov, R. V., Astok, V. K., Gesentswei, A. N., and Yukhat, M. K. 'Statistical properties of concentration fields of a passive pollutant artifically introduced in the sea', *Atm. Ocean. Phys.*, 1971, **7** (9), 963–973 (in Russian).
88. Ozmidov, R. V., Gezentswei, A. N., and Karabashev, G. S. 'New data on the admixture diffusion in the sea', *Atm. Ocean. Phys.*, 1969, **5** (11), 1191–1204 (in Russian).
89. Ozmidov, R. V., Gezentswey, A. V., Le-Kuang-Toai *et al.* 'Experimental studies on the diffusion of contaminants introduced into the sea', In *Studies on Ocean Turbulence*, Moscow, Nauka, 1973 (in Russian).
90. Ozmidov, R. V. and Popov, N. I. 'The study of the vertical water exchange in the ocean using Strontium-90 distribution data', *Atm. Ocean. Phys.*, 1966, **2** (2), 183–190, 64–78 (in Russian).
91. *Oceanology, vol.* 1, *Hydrophysics of the Ocean*. Edited by V. M. Kamenkovich and A. S. Monin, Moscow, Nauka, 1978 (in Russian).
92. Okubo, A., Ozmidov, R. V. 'An empirical relationship between the horizontal eddy diffusivity in the ocean and the scale of the phenomenon', *Atm. Ocean. Phys.*, 1980, **6** (5), 534–536 (in Russian).
93. Panchev, S. 'On the existence of power dependences in the spectra of oceanic turbulence', *Atm. Ocean. Phys.*, 1975, **11** (6), 620–625 (in Russian).
94. Panchev, S. 'Spectra of large-scale geophysical turbulence in inertial ranges', *Blgarsko Geofizichno Spisanie*, 1976, **2** (2), 3–11 (in Russian).
95. Pivovarov, A. A. *Thermodynamics of the Ocean*. Published by MGU, Moscow, 1979 (in Russian).
96. Ryabchenko, V. A. 'Numerical modelling of dissolved oxygen distribution in the World Ocean', *Oceanology*, 1977, **17** (6), 1004–1009 (in Russian).
97. Simonov, A. I., Nazarov, I. M., Gruzinov, V. M. *et al.* 'The balance of petroleum hydrocarbons in the World Ocean (exemplified by the North Atlantic)', *Meteorology and Hydrology*, 1984, **3**, 64–72 (in Russian).
98. Styro, B. I. *Self-Clearing of the Atmosphere from the Radioactive Contaminations*. Gidrometeoizdat, Leningrad, 1968 (in Russian).
99. *Currents in Baikal*. Edited by G. I. Galazii, Nauka, Novosibirsk, 1977 (in Russian).
100. *Currents and Diffusion of the Waters in Baikal*. Trudy Limnologicheskogo Inst. SO Ak. Nauk SSSR, v. 14/34. Nauka, Leningrad, 1970 (in Russian).
101. Tikhonov, A. N. and Samarskii, A. A. *Equations of Mathematical Physics*. Gos. izd. tekhn.-teor. lit., Leningrad, 1951 (in Russian).
102. Toompu, A. 'Non-self-similarity of concentration variance in turbulent diffusion in the sea', *Oceanology*, 1982, **22** (5), 738–743 (in Russian).
103. *Trudy Inst. Okeanologii Akad. Nauk SSSR*, 1946, v. 1, Izd. Ak. Nauk SSSR (in Russian).
104. Frenkiel, F. N. 'Turbulent diffusion: mean concentration distribution in a flow field of homogeneous turbulence', In *Advances in Appl. Mech.*, 1953, v. 3.
105. *Chemistry in the Ocean*. v. 1. Edited by O. K. Bordovskii and V. N. Ivanenkov, Nauka, Moscow, 1979 (in Russian).
106. Shkudova, G. Y. and Djoiyev, T. Z. 'Numerical model of the waters circulation and spreading of contaminant in a deep barocline sea (exemplified by the Black Sea)', *Trudy Gosudarstennyi Okeanograficheskii Inst.*, 1975, **126**, 92–103 (in Russian).
107. Shtokman, V. B. 'Turbulent exchange off the Agrahanskii peninsula in the Caspian Sea', *Zhurnal geofiziki*, 1936, **6** (4), 340–388 (in Russian).
108. Shtokman, V. B. 'On turbulent exchange in the middle and southern parts of the Caspian Sea', *Izv. Ak. Nauk SSSR, ser. geogr. geofiz.*, 1940, **4**, 569–592 (in Russian).
109. Shtokman, V. B. 'On the fluctuations of horizontal velocity components in the sea due to large-scale turbulence', *Izv. Ak. Nauk SSSR, ser. geogr. geofiz.*, 1941, **4–5**, 476–486 (in Russian).

110. Shtokman, V. B. 'Principles of the T,S-curves theory as a method of investigation of the sea-waters masses mixing and transformation', *Problemy Arktiki*, 1943, **1**, 32–71 (in Russian).
111. Yakushev, E. V. 'Mathematical modelling of phosphate distribution in the meridional plane of the southern part of the World Ocean', *Oceanology*, 1984, **24** (2), 277–284 (in Russian).
112. Yakushev, E. V. 'Numerical simulation of distribution and variability of phosphorus compounds in the ocean', Cand. Thesis, Inst. Okeanologii Ak. Nauk SSSR, Moscow, 1984 (in Russian).
113. Assaf, G., Gerard, R., and Gordon, H. 'Some mechanisms of oceanic mixing revealed in aerial photographs', *J. Geoph. Res.*, 1971, **76**, 6550–6572.
114. Batchelor, G. K. 'The application of the similarity theory of turbulence to atmospheric diffusion', *Quart. J. Roy. Met. Soc.*, 1950, **76** (328), 133–146.
115. *Beitrage zur Meereskunde*. Akademie-Verlag. Berlin, 1972, Heft 30/31.
116. Bowden, K. F., 'Horizontal mixing in the sea due to a shearing current', *J. Fluid Mech.*, 1965, **21** (2), 83–95.
117. Brosin, H.-J., Kremser, U., Muraviev, S. S., and Zeidler, R. 'Parametric description of mean concentration fields of passive additives in the sea', *Reporty MJR, Ser. R.*, Gdynia, 1976, No. 2a, 139–151.
118. Chandrasekhar, S. 'The theory of axisymmetric turbulence', *Phil. Trans. Roy. Soc.*, 1950, **242** (855), 557–577.
119. Corrsin, S. 'Heat transfer in isotropic turbulence', *J. Appl. Phys.*, 1952, **23** (1), 113–118.
120. Corrsin, S. 'Theories of turbulent dispersion. Mechanics of turbulence', 1962, Proc. Intern. Symp. National Science Research Center, Marseille, 1961, 27–52.
121. Csanady, G. T. *Turbulent Diffusion in the Environment*. D. Reidel Publishing Comp., Dordrecht-Boston. 1973, 248.
122. Fennel, W. and Lass, H. U. 'Phytoplankton patchiness and advection diffusion models', *Beitrage zur Meereskunde*, 1982, H. 47, 95–103.
123. Fjeldstad, J. E. 'Wärmeleiting im Meer', *Geophys. Publik.*, 1933, **10** (7), 1–20.
124. Fox, D. G. 'Forced plume in a stably stratified fluid', *J. Geophys. Res.*, 1970, **75** (33), 6818–6835.
125. Gyllander, C. 'Water exchange and diffusion processes in Tvären, a Baltic bay', *Proc. Symp. Int. Atom. Energy Agency*, 1966, Vienna, 207–220.
126. Hela, J. and Viopio, A. 'Tracer dyes as a means of studying turbulent diffusion in the sea', *Annales Acad. Sci. Fennica*, 1960, *ser. A*, **6** (69).
127. Howard, L. M. 'Note on a paper of John W. Miles', *J. Fluid. Mech.*, 1961, **10** (4), 509–512.
128. Ichiye, T. 'Upper ocean boundary-layer flow determined by dye diffusion', *Phys. Fluids*, 1967, **10** (2), part 2, 270–277.
129. Joseph, J. and Sendner, H. 'Über die horizontale Diffusion in Meer', *Deutsche Hydr. Z.*, 1958, **11** (2), 49–77.
130. Joseph, J., Sendner, H., and Weideman, H. 'Untersuchungen über die Horizontale Diffusion in der Nordsee', *Deutsche Hydr. Zeitsch.*, 1964, **17** (2), 57–75.
131. Katz, B., Gerard, R., and Costin, M. 'Response of dye tracer to sea surface conditions', *J. Geophys. Res.*, 1965, **70** (22) 5505–5513.
132. Kraichnan, R. H. 'Diffusion by a random velocity field', *Phys. Fluids*, 1970, **13** (1), 22–31.
133. Krauss, W. 'Theorie des Triftstromes und der vertuellen Reibung im Meer', *Deutsch. Hydr. Zeitsch.*, 1965, **18** (5), 193–210.
134. Kremser, U. and Brosin, H.-J. 'Untersuchungen über vertikale Austauschkoeffizienten und Richardson'sche Zahlen in der mittleren Ostsee', *Beitrage zur Meeresk.*, 1969, No. 26, 5–10.
135. Kremser, U., and Matthäus, W. 'Grundlagen und Methoden zur Berechnung mittlerer vertikaler Warmeaustauschkoeffizienten in der Ostsee', *Gerlands Beitr. Geophys.*, 1973,

82 (2), 128–134.
136. Levy, E. M., Ehrhardt, M., Kohnke, D., Sobtchenko, E., Suzuoki, T., and Tokuhiro, A. *Global Oil Pollution.* Intergovernment Oceanographic Commission UNESCO, Paris, 1981.
137. Lin, J. T. 'Relative dispersion in the enstrophy cascading inertial range of homogeneous two-dimensional turbulence', *J. Atmos Sci.*, 1972, **29** (2), 394–395.
138. McEwen, G. F. 'The dynamics of large horizontal eddies (axes vertical) in the ocean off Southern California', *J. Mar. Res.*, 1948, **7** (3), 188–216.
139. Miles, J. M. 'On the stability of a heterogeneous shear flow', *J. Fluid Mech.*, 1961, **10** (4), 496–508.
140. Munk, W. H., Ewing, C. C., and Revelle, R. R. 'Diffusion in Bikini Lagoon', *Trans. Amer. Geophys. Union*, 1949, **30** (1), 59–66.
141. Nichoul, J. (ed.) *Remote Sensing of Shelf Sea Hydrodynamics.* 1984, Elsevier, Amsterdam, 354 p.
142. Obukhov, A. M. 'Description of turbulence in terms of Lagrangian variables', *Adv. in Geophys.*, 1959, **6**, 113–116.
143. Okubo, A. 'A review of theoretical models for turbulent diffusion in the sea', *J. Oceanogr. Soc. Japan*, 1962, 20th Anniversary v., 286–320.
144. Okubo, A. 'A note on horizontal diffusion from an instantaneous source in a nonuniform flow', *J. Oceanogr. Soc. Japan*, 1966, **22** (2), 36–40.
145. Okubo, A. 'The effect of shear in an oscillatory current on horizontal diffusion from an instantaneous source', *Oceanology and Limnology*, 1967, **1** (3), 194–204.
146. Okubo, A. 'Oceanic diffusion diagrams', *Deep.-Sea Res.*, 1971, **18**, 789–802.
147. Okubo, A., Andreasen, V., and Mitchel, J. 'Chaos-induced turbulent diffusion', *Physics letters*, 1984, **105A** (4–5), 169–172.
148. Okubo, A. and Carter, H. H. 'An extremely simplified model of the "shear effect" on horizontal mixing in a bounded sea', *J. Geoph. Res.*, 1966, **71** (22), 5267–5270.
149. Okubo, A., Ebbesmeyer, C. C., and Sanderson, B. G. 'Lagrangian diffusion equation and its application to oceanic dispersion', *Journal Oceanogr. Soc. Japan*, 1983, **39** (5), 259–266.
150. Ozmidov, R. V. and Zatz, V. I. 'Study of mesoscale processes in the shelf zone of the Black Sea using remote techniques', In *Remote Sensing of Shelf Sea Hydrodynamics.* 1984. Ed., J. Nichoul, Elsevier, Amsterdam, 149–157.
151. Pritchard, D. W., Okubo, A., and Carter, H. *Observation and Theory of Eddy Movement and Diffusion of an Introduced Tracer Material in the Surface Layer of the Sea. Disposal Radioactive Wastes Seas, Oceans and Surface Waters.* 1966, Vienna.
152. Richardson, L. F. 'Atmospheric diffusion shown on a distance-neighbor graph', *Proc. Roy. Soc.*, 1926, *ser. A*, **110** (753), pp. 709–737.
153. Richardson, L. F. and Stommel, H. 'Note on eddy diffusion in the sea', *J. Meteorol.*, 1948, **5** (5), 238–240.
154. Saffman, P. G. 'The effects of wind shear on horizontal spread from instantaneous ground source', *Quart. J. Roy. Meteorol. Soc.*, 1962, **88** (378), 382–393.
155. Schmidt, W. 'Der Massenaustausch bei der ungeordneten Strömung in freier Luft und seine Folgen', *Sitzungsber. Akad. Wiss.* Wien, Math.-nat. Kl. (2a), 1917, **126** (6), 757–804.
156. Stommel, H. 'Horizontal diffusion due to oceanic turbulence', *J. Mar. Res.*, 1949, **8** (3), 199–225.
157. Sutton, O. G. 'The theory of eddy diffusion in the atmosphere', *Proc. Roy. Soc.*, 1932, *ser. A*, **135** (826), 143–165.
158. Sverdrup, H. U. 'Lateral mixing in the deep water of the south Atlantic ocean', *J. Mar. Res.*, 1939, **2** (3), 195–207.
159. Sverdrup, H. U. and Fleming, R. 'The waters of the coast of Southern California', *Bull. Scripps. Inst. Oceanograph. Univ. Calif.*, 1941, **4** (10), 261–378,
160. Taylor, G. I. 'Eddy motion in the atmosphere', *Phil. Trans. Roy. Soc.*, 1915, *ser. A*, **215**, 1–26.
161. Taylor, G. I., 'Diffusion by continuous movements', *Proc. Lond. Math. Soc.*, 1921,

ser. A, **20**, 196–211.

162. Taylor, G. I., 'Dispersion of soluble matter in solvent flowing slowly through a tube', *Proc. Roy. Soc. London*, 1953, *ser. A*, **219** (1137), 186–203.
163. Taylor, G. I., 'The dispersion of matter in turbulent flow through a pipe', *Proc. Roy. Soc. London*, 1954, *ser. A*, **223** (1155), 446–468.
164. Townsend, A. A. 'The diffusion of heat spots in isotropic turbulence', *Proc. Roy. Soc.*, 1951, *ser. A*, **209**, 418–430.
165. Williams, R. W. and Gibson, C. H. 'Direct measurement of turbulence in the Pacific Equatorial Undercurrent', *J. Phys. Oceanogr.*, 1974, **4** (1), 104–108.

Subject index

Oceanographic Sciences Library

1. V.M. Litvin: *The Morphostructure of the Atlantic Ocean Floor*. Its Development in the Meso-Cenozoic. 1984 ISBN 90-277-1933-0
2. E.C. Monahan and G. Mac Niocaill (eds.): *Oceanic Whitecaps*. And Their Role in Air-Sea Exchange Processes. 1986 ISBN 90-277-2251-X
3. G.I. Marchuk and B.A. Kagan: *Dynamics of Ocean Tides*. 1989 ISBN 90-277-2552-7
4. R.V. Ozmidov: *Diffusion of Contaminants in the Ocean*. 1990 ISBN 0-7923-0611-2

KLUWER ACADEMIC PUBLISHERS – DORDRECHT / BOSTON / LONDON

MIX
Papier aus verantwortungsvollen Quellen
Paper from responsible sources
FSC® C105338

If you have any concerns about our products,
you can contact us on
ProductSafety@springernature.com

In case Publisher is established outside the EU,
the EU authorized representative is:
Springer Nature Customer Service Center GmbH
Europaplatz 3, 69115 Heidelberg, Germany

Printed by Libri Plureos GmbH
in Hamburg, Germany